THE FULLERENES

Pergamon Journals of Related Interest

Applied Superconductivity
Carbon
Journal of Physics & Chemistry of Solids
Nanostructured Materials
Polyhedron
Solid State Communications
Tetrahedron
Tetrahedron Letters

THE FULLERENES

Edited by

HAROLD W. KROTO
University of Sussex, UK

JOHN E. FISCHER
University of Philadelphia, USA

DAVID E. COX
Brookhaven National Laboratory, USA

PERGAMON PRESS

Oxford · New York · Seoul · Tokyo

U.K.	Pergamon Press Ltd, Headington Hill Hall, Oxford OX3 0BW, England
U.S.A.	Pergamon Press, Inc., 660 White Plains Road, Tarrytown, New York 10591-5153, U.S.A.
KOREA	Pergamon Press Korea, K.P.O. Box 315, Seoul 110-603, Korea
JAPAN	Pergamon Press Japan, Tsunashima Building Annex, 3-20-12 Yushima, Bunkyo-ku, Tokyo 113, Japan

First edition 1993

ISBN 0-08-042152-0

Library of Congress Cataloging in Publication Data
A catalogue record for this book is available from the Library of Congress.

British Library Cataloguing in Publication Data
A catalogue record for this book is available from the British Library.

Reprinted from:
Carbon, Vol. 30 (8)
Journal of Physics & Chemistry of Solids, Vol. 53 (11)

Printed and bound in Great Britain by BPCC Wheatons Ltd, Exeter

CONTENTS

Contents

Part One
The Fullerenes

INTRODUCTION

There are several fascinating strands in the story of the birth of fullerene chemistry, physics, and materials science. The earliest record of the molecule in the literature is contained in an article by Eiji Osawa in *Kagaku* (in Japanese) in 1970[1]. Osawa conjectured that such a molecule would be stable and the following year he and Yoshida described it more fully in a book on aromatic molecules—again in Japanese[2]. A little later Bochvar and Gal'pern published a Hückel calculation on C_{60}[3], and in 1980 Davidson applied general group theoretical techniques to a range of highly symmetric molecules, one of which was C_{60}[4]. However, somewhat earlier in a remarkable article in 1966, David Jones (writing under the pseudonym of Daedalus in the *New Scientist*) discussed the possibility of creating graphite balloons[5,6] similar to geodesic cages—these objects are essentially giant fullerenes[7,8]. The molecule forms spontaneously under our noses when 60 or so carbon atoms are allowed to aggregate during gas-phase nucleation. In general, such conditions may occur relatively frequently, but they appear never previously to have been probed carefully enough. Often when such a situation has arisen many other reactions take place as well, and so the fleeting, shadowy existence of this third member of the carbon family of allotropes has been obscured. So the discovery had to wait until the right experiment could be carried out at the right time with the right instrument. In September 1985 this conjunction took place [9–11].

Some long carbon chain molecules (the cyanopolyynes), which had originally been discovered in space during a Sussex/NRC (Ottawa) radioastronomy collaboration (with David Walton, Takeshi Oka, and Canadian astronomers)[12] were, by the early 80s, found to be streaming out of red giant carbon stars. At just this moment the laser vaporisation cluster beam technique developed by Rick Smalley and coworkers at Rice University[13] was beginning to revolutionise the study of refractory clusters. This technique seemed to offer the perfect way to simulate stellar nucleation conditions if graphite were vaporised. Graphite vaporisation had actually been carried out somewhat earlier by Rohlfing *et al.*[14] and Bloomfield *et al.*[15] under conditions that had not highlighted the special nature of C_{60}—so the molecule was overlooked. A collaboration with Robert Curl and Rick Smalley at Rice University (Houston, TX) was initiated in 1985 and the experiments performed with graduate students Jim Heath, Sean O'Brien, and Yuan Liu. The experiments immediately confirmed that long cyanopolyynes, including those detected in space, are formed in a plasma produced by a laser focused on a graphite target[16,17]. During these experiments the serendipitous discovery was made that the C_{60} molecule was remarkably stable[9].

The result was rationalised on the basis of a closed fullerene cage structure consisting of 12 pentagons and 20 hexagons with the symmetry of a soccerball [9]. Geodesic and chemical factors inherent in such a structure seemed to account for the stability of the molecule, and it was named Buckminsterfullerene after the designer/inventor of the geodesic domes. During the period 1985–90 unequivocal evidence for the exceptional stability of C_{60} was amassed, which provided convincing evidence for the cage structure proposal[10,11,18]. Endohedral complexes were detected[19] and reactivities studied[20]. Other groups, such as that of Busmann *et al.*[21–23], showed that the fragmentation behaviour was consistent with the fullerene proposal. O'Keefe *et al.*[24–28] obtained important electronic and dynamic information—including ionisation potentials, and UPS spectra of the negative ion carbon clusters were measured at Rice[29–30]. All these experimental studies added considerable weight to the fullerene structural proposal and Robert Curl, in this volume, describes the pre-1990 evidence. Cambell and Hertel complement this study with their article on beam studies of carbon clusters. A large number of theoretical papers also lent considerable weight to the proposal. Particularly important were those by Tom Schmalz *et al.*[31,32], who among other things pointed out that carbon cage structures in which the pentagons are isolated are likely to be more stable than structures in which they abut. The requirement that pentagons do not abut was subsequently recognised as having quantitative significance and the *pentagon isolation principle* formulated[33,34], which predicted the magic numbers observed experimentally—in particular, 60 and 70. Indeed, using a generalised form of this rule, together with symmetry principles, a whole family of closed carbon cages was predicted to exist with $20 + 2n$ atoms($n = 0,2,3,4 \ldots$)[33]. The study led to the conclusion that C_{28} should be very special and that the tetravalent analogues such as $C_{28}H_4$ should be particularly stable[33,35]. Fowler and Woolrich discovered that fullerene cages, C_n, with $n = 60 + 6k$ ($k = 0,2,3,4 \ldots$) should be electronically stable[36]—a fascinating analogue of Hückel's famous $4n + 2$ rule for aromatic compounds. Giant fullerenes with $n = 240, 560, 960$, etc., were studied and discovered to have beautiful quasi-icosahedral shapes[7].

A possible mechanism for the formation of C_{60} was proposed, and included a suggestion that it might be a by-product of soot formation[20]. This prediction was verified when Gerhardt et al. in 1987 found that C_{60} was indeed detectable during the soot formation process[37]. The relationship between soot and C_{60} generation is an interesting issue and the subject of some debate—in this special issue of *Carbon,* Jack Howard and co-workers discuss some aspects of it.

The original detection of C_{60} was made on minute amounts of the species in a beam using that most sensitive of detection devices—the mass spectrometer. During 1989–90 Wolfgang Krätschmer, Lowell Lamb, Kostas Fostiropoulos and Don Huffman (Max Planck Institut, Heidelberg, and University of Arizona, Tucson) put two camel humps together and got sixty[38,39]. They first presented a remarkable paper at an astronomy conference[39] in which they suggested that they might actually have detected the IR absorption signal of C_{60} in carbon dust deposited by carbon arc struck under an inert gas. In August 1990 any remaining scepticism surrounding the assignment of these IR features, and of course the original fullerene-60 cage concept[9], was finally and emphatically laid to rest by their announcement that they had extracted macroscopic amounts of a soluble red material containing C_{60} and that the molecule was indeed a round pure carbon ball[38]. Similar experiments were being carried out at Sussex (with Jonathan Hare and Ala'a Abdul Sada), where it was also discovered independently that the arc-processed material showing the IR features gave rise to a prominent mass spectrometric line at 720 amu, consistent with C_{60} and that, furthermore, a soluble red extract could be obtained from this material[40,35]. From a solution of the extract, Krätschmer et al. obtained crystals shown by x-ray analysis to consist of arrays of spherical molecules some 7Å in diameter—a result beautifully and unequivocally consistent with expectation for C_{60}[38]. At Sussex Roger Taylor showed that the soluble extract was chromatographically separable into two components, C_{60} and C_{70}. C_{60} gave an exquisitely delicate magenta solution. ^{13}C NMR analysis yielded a single resonance[40] which indicated that all the C atoms in the molecule are equivalent—a result perfectly commensurate with the proposed cage structure. Elegant further support for the fullerene proposal came from the ^{13}NMR spectrum of the C_{70} component (red in solution), which exhibited five lines, again in perfect agreement with expectation and, most importantly, proof of the existence of other members of the fullerene family. In a parallel investigation, Meijer and Bethune[41,42] were able to identify C_{60} by mass spectrometry in the solid material deposited from a Smalley-type laser vaporisation nozzle. Subsequent studies by Johnson et al.[43] confirmed the NMR measurement on C_{60} and Ajie et al.[44] confirmed that the material could be chromatographically separated.

The fact that this new third form of carbon may be created spontaneously and has been under our noses since time immemorial is almost unbelievable. In their article for *Carbon,* Wolfgang Krätschmer and Don Huffman describe the exciting events that led to their momentous breakthrough. Bob Curl in his article describes the pre-isolation evidence for C_{60}—in particular, the considerable quantity of evidence amassed in cluster beam studies at Rice. The article written by Eleanor Campbell and Ingolf Hertel focuses on complementary carbon cluster studies. Fullerene-60 appears to make up some $10\% \pm 5\%$ of the material produced by a carbon arc under an inert gas, and various aspects of the production procedure are described in the article by Deborah Parker et al. Just as conjectured by Zhang et al. in 1986[20], C_{60} does indeed form in a sooting flame. In their article, Jack Howard et al. describe the conditions in detail. Since the fullerenes have been available in bulk quantities, the field has exploded and the chemical and physical properties of the new materials are now being investigated in minute detail. Already there are exciting results: The molecule undergoes a vast new range of novel reactions; it readily accepts and donates electrons, suggesting battery applications; it forms high-temperature superconducting compounds; complexes exhibit ferromagnetic behaviour. This volume has brought together reviews from groups that show how wide are the physico/chemical implications of the new material. George Olah et al. describe aspects of fullerene reactivity and functionalization. The article from du Pont by Paul Fagan et al. also deals with the reactivity, focusing on techniques for structure characterisation. From these two papers it is clear that fullerene chemistry presents a real and exciting analytical challenge. Lawrence Dunne and his co-workers discuss the implications of the fact that C_{60} forms spontaneously on our understanding of the structure of graphitic materials. Patrick Fowler et al. discuss the fascinating problem of isomerisation, which is vital for a basic understanding of fullerene family structure and behaviour. Humberto Terrones and Alan MacKay in their article go beyond totally closed structures and expand our horizons to describe open, extended, carbon networks that may be fabricated in the future. Karoly Holczer and Robert Whetten describe the remarkable discovery that metal doped fullerenes exhibit high-temperature superconductivity. Finally, the results of some solid state neutron scattering studies obtained during a collaborative investigation between the Sussex group (Kosmas Prassides, myself, Roger Taylor, and David Walton, working with Bill David and John Tomkinson at ISIS and Mat Rosseinsky and Don Murphy at AT&T Bell Labs) are presented.

There is food for thought in the fact that this fascinating, new round world of synthetic chemistry and materials science was discovered as a consequence of the desire to understand the role of carbon in space and stars, rather than the probing of carbon's possible material applications. It was, however, dependent on the development of apparatus designed to probe clusters and develop an understanding of the ways in which atomic and molecular properties correlate with those of the bulk. Finally, it is interesting to reflect on what new light the terrestrial experiments shed on the carbon content of space and stars. Although C_{60} appears to be unstable in the environment, it appears to be stable in beams, where the conditions are similar to those in space. Thus it seems very likely that C_{60} is formed in those regions of space where polyynes and carbon dust are formed, and there is reason to believe that it may be stable and may be a major constituent of the space between the stars[45,46]. It would certainly be a most fitting last chapter in the story of the discovery if it turned out that C_{60} does indeed play a mysterious key role in space.

Acknowledgements—I am most grateful to the many co-workers and colleagues, particularly those named in the text who were most directly involved in the many aspects of the studies covered above. In addition to those mentioned in this short article, I thank Michael Jura and Steve Wood. I also wish to thank the authors of the articles contained here. The contributors reflect the interdisciplinary breadth of fullerene research and, in particular, show how these new materials will expand the type of science that *Carbon* is likely to publish in the future. I thank David Walton for advice.

HAROLD KROTO
Guest Editor

REFERENCES

1. E. Osawa, *Kagaku* (Kyoto) **25**, 854–863 (in Japanese) (1970). *Chem. Abstr.* **74**, 75698v (1971).
2. Z. Yoshida and E. Osawa, *Aromaticity,* pp. 174–178. Kagakudojin, Kyoto (1971) (in Japanese).
3. D. A. Bochvar and E. G. Gal'pern, *Dokl. Akad. Nauk SSSR* **209**, 610–612 (1973) (English translation *Proc. Acad. Sci. USSR* **209**, 239–241 1973).
4. R. A. Davidson, *Theor. Chim. Acta.* **58**, 193–195 (1981).
5. D. E. H. Jones, *New. Sci.* **32**, 245 (1966).
6. D. E. H. Jones, *The Inventions of Daedalus* pp. 118–119. Freeman, Oxford (1982).
7. H. W. Kroto and K. G. McKay, *Nature* (London) **331**, 328–331 (1988).
8. H. W. Kroto, *Chem. Brit.* **26**, 40–45 (1990).
9. H. W. Kroto, J. R. Heath, S. C. O'Brien, R. F. Curl, and R. E. Smalley, *Nature* (London) **318**, 162–163 (1985).
10. H. W. Kroto, *Science* **242**, 1139–1145 (1988).
11. R. F. Curl and R. E. Smalley, *Science* **242**, 1017–1022 (1988).
12. H. W. Kroto, *Chem. Soc. Rev.* **11**, 435–491 (1982).
13. T. G. Dietz, M. A. Duncan, D. E. Powers, and R. E. Smalley, *J. Chem. Phys.* **74**, 6511–6512 (1981).
14. E. A. Rohlfing, D. M. Cox, and A. Kaldor, *J. Chem. Phys.* **81**, 3322–3330 (1984).
15. L. A. Bloomfield, M. E. Geusic, R. R. Freeman, and W. L. Brown, *Chem. Phys. Lett.* **121**, 33–37 (1985).
16. J. R. Heath, Q. Zhang, S. C. O'Brien, R. F. Curl, H. W. Kroto, and R. E. Smalley, *J. Am. Chem. Soc.* **109**, 359–363 (1987).
17. H. W. Kroto, J. R. Heath, S. C. O'Brien, R. F. Curl, and R. E. Smalley, *Astrophys. J.* **314**, 352–355 (1987).
18. H. W. Kroto, S. P. Balm, and A. W. Allaf, *Chem. Revs.* **91**, 1213 (1991).
19. J. R. Heath, S. C. O'Brien, Q. Zhang, Y. Liu, R. F. Curl, H. W. Kroto, and R.E. Smalley, *J. Am. Chem. Soc.* **107**, 7779–7780 (1985).
20. Q. L. Zhang, S. C. O'Brien, J. R. Heath, Y. Liu, R. F. Curl, H. W. Kroto, and R. E. Smalley, *J. Phys. Chem.* **90**, 525–528 (1986).
21. E. E. B. Campbell, G. Ulmer, B. Hasselberger, H. G. Busmann, and I. V. Hertel, *J. Chem. Phys.* **93**, 6900–6907 (1990).
22. B. Hasselberger, H. G. Busmann, and E. E. B. Campbell, *Appl. Surf. Sci.* **46**, 272–278 (1990).
23. E. E. B. Campbell, G. Ulmer, H. G. Busmann, and I. V. Hertel, *Chem. Phys. Lett.* **175**, 505–510 (1990).
24. A. O'Keefe, M. M. Ross, and A. P. Baronavski, *Chem. Phys. Lett.* **130**, 17–19 (1986).
25. S. W. McElvany, H. H. Nelson, A. P. Baronavski, C. H. Watson, and J. R. Eyler, *Chem. Phys. Lett.* **134**, 214–219 (1987).
26. S. W. McElvany, B. I. Dunlap, and J. O'Keefe, *J. Chem. Phys.* **86**, 715–725 (1987).
27. J. A. Zimmerman, J. R. Eyler, S. B. H. Bach, and S. W. McElvany, *J. Chem. Phys.* **94**, 3556–3562 (1991).
28. S. W. McElvany, *Int. J. Mass Spectrom. Ion Process* **102**, 81–98 (1990).
29. S. H. Yang, C. L. Pettiette, J. Conceicao, O. Cheshnovsky, and R. E. Smalley, *Chem. Phys. Lett.* **139**, 233–238 (1987).
30. O. Cheshnovsky, S. H. Yang, C. L. Pettiette, M. J. Craycraft, Y. Liu, and R. E. Smalley, *Chem. Phys. Lett.* **138**, 119–124 (1987).
31. T. G. Schmalz, W. A. Seitz, D. J. Klein, and G. E. Hite, *Chem. Phys. Lett.* **130**, 203–207 (1986).
32. D. J. Klein, T. G. Schmalz, G. E. Hite, and W. A. Seitz, *J. Am. Chem. Soc.* **108**, 1301–1302 (1986).
33. H. W. Kroto, *Nature* (London) **329**, 529–531 (1987).
34. T. G. Schmalz, W. A. Seitz, D. J. Klein, and G. E. Hite, *J. Am. Chem. Soc.* **110**, 1113–1127 (1988).
35. H. W. Kroto, *Angewandte Chemie* **31**, 111–129 (1991).
36. P. W. Fowler and J. Woolrich, *Chem. Phys. Lett.* **127**, 78–83 (1986).
37. W. Krätschmer, L. D. Lamb, K. Fostiropoulos, and D. R. Huffman, *Nature* (London) **347**, 354–358 (1990).
38. P. Gerhardt, S. Loeffler, and K. Homann, *Chem. Phys. Lett.* **137**, 306–310 (1987).
39. W. Krätschmer, K. Fostiropoulos, and D. R. Huffman, *Dusty Objects in the Universe,* (Edited by E. Bussoletti and A. A. Vittone). Kluwer, Dordrecht (1990) (conference in 1989).
40. R. Taylor, J. P. Hare, A. K. Abdul-Sada, and H. W. Kroto, *J. Chem. Soc. Chem. Commun.* 1423–1425 (1990).
41. G. Meijer and D. S. Bethune, *Chem. Phys. Lett.* **175**, 1–2 (1990).
42. G. Meijer and D. S. Bethune, *J. Chem. Phys.* **93**, 7800–7802 (1990).
43. R. D. Johnson, G. Meijer, and D. S. Bethune, *J. Am. Chem. Soc.* **112**, 8983–8984 (1990).
44. H. Ajie, M. M. Alvarez, S. J. Anz, R. D. Beck, F. Diederich, K. Fostiropoulos, D. R. Huffman, W. Kraetschmer, Y. Rubin, K. E. Schriver, K. Sensharma, and R. L. Whetten, *J. Phys. Chem.* **94**, 8630–8633 (1990).
45. H. W. Kroto and M. Jura, *Astron. Astrophys.* (in press).
46. J. P. Hare and H. W. Kroto, *Acc. Chem. Res.* **25**, 106–112 (1992).

tinction properties. What we wanted to achieve was to make graphite particles as small and as monodispersed as we possibly could—preferably spherical in shape and with a high degree of atomic order. In retrospect, we can state that we succeeded beyond our fondest expectations in creating copious quantities of C_{60} molecules; however, the mysteries of the astronomical bands still remain.

At first we repeated the older smoke experiments, in which graphite rods were evaporated in an atmo-sphere of helium. A conventional bell-jar carbon evaporator was used for this purpose. The carbon was vaporized by resistive heating (i.e., by driving a current through a pair of contacted graphite rods). Pressures ranging between a few and about 20 torr of helium were used to cool the carbon vapor and promote nucleation and aggregation of carbon atoms into small solid particles. The smoke was collected at different positions in the bell jar, measured spectroscopically in the UV-VIS range, and examined with a

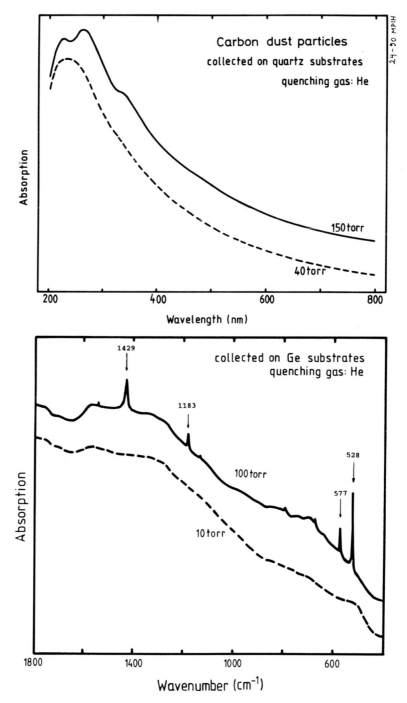

Fig. 2. Extinction spectra of soot particles produced by evaporation of graphite under different helium quenching gas pressures. Notice the occurrence of additional absorptions in the UV-VIS and IR at elevated helium pressures. These features turned out to originate from C_{60}—Buckminsterfullerene.

transmission electron microscope. This work and some later works on smaller carbon clusters were part of the Diplomarbeit of Norbert Sorg. We observed that the degree of structural disorder of the graphite and the size of the particles seemed to depend considerably on the helium quenching gas pressure. However, in all the recorded spectra of the soot samples, the discrepancy between the widths of the measured extinction curves and calculated (i.e., interstellar-like) curves remained. We attributed this to unavoidable clumping of our laboratory-produced particles, which to this day is a major experimental problem. Clumping changes the optical properties of the individual particles and makes it rather difficult to compare the extinction spectra of laboratory-produced and interstellar dust grains.

Even though the Heidelberg smoke experiments were not very conclusive with respect to interstellar dust, there occurred an interesting effect that caught our attention. At our standard 20 torr pressure of helium quenching gas we occasionally observed, besides the "regular" soot absorption feature centered at about 230 nm, three additional features showing up out of the continuum (see Fig. 2), sometimes stronger, sometimes weaker, and sometimes absent. Having no explanation, we whimsically called these features "camel humps," and the samples bearing them "camel samples" (actually the very similar German word was used). Only a few things could be found out about the carrier of the camel humps: The samples showed a peculiar raman spectrum[2] and the humps disappeared when the sample was heated in air. After much discussion we concluded that the carrier must be some kind of "junk." This did not appear unlikely, since the soot with its high specific surface area should be able to sweep up various contaminants, such as pump oil or vacuum grease used in the bell-jar evaporator. Our attention returned to properties of other small particles and to the low temperature matrix isolation of small carbon clusters.

3. INTERMISSION

In the summer of 1984, a year after our joint work at Heidelberg had ended, the two of us got together again at the Third ISSPIC cluster conference at the Freie Universität in Berlin. We had jointly submitted two papers—one on our matrix isolated carbon cluster work[11] and one on our carbon smoke-making experiments that had revealed the camel samples. The carbon smoke paper was rejected.

To us, one of the conference highlights was the impressive mass spectrum of carbon clusters obtained by the Exxon group and presented by Andrew Kaldor in a poster session[12]. He showed the wide distribution of even-numbered clusters (now known as fullerenes), and we were among the rest who did not notice the slight overabundances of C_{60} and C_{70} in the mass spectra. Since we were just starting to work on clusters like C_3, C_4, C_5 up to perhaps C_9, it became

clear that we had only reached the first islands of the huge continent of carbon molecules.

In the fall of 1985 we read with interest about the C_{60} (Buckminsterfullerene) discovery by Harry Kroto, Richard Smalley, and co-workers[13], and about the soccerball proposal for its molecular structure. Many theorists began to predict (among other properties) its ultraviolet, infrared, and raman spectra. A remarkable result was that—for symmetry reasons—the infrared spectrum of soccerball C_{60} should exhibit only four absorptions[14], two grouped at around $1300\ cm^{-1}$, and two more grouped at around $600\ cm^{-1}$.

One of us (DRH) began to consider that our camel hump carrier might possibly be due to the elusive Buckminsterfullerene. In 1987 a patent disclosure was even submitted to appropriate channels of the University of Arizona, but was withdrawn when the production technique proved to be too unreliable to reproduce regularly. During an IAU symposium on interstellar dust held in Santa Clara, California, in the summer of 1988, the two of us discussed the evidence for having C_{60} in our soot. In case the camel hump carrier is not "junk," what else could it be other than C_{60}? Although intriguing, the evidence was rather circumstantial. The main argument in favour of C_{60} was that the UV spectrum theoretically predicted by the group of Arne Rosén[15] was very similar to our camel hump pattern. The counter-argument, however, appeared quite substantial. Depletion spectroscopy of van der Waals complexes of C_{60} and other small molecules seemed to show that the free Buckminsterfullerene would exhibit only one weak absorption band located at 386 nm[16]. This result misled us for quite a while. It turned out later that the C_{60} molecule in fact has an absorption at 386 nm; however, there are much stronger lines in the UV region of the spectrum, which apparently remain undetected by this kind of spectroscopy.

In Santa Clara we both agreed to start new efforts in producing and investigating the camel hump soot. With the help of Bernd Wagner (in Heidelberg) and Lowell D. Lamb (in Tucson), it was found that the helium pressure is the key parameter for producing camel samples. Between 100 and 200 torr of helium the best yields were obtained, and we continued to use such pressures in future work.

We in Heidelberg had purchased a very sensitive infrared Fourier-transform spectrometer to calibrate filters for the ISO (Infrared Space Observatory) project. The camel samples were soon studied and, to our surprise, showed besides an intense continuum of regular soot, four stronger line absorptions very close to the positions predicted by theory for soccerball C_{60} (see Fig. 2). There was a lot of excitement, since this would imply that our soot contained C_{60} in considerable quantities. We estimated a concentration in the order of percent. However, the euphoria soon was followed by considerable concern. We found that our diffusion pump oil has a rather intense absorption line at around $1430\ cm^{-1}$, rather close to one of the four soot features. We thus still could not rule out that the IR lines of the camel carrier was "junk," as we had believed before.

4. SUCCESS

As part of his thesis work, Konstantinos Fostiropoulos in Heidelberg investigated the effects of possible soot contaminations and employed much cleaner conditions and procedures than we had used before. The appearance of the camel hump was not affected. With slightly renewed confidence a paper entitled "Search for the UV and IR Spectra of C_{60} in Laboratory Produced Carbon Dust"[17] was submitted and presented at a workshop on interstellar matter, which took place in September, 1989, in Capri, Italy. We later learned that Mike Jura, who attended the workshop, sent our paper directly to Harry Kroto with the written remark, "Do you believe this?".

In a final attack on the problem "junk" or not, we in Heidelberg succeeded in producing carbon rods sintered from isotopically enriched (99% ^{13}C) powder. In order not to dilute the ^{13}C with regular carbon, no glue or binder was used. The rods were successfully evaporated and the valuable smoke collected. In the IR spectrum of the soot, we observed precisely the lineshifts expected for pure carbon, while the UV spectrum remained unchanged. The conclusion was now clear: The carrier of the mysterious UV and IR lines had to be a pure carbon molecule (i.e., not "junk"). This and some additional results were writ-

ten up and submitted for publication in Chemical Physics Letters[18]. When the work was published by the beginning of July, we already had in hand the ultimate product of our research: crystalline C_{60}.

In the early months of 1990 work had continued in both Heidelberg and Tucson, with steady contact by mail, FAX, and phone. There was now more urgency to our work, since the Sussex group had succeeded in reproducing the camel soot and confirmed that it exhibited four IR lines. We got into contact with W. Schmidt, a chemist we knew from his work on interstellar dust and polycyclic aromatic hydrocarbon (PAH) molecules. Following his suggestions, in May 1990 we achieved two important successes leading to a separation of the C_{60} from the soot. First, there was the discovery that heating of the soot to about 500°C in vacuum or in an inert atmosphere would result in sublimed coatings with all the absorption features of the camel samples, but with no background of the soot continuum discernable. It was further observed that the coatings dissolved in benzene, toluene, and various other nonpolar solvents. This finding naturally led to the method of separating soot and fullerenes by simply washing the fullerenes out of the insoluble soot. A drop of the concentrated solution, dried on a microscope slide, provided the first view of the crystals of this new form of carbon, as shown in Fig. 3. The thin flake-like crystals were ideally suited for electron diffraction studies, and the powder could be subjected to x-ray diffraction analysis. Mass spectra confirmed that the separated ma-

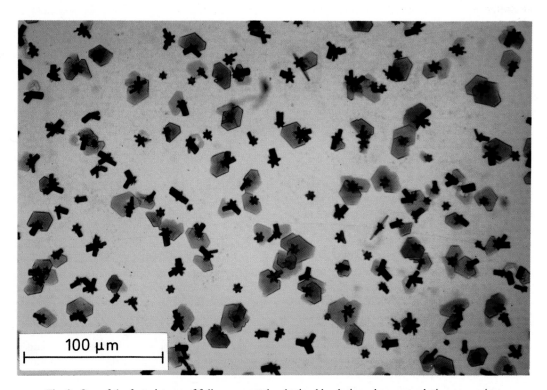

Fig. 3. One of the first pictures of fullerene crystals, obtained by drying a benzene solution on a microscope slide. The picture was taken in transmitted light. The crystals consist of C_{60} with a small admixture of C_{70} and traces of solvent molecules. We suggested calling such solids "fullerites."

terial consisted of C_{60} with a few percent C_{70}. Although we were a bit confused over whether to index the diffraction pattern as fcc or hcp, it was clear that the molecules form a lattice according to the close packing of spheres and with a nearest neighbor distance of about 1 nm. This is the distance expected for the soccerball shaped C_{60}, when a reasonable extension of the pi-electron cloud is taken into account.

By this time it was clear that we had produced, for the first time, a brand new crystalline form of pure carbon, and that the molecular crystals are composed of a close packing of the long-sought spherical molecule C_{60}, Buckminsterfullerene. For this reason we suggested the special term "fullerites" to name such solids.

In addition, we had provided a method for making copious supplies of this molecule, discovered in 1985, but never before made in sufficient quantity to see. In the work following our first studies, many other groups used the new material, and soon it became proved beyond any doubt that the proposal by Kroto, Smalley, and co-workers of a soccerball-shaped C_{60} molecule was correct.

At the time of this writing, hundreds of papers have been published reporting measurements made using C_{60} and C_{70}, and the other "magic" fullerenes prepared and extracted by the method we discovered. Through the efforts of many investigators, it is now becoming clear that these fullerenes, extracted with benzene or toluene, constitute only a fraction of the fullerenes apparently produced in the smoke-making process. The soot generated by evaporating carbon in an atmosphere of helium, which had already begun in the early 1970s and which the authors refined during their collaborative efforts in 1982 and in the following years, now seems to contain a wondrous concoction of closed-cage molecules—round, elongated, tube-like, helical, nested, to name some of their known features. In our view, the challenge is to work out how to extract, separate, and crystallize all these new molecules, determine their properties, and finally find uses for them.

REFERENCES

1. W. Krätschmer, L. D. Lamb, K. Fostiropoulos, and D. R. Huffman, *Nature* **347**, 354–358 (1990).
2. D. R. Huffman and W. Krätschmer, In *Symp. Proc.* (Edited by R. S. Averback, J. Bernholc, and D. L. Nelson), Vol. *206*, p. 601 Mater. Res. Soc., Pittsburgh, PA (1991).
3. W. Krätschmer, *Z. Phys. D.,* **19**, 405 (1991).
4. D. R. Huffman, *Physics Today,* p. 22 (November 1991).
5. D. R. Huffman, *Adv. Phys.* **26**, 129 (1977).
6. G. H. Herbig, *Astrophys. J.* **196**, 129 (1975).
7. T. P. Stecher, *Astrophys. J.* **157**, L125 (1969).
8. K. L. Day and D. R. Huffman, *Nature, Phys. Sci.,* **243**, 50 (1973).
9. C. F. Bohren and D. R. Huffman, *Absorption and Scattering of Light by Small Particles.* John Wiley & Sons (1983).
10. W. Krätschmer and D. R. Huffman, *Astrophysics and Space Science,* **61**, 195 (1979).
11. W. Krätschmer, N. Sorg, and D. R. Huffman, *Surface Science* **156**, 814 (1985).
12. E. A. Rohlfing, D. M. Cox, and A. Kaldor, *J. Chem. Phys.* **81**, 3322 (1984).
13. H. W. Kroto, J. R. Heath, S. C. O'Brien, R. F. Curl, and R. E. Smalley, *Nature* **318**, 162–163 (1985).
14. D. E. Weeks and W. G. Harter, *J. Chem. Phys.* **90**, 4744 (1989).
15. S. Larsson, A. Volosov, and A. Rosén, *Chem. Phys. Lett.* **137**, 501 (1987).
16. J. R. Heath, R. F. Curl, and R. E. Smalley, *J. Chem. Phys.* **87**, 4236 (1987).
17. W. Krätschmer, K. Fostiropoulos, D. R. Huffman, In *Dusty Objects in the Universe* (Edited by E. Bussoletti and A. A. Vittone), pp. 89–93. Norwell: Kluwer Academic Publishers (1990).
18. W. Krätschmer, K. Fostiropoulos, and D. R. Huffman, *Chem. Phys. Lett.* **170**, 167 (1990).

PRE-1990 EVIDENCE FOR THE FULLERENE PROPOSAL

R. F. CURL

Chemistry Department and Rice Quantum Institute, Rice University, Houston, TX 77251, U.S.A.

(*Accepted in final form* 28 *April* 1992)

Abstract—During the period 1985 to 1990, the closed cage carbon molecules known as the fullerenes could be observed only in the mass spectrometer. Through a number of mass spectrometric and theoretical investigations, a considerable amount of evidence was obtained supporting the hypothesis that closed cage carbon clusters are formed spontaneously in condensing carbon vapor, with the most stable one being the highly symmetrical, truncated icosahedron C_{60} molecule. From these studies a coherent description of the structure and properties of the fullerenes, especially C_{60} and C_{70}, emerged which has been validated and greatly extended by investigations by many workers on macroscopic samples of these materials during the last two years.

Key Words—Fullerenes, metalation, photofragmentation, mass spectroscopy.

1. INTRODUCTION

Almost five years elapsed between our proposal[1] in 1985 that the highly symmetric, closed cage carbon molecule C_{60} is formed spontaneously in condensing carbon vapor and the isolation of macroscopic samples of this molecule by Krätschmer *et al.*[2] in 1990. During that period, C_{60} and the other fullerenes were known and investigated experimentally only as species in a mass spectrometer. The purpose of this article is to review, incompletely, mass spectrometer experiments probing of the carbon clusters between 1985 and 1990. As these experiments progressed, an almost incredible picture emerged: out of the chaos of carbon vapor condensing at high temperature, not only C_{60}, but a whole family of closed cage carbon molecules, the fullerenes, are synthesized. Each new experiment fit beautifully into the picture revealing some new aspect until by 1990 it was almost impossible to doubt this fullerene hypothesis.

2. LASER VAPORIZATION CLUSTER SUPERSONIC BEAM MASS SPECTROMETRY

These experiments are based upon the principle that any material no matter how refractory can be vaporized when placed at the focus of an intense laser pulse such as the $10–100^+$ mJ energy, 5-ns duration, frequency-doubled output of a Q-switched Nd:YAG laser. The vaporized material is entrained into an inert gas pulse flowing over the surface during vaporization as shown in Fig. 1. The hot plasma jet of refractory material is quenched by mixing with the gas stream and condenses back into clusters containing from a few to several hundred atoms. Such clustering is permitted to take place for 30–200 μs, then the gas stream undergoes supersonic expansion into a vacuum terminating further clustering and cooling the clusters to a few degrees Kelvin. The supersonic jet is skimmed into a molecular beam and probed by mass spectrometry. Either residual positive or negative

ions from the vaporization plasma may be observed, or neutrals in the beam can be UV excimer laser ionized and the resulting positive ions observed. Four different kinds of mass spectrometry experiments carried out on fullerenes are discussed in this brief review.

3. DISCOVERY OF C_{60} AND THE FULLERENES

In the simplest experiment, cluster ions were pulse extracted at a right angle from the molecular beam and observed by time-of-flight mass spectrometry. Rohlfing *et al.*[3] first observed the distribution of carbon cluster sizes formed by laser vaporization of graphite in this way in 1984 obtaining the spectrum shown in Fig. 2. As can be seen it appears to consist of two distributions: one in which both odd and even clusters are represented ending at about 20 atoms, and another beginning at about 40 atoms consisting of only even clusters. We now know that the high mass, even clusters are the closed cage fullerenes.

This distribution where C_{60} is only slightly more prominent than its neighbors is characteristic of that obtained when the clustering process is quenched at a fairly early stage. When clustering is allowed to proceed to a greater extent, a distribution is obtained[1] with C_{60} far more prominent than its neighbors as shown in Fig. 3. A closed cage, highly symmetric truncated icosahedron structure was proposed[1] as a unique structure explaining the prominence of C_{60} in Fig. 3, and the molecule was named "buckminsterfullerene" as the geodesic domes of R. Buckminster Fuller guided the search for a closed structure. This structure was not proposed merely because it is highly symmetric and therefore quite beautiful. In Fig. 4, its chemical structure is depicted. To chemists, this structure seems uniquely satisfactory: all the carbon valences are satisfied, the six-membered rings are stabilized by aromaticity, and the strains of bending the polyaromatic system out of planarity are uniformly

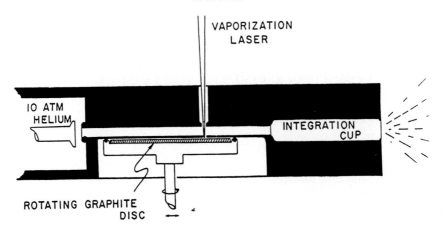

Fig. 1. Laser vaporization disk source. The sequence of events is that a pulsed valve opens flowing He gas over the surface. Near the middle of the pulse, the laser pulse strikes the surface vaporizing material to atoms and ions. The material is entrained in the He carrier gas and clusters condense. After clustering, the gas mixture undergoes supersonic expansion into a vacuum with resultant cooling to a few K. Reprinted by permission for *Nature* vol. 318, pp. 162–163. Copyright © 1985 MacMillan Magazines Limited.

distributed over the surface. Moreover, many Kekulé structures exist for this molecule (Klein *et al.*[4] determined that there are 12,500) offering the promise of further resonance stabilization.

Such a molecule should be chemically stable and relatively unreactive. To test this, in a second kind of experiment[5], a reaction tube was added to the end of the clustering chamber and reagents such as NO, H_2, CO, SO_2, O_2, and NH_3 were injected into the gas stream. When this was done, no effect on the C_{60} mass peak which could be attributed to chemical reaction was observed, establishing that C_{60} is chemically rather inert as expected from the structure.

Symmetry, chemical inertness, and the existence of many resonance forms are not sufficient to establish the correctness of the truncated icosahedron

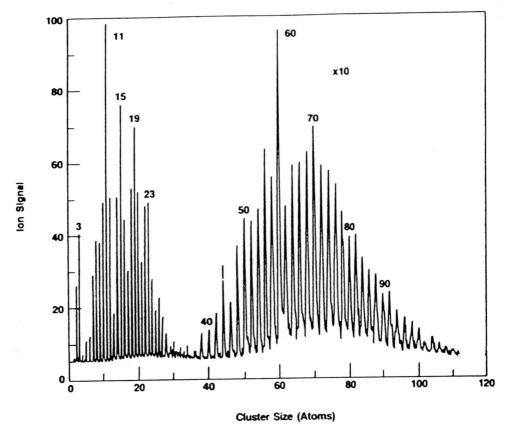

Fig. 2. Distribution of C_n^+ ions obtained with mild clustering conditions. Reprinted with permission from Rohlfing *et al.*[3]. Copyright American Institute of Physics.

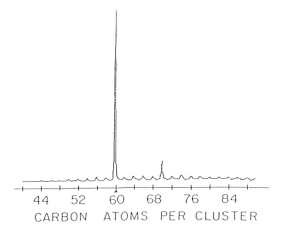

Fig. 3. Distribution of Cn^+ ions obtained under extreme clustering conditions. Reprinted by permission for *Nature* vol. 318, pp. 162–163; copyright © 1985 MacMillan Magazines Limited.

buckminsterfullerene structure. Haymet[6] pointed out that there is more than one closed cage structure for C_{60} satisfying Euler's rule[7] even if the structures are restricted to those containing five- and six-membered rings. He compared Hückel calculations on the truncated icosahedron with a D_{6h} symmetry cage he termed "graphitene" which has even more Kekulé structures (Schmalz *et al.*[8] determined that there are 12,688) than buckminsterfullerene. However, the Hückel stabilization energy of buckminsterfullerene is marginally greater than that of graphitene[6], and Haymet noted that the connected pentagonal rings of graphitene introduce more strain. Schmalz *et al.*[8] used both Hückel calculations and Herndon-Simpson (conjugated circuit) theory to investigate the relative stability of buckminsterfullerene, graphitene, and three other closed cage structures for C_{60}, and concluded that buckminsterfullerene is the most stable isomer. They noted that abutting pentagons raise

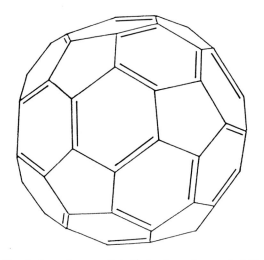

Fig. 4. Lowest energy Kekulé structure of soccer ball C_{60}. Reprinted by permission for *Nature* vol. 329, pp. 529–531; copyright © 1987 MacMillan Magazines Limited.

the π-energy because destabilizing conjugated 8-circuits can occur. This prescription that abutting pentagons are to be avoided has become known as the "pentagon rule."

Remarkably, not only C_{60}, but all the even numbered carbon clusters in the 40–80 carbon atom range were resistant to chemical attack[5]. According to Euler's rule for any C_n cluster with n even and greater than 22, at least one (and for n large many different) closed hollow cages consisting of 12 pentagons and $(n\text{-}20)/2$ hexagons can be constructed. Such closed cages will be aromatic molecules without dangling chemical bonds, and therefore are expected to be chemically rather inert as observed. Therefore, it was proposed[5] that the entire high mass distribution of even numbered carbon clusters consists of these closed cage "fullerene" molecules. Such hollow carbon cage molecules (now known as fullerenes) were originally conceived by Jones[9]; the truncated icosahedron buckminsterfullerene C_{60} appears to have been originally conceived by Osawa[10]. However, no one ever imagined that such molecules would form spontaneously in condensing carbon vapor.

What structures should be preferred for these other cage molecules? In particular, what is the structure of the next most prominent mass spectrometer species C_{70}? An egg-shaped structure based on modifying C_{60} by introducing a band of hexagons was proposed[11]. Schmalz *et al.*[12] found "upon making a diligent search" that this C_{70} structure is the next largest after the buckminsterfullerene structure that avoids abutting pentagons. Since adjacent pentagons are destabilizing[8], this provides an explanation for why C_{70} should be the next prominent mass peak. For clusters with less than 60 atoms, it is impossible to avoid adjacent pentagons. On the basis that the number of abutting pentagons should be minimized, Kroto[13] provided a nice explanation of the special prominence of C_{50} in some mass spectra, and predicted special stability for 24, 28, 32, and 36 atom carbon clusters.

This emerging picture was self-consistent and reasonable. It explained both the chemical inertness of these carbon clusters and through the pentagon rule the special prominence of C_{60}, C_{70}, and C_{50}. However, it did not constitute proof of the buckminsterfullerene structure for C_{60} or of the fullerene hypothesis. A direct determination of structure by rotationally resolved high resolution spectroscopy or by diffraction methods would have been proof of structure. Unfortunately, direct structural investigation by mass spectrometric cluster beam methods is mostly limited to small clusters of a two to four atoms using rotationally resolved resonant two photon ionization spectroscopy. This technique is very difficult to apply to clusters of 40+ atoms and has not at this date yet been applied to C_{60}. However, in science, proof of any hypothesis is elusive; usually a hypothesis is accepted when it explains so many observations that it is more difficult to work without it than it is to believe it. Thus, in search of this sort of "proof" or perhaps contradiction, a series of experiments were carried out. The results of these experiments provided indirect but compelling evidence for the correctness of the fullerene picture.

4. ENDOHEDRAL METAL COMPLEXES

Buckminsterfullerene C_{60} should be hollow with ample space to contain an atom encapsulated by the cage. In an effort to make such species, a disk of low density graphite was soaked in a water solution of $LaCl_3$, dried, and used as a laser target. The resulting[11] mass distribution under two sets of conditions is shown in Fig. 5. The lower trace is taken with a low 193-nm ionizing fluence which should produce little fragmentation upon ionization. This spectrum is poorly resolved with an unresolved background probably caused by clusters containing Cl atoms. In the upper panel, the ionizing laser has been turned up. This higher fluence should have caused extensive photofragmentation with only the most stable ions surviving. As can be seen, this mass spectrum is much cleaner. It shows a prominent $C_{60} \cdot La^+$ peak but no $C_{60} \cdot La_2^+$ peak. Both the survival of $C_{60} \cdot La^+$ and the absence of $C_{60} \cdot La_2^+$ are strong indicators that the

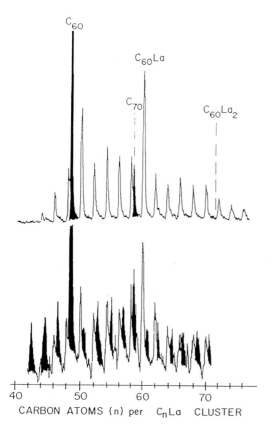

Fig. 5. Time-of-flight mass spectrum of C_nLa complexes (open peaks) and C_n clusters (blackened peaks) produced by laser vaporization of a $LaCl_3$ impregnated graphite disk in a pulsed supersonic nozzle. In the bottom trace, the fluence of the ionizing ArF excimer laser is <0.01 mJ/cm^2; in the upper trace, the ArF fluence has been turned up to 1–2 mJ/cm^2 Copyright 1985 American Chemical Society. Used with permission.

$C60 \cdot La^+$ ion seen in this spectrum has the La atom inside the cage ("endohedral"). Note that several peaks with only one La atom attached can be seen with $C_{60} \cdot La^+$ being the most prominent as would be expected because the C_{60} cage is especially stable. All these ions have remarkable photoresistivity. It is now believed that they are all endohedral La complexes. Recently[14], the notation $La@C_{60}$ has been proposed to indicate such endohedral complexes.

5. PHOTOFRAGMENTATION

The photofragmentation of the C_n^+ ions was explored[15] in a third kind of mass spectrometric experiment using a tandem time-of-flight mass spectrometer. This apparatus is shown in Fig. 6. In brief, a mass gate has been added at the end of the drift region of the first time-of-flight permitting the selection of a single mass ion for further interrogation by a photofragmentation laser. About two to three microseconds after excitation by this laser, the fragment ions are accelerated into the second time-of-flight drift region giving the mass spectrum of the fragment ions. The fragmentation of several even C_n^+ ions with $n \geq 60$ was explored.

High resistivity to photofragmentation is to be expected for these spheroidal carbon shell molecules whether empty or endohedral complexes. The fluence dependence of the fragmentation of buckminsterfullerene was given special attention because this ion should be especially resistant to photofragmentation since the species has an especially stable structure and all the carbon atoms are equivalent; thus there is no weak point in the cage. In these experiments, C_{60}^+ indeed proved to be highly photoresistive. Our best estimate is from studying the fluence dependence of the fragmentation was that at least 19 eV of energy is required for significant fragmentation of C_{60}^+ in 3 μs.

Figure 7 shows the pattern of fragment ions produced by C_{60}^+ at several fluences of 266-nm radiation. Similar patterns exhibiting loss of an even number of carbon atoms were found when C_{68}^+, C_{74}^+, and C_{80}^+ were photofragmented. Note in Fig. 7 that only fragment ions containing an even number of carbon atoms are observed down to C_{32}^+ where the pattern abruptly breaks off. This pattern was simply interpreted in a way that provided strong support for the cage hypothesis. When C_{60}^+ fragmented by loss of an even number of carbon atoms, the process occurred by a mechanism which left a closed cage ion product. Figure 8 gives some simple pictures showing how this can happen. As the cage shrinks by fragmentation, the steric strains in these even smaller fullerenes increase until at C_{32}^+ the cage ruptures.

6. "SHRINK-WRAPPING" METAL CONTAINING C_{60}

These observations led to a series of photofragmentation experiments[16] on $C_{60}K^+$ and $C_{60}Cs^+$. In

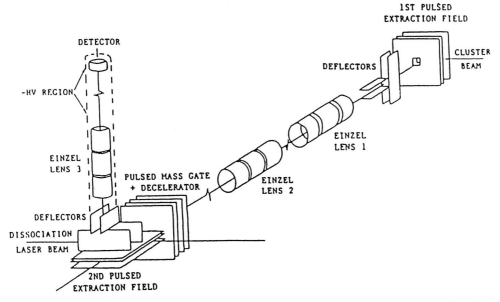

Fig. 6. Overview of the tandem time-of-flight mass spectrometer. The cluster ions are extracted from the molecular beam by the first pulsed extraction field and the selected ion is chosen by the mass gate, decelerated, and probed by the dissociation laser beam. The resulting photofragment ions are extracted by the second pulsed extraction field and mass analyzed by time of flight. Copyright American Institute of Physics. Used with permission.

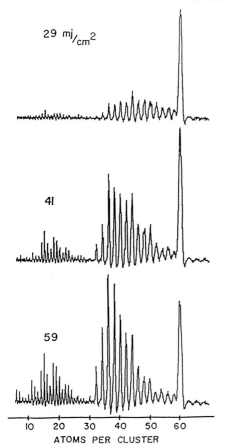

Fig. 7. Mass spectra of photofragment ions from C_{60}^+ observed with three different fluences of 353-nm pulses. The detection is optimized for fragment ions near C_{30}^+. The abrupt change in fragment pattern at C_{32}^+ indicates that it is the last stable cage ion. Copyright American Institute of Physics. Used with permission.

this fourth kind of mass spectrometer experiment, a supersonic beam of cluster ions is prepared as described previously, but instead of mass analyzing by time of flight, the beam is injected into a fourier transform ion cyclotron resonance cell. The ions can be trapped in the cell for several minutes and their cyclotron resonance frequencies driven by an applied radio frequency field. By using an appropriate range of rf frequencies, all ions but those in a small mass range can have their cyclotron orbits driven out of the observation region. The remaining ions can be subjected to a sequence of laser pulses and the masses of the resulting photofragment ions observed by cyclotron resonance. In the investigations of present interest, the ions $C_{60}K^+$ and $C_{60}Cs^+$ were selected and studied.

Figure 9 compares the mass spectra of the resulting fragment ions. The separation of $C_{60}K^+$ from C_{64}^+ and of $C_{60}Cs^+$ from C_{72}^+ is incomplete so that the peaks marked as C_n^+ are to be ignored. (If fragmentation of the metal containing species into a bare carbon cluster and the alkali metal occurs, the positively charged species should be the metal ion.) The first thing to note is that fragmentation of $C_{60}M^+$ is proceeding by loss of an even number of carbon atoms as described above for the bare clusters. This is strong evidence that the metal atom is inside the cage: loss of M^+ is to be expected in the unimolecular decay of a metal carbon complex with the metal stuck on the outside because the M^+–C_n bond is the weakest.

When the cage contains a metal atom, the strain on a small cage should be much greater because of the repulsive force between the hard core of the metal ion electrons and the carbon atom cores. Thus it is expected that cluster size where the apparent C_2 loss

16

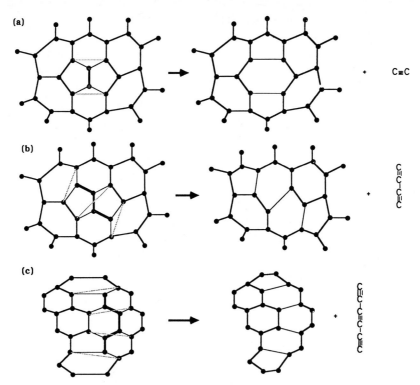

Fig. 8. Possible mechanisms for fragmentation of a spheroidal carbon shell to lose (a) C_2, (b) C_4, or (c) C_6 while maintaining the integrity of the cage. The individual bond centers may undergo a concerted process as the carbon chain is unzipped from the cage. Copyright American Institute of Physics. Used with permission.

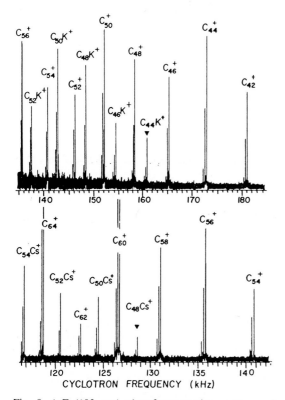

Fig. 9. ArF (193 nm) photofragmentation patterns of $C_{60}K^+$ (top) and $C_{60}Cs^+$ (bottom) detected by FT–ICR. The C_nK^+ fragmentation breaks off at $C_{44}K^+$ while the C_nCs^+ fragmentation breaks off at $C_{48}Cs^+$. The bare C_n^+ clusters seen in this spectra are fragments from C_{64}^+ in the potassium case and fragments from C_{72}^+ in the cesium case and are present because the prephotolysis mass selection is incomplete. Copyright 1988 American Chemical Society. Used with permission.

fragmentation breaks off should be larger when the cage contains a metal than when it is bare and that this breakoff point should depend upon the size of the contained metal ion occurring at larger sizes for larger metals. The carbon cage is being "shrink-wrapped" around the metal ion by photofragmentation until it bursts. This is precisely what is seen in Fig. 9. Fragmentation breaks off at $C_{44}K^+$ and at $C_{48}Cs^+$ as compared with C_{32}^+ for the bare clusters. These breakoff points agree well with a simple model using the van der Waals radii of carbon and the metal species involved. In our minds, these experiments provided definitive proof that these carbon clusters containing more than about forty carbon atoms are spheroidal carbon shells (the fullerenes) and that a metal atom could be introduced into the spheroidal cavity they possess.

7. PRE-1990 MASS SPECTROMETRY AND RECENT ADVANCES

Since macroscopic samples of C_{60} and C_{70} became available in September 1990[2], many advances have been made. The structures proposed for C_{60} and C_{70} have been verified in a variety of ways. Macroscopic amounts of materials containing endohedral metal fullerene complexes have been produced[14]. A totally unexpected development is the isolation of new stable carbon cages, such as C_{76}, C_{78}, and C_{84}, which did not appear to have special prominence in the mass spectrometer, but which have been found[17] in macroscopic samples. Pre-1990 mass spectrometer experiments showed[18] that C_{60} has a large electron affinity and the triply degenerate LUMO of C_{60} could contain up to six electrons so that a rich negative ion electrochemistry of C_{60} was expected[19] before it was observed[20], and the existence of buckide salts of $C_{60}{}^{n-}$ with extrahedral metals was anticipated. However, the observation[21] that some such salts become superconducting at relatively high temperature could not have been predicted. At this moment, research on fullerene systems is flourishing with exciting developments and new surprises almost daily.

Acknowledgement—This work was supported by the Robert A. Welch Foundation.

REFERENCES

1. H. W. Kroto, J. R. Heath, S. C. O'Brien, R. F. Curl, and R. E. Smalley, *Nature* **318**, 162–163 (1985).
2. W. Krätschmer, L. D. Lamb, K. Fostiropoulos, and D. R. Huffman, *Nature* **347**, 354–358 (1990).
3. E. A. Rohlfing, D. M. Cox, and A. Kaldor, *J. Chem. Phys.* **81**, 3322–3330 (1984).
4. D. J. Klein, T. G. Schmalz, T. G. Hite, and W. A. Seitz, *J. Am. Chem. Soc.* **108**, 1301–1302 (1986).
5. Q. L. Zhang, S. C. O'Brien, J. R. Heath, Y. Liu, R. F. Curl, H. W. Kroto, and R. E. Smalley, *J. Phys. Chem.* **90**, 525–528 (1986).
6. A. D. J. Haymet, *J. Am. Chem. Soc.* **108**, 319–321 (1986).
7. L. Euler, *Elementa Doctrinae Solidorum, Novi corumentarii academie Petropolitanae* **4**, 109 (1752/3) (1758).
8. T. G. Schmalz, W. A. Seitz, D. J. Klein, and G. E. Hite, *Chem. Phys. Lett.* **130**, 203–207 (1986).
9. D. E. H. Jones, *New Scientist* **32**, 245 (1966). D. E. H. Jones, *The Inventions of Daedalus,* pp. 118–119. W. H. Freeman, Oxford and San Francisco (1982).
10. E. Osawa, *Kagaku* (Kyoto) **25**, 854–863 (1970).
11. J. R. Heath, S. C. O'Brien, Q. Zhang, Y. Liu, R. F. Curl, H. W. Kroto, and R. E. Smalley, *J. Am. Chem. Soc.* **107**, 7779–7780 (1985).
12. T. G. Schmalz, W. A. Seitz, D. J. Klein and G. E. Hite, *J. Am. Chem. Soc.* **110**, 1113–1127 (1988).
13. H. W. Kroto, *Nature* **329**, 529–531 (1987).
14. Y. Chai, T. Guo, C. Jin, R. E. Haufler, L. P. F. Chibante, J. Fure, L. Wang, J. M. Alford, and R. E. Smalley, *J. Phys. Chem.* **95**, 7564–7568 (1991).
15. S. C. O'Brien, J. R. Heath, R. F. Curl, and R. E. Smalley, *J. Chem. Phys.* **88**, 220–230 (1988).
16. F. D. Weiss, S. C. O'Brien, J. L. Elkind, R. F. Curl, and R. E. Smalley, *J. Am. Chem. Soc.* **110**, 4464 (1988).
17. F. Diederich, R. Ettl, Y. Rubin, R. L. Whetten, R. Beck, M. Alvarez, S. Anz, D. Sensharma, F. Wudl, K. C. Khemani, and A. Koch, *Science* **252**, 548–551 (1991).
18. S. Yang, C. L. Pettiette, J. Conceicao, O. Cheshnovsky, and R. E. Smalley, *Chem. Phys. Lett.* **139**, 233–238 (1987).
19. R. C. Haddon, L. E. Brus, and K. Ragavachari, *Chem. Phys. Lett.* **125**, 459–464 (1986).
20. R. E. Haufler, J. Conceicao, L. P. F. Chibante, Y. Chai, N. E. Byrne, S. Flanagan, M. M. Haley, S. C. O'Brien, C. Pan, Z. Xiao, W. E. Billups, M. A. Ciufolini, R. H. Hauge, J. L. Margrave, L. J. Wilson, R. F. Curl, and R. E. Smalley, *J. Phys. Chem.* **94**, 8634–8636 (1990).
21. A. F. Hebard, M. J. Rosseinsky, R. C. Haddon, D. W. Murphy, S. H. Glarum, T. T. M. Palstra, A. P. Ramirez, and A. R. Kortan, *Nature* **350**, 600 (1991).

MOLECULAR BEAM STUDIES OF FULLERENES

Eleanor E. B. Campbell and Ingolf V. Hertel

Fakultät für Physik and Freiburger Materialforschungszentrum, Albert-Ludwigs Universität, Hermann-Herder-Str. 3, W-7800 Freiburg, Germany

(*Accepted* 1 *June* 1992)

Abstract—A number of molecular beam experiments with the fullerenes are described. Such experiments allow a detailed probing of the interactions between fullerenes and other species and among the fullerenes themselves. Emphasis is placed on work involving photophysics and collisional interactions of the fullerenes.

Key Words—Fullerenes, C_{60}, molecular beams, collisions, photophysics.

1. INTRODUCTION

Molecular beam experiments have played an extremely important role in the discovery and investigations of C_{60}. The first experiments that showed evidence of the particular stability of Buckminsterfullerene were carried out in a molecular beam apparatus designed by Rick Smalley and his group at Rice University[1]. This group had developed a laser vaporisation source to produce beams of clusters (aggregates of up to a few hundred atoms) of nonvolatile materials like metals and semiconductors, which could be produced as ions (or as neutral species that were then photoionised) and detected by time-of-flight mass spectrometry. Harry Kroto, who visited Rice University in 1984, persuaded the group to investigate carbon clusters and, as they say, "the rest is history." A number of excellent accounts have been given of this discovery period[2–4], all written from a somewhat different perspective. The highly original and imaginative letter published by Kroto *et al.* in *Nature* in 1985[1] not only suggested the "soccer ball" structure and the name for the C_{60} molecule, but also predicted some of the exciting developments that we are now witnessing.

The molecular beam experiments on the fullerenes carried out in the pre-Krätschmer/Huffman era were dominated by the Rice University group, and have been described in various review articles, including the article by Robert Curl in this special issue[5]. In the next section we will only briefly mention these results when relevant to the discussion of results from other groups, and will concentrate on the kind of experiments carried out at Freiburg University in the period leading up to the discovery of the Krätschmer/Huffman fullerene production method[6].

In sections 3 to 6—the post-Krätschmer/Huffman period—we will also concentrate on work that especially interests us in the physics department at Freiburg. This falls into basically two categories: photophysics either with lasers or with synchrotron radiation, and collision experiments in the gas phase or at surfaces.

2. STABILITY OF THE FULLERENES: METASTABLE FRAGMENTATION OF "HOT" BEAMS

We became involved in fullerene research more or less by accident in 1989 while investigating UV laser ablation of polymers. A few years earlier Srinivasan had discovered that when pulsed, ultraviolet laser radiation falls on the surface of an organic polymer or biological tissue, the material at the surface is spontaneously etched away with no thermal damage caused to the substrate[7]. The products from this so-called "ablative photodecomposition" or "photoablation" are ejected from the surface at supersonic velocities. Many attractive applications for this technique were proposed, and are still being developed, mainly for photolithography in semiconductor processing and surgical applications[8]. We became interested in trying to understand the fundamental physics and chemistry of the ablation process by looking at the velocity and angular distributions of the charged and neutral products[9,10] by using a reflectron time-of-flight mass spectrometer. Polyimide (perhaps better known as "Kapton") was chosen to be the main object of our investigations. This polymer is technologically very important because of its mechanical, thermal, and electrical properties; but because it is resistant to chemical attack, it is extremely difficult to treat using standard "wet" etching techniques. To our great surprise (and delight) we discovered that, in the supersonic expansion of the small polymer fragments into the vacuum, aggregation was occuring to produce many larger species, including the fullerenes[11]. A typical time-of-flight mass spectrum of the positive ions produced from laser ablation of polyimide (C:H:O:N = 22:10:5:2) is shown in Fig. 1. Similar spectra are also obtained from other aromatic polymers[12]. This mass spectrum can be divided into three regions: the low mass range (top part of Fig. 1), which consists mainly of hydrocarbon chains and depends in detail on the structure of the polymer being ablated; the middle mass range (second from top in Fig. 1), where very interesting oscillatory structure is seen with minima in intensity for

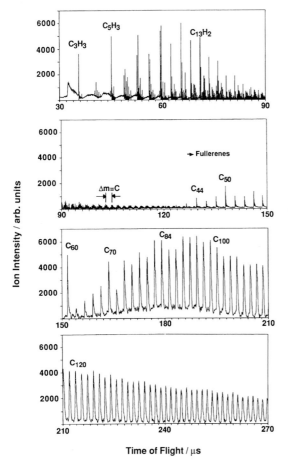

Fig. 1. Positive ion mass spectrum produced from 308 nm laser ablation of polyimide.

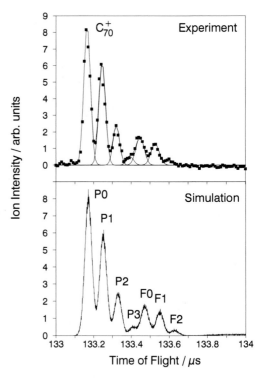

Fig. 2. Upper part: Isotope distribution of C_{70}^+ from 308 nm laser ablation of polyimide (from ref. [13]). Lower part: Simulated intensity distribution calculated using the natural isotope ratio for carbon (1.1% ^{13}C). Peaks P0 ($^{12}C_{60}$), P1 ($^{12}C_{59}\,^{13}C$), P2 ($^{12}C_{58}\,^{13}C_2$), and P3 ($^{12}C_5\,^{13}C_3$) are due to unfragmented clusters. Peaks F0–F2 due to metastable fragmentation of C_2 from C_{72}.

pure carbon species and maxima for $C_nH_{6-8}^+$, which has still to be understood; and finally the fullerenes with the characteristic C_2 mass difference between peaks, extending to C_{400}^+ and beyond (the limit of the mass spectrometer detection range). The surprising aspect for us was that a polymer containing four different atomic species could be such an efficient source of pure carbon "clusters." Proof that they are in fact pure carbon is obtained on looking more closely at the individual mass peaks. Figure 2 shows C_{70}^+ on a greatly expanded time scale to enable the isotope distribution to be resolved. Natural carbon has 1.1% ^{13}C so that it is a fairly easy exercise to calculate the percentage of C_{70} that will have no ^{13}C atoms, one ^{13}C atom, etc., using the binomial distribution. This has been done in the time-of-flight simulation in the lower part of Fig. 2 (peaks P0–P3), and gives excellent agreement with the measured distribution, showing that no other species can be present. Similar comparisons were carried out for all C_n^+ in the range n = 50 to 140, with the same result[13]. At this time in the fullerene story (late 1989, early 1990), there was still some controversy over the structure of the large carbon clusters, with some groups tending to be opposed to the idea of the fullerene "cages"[14,15]. Although on the side of the "believers" from the beginning, we finally became completely convinced of the fullerene hypothesis after analysing the mass spectra obtained from polyimide.

Figure 2 also shows some other peaks arriving slightly later than C_{70}^+, which can be identified as products from metastable fragmentation of hot C_{72}^+ in the field-free region of the mass spectrometer[13]. The carbon clusters produced by the ablation have large internal energies and cool by emitting C_2 molecules:

$$C_n^+ \rightarrow C_{n-2}^+ + C_2; \qquad (1)$$
$$C_{n-2}^+ \rightarrow C_{n-4}^+ + C_2 \text{ etc.}$$

The reason for loss of C_2 is a combination of the high stability of the even-numbered clusters and the relatively high binding of energy of C_2 (3.6 eV), which makes it energetically more favourable to lose the molecule than two individual carbon atoms. A mechanism for such a fragmentation process has been suggested by O'Brien et al.[16], and is shown schematically in Fig. 3. The first step is rearrangement of the carbon atoms to produce two adjacent pentagons followed by C_2 expulsion, keeping the number of pentagons constant, as needed for a closed shell structure, but reducing the number of hexagons by one. Obviously, since rearrangement has to take place and four carbon bonds need to be broken, one requires an activation energy for the fragmentation that is consid-

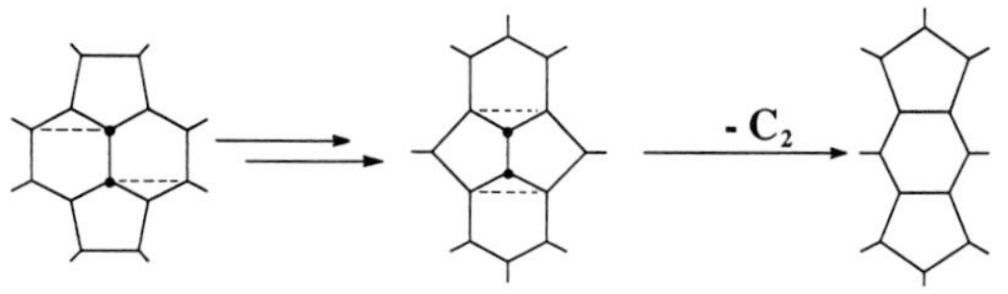

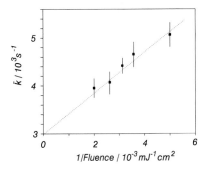

Fig. 3. Possible scheme for C_2 loss from the fullerenes.

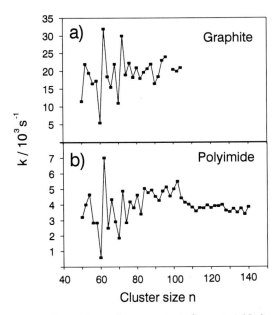

Fig. 5. Measured average fragmentation rate constant for $70 < n < 100$ as a function of ablating laser fluence for ions of the same average velocity obtained from laser ablation of polyimide (from ref. [11]).

erably larger than the fragmentation energy obtained from kinetic energy release measurements (4.6 eV for fragmentation of C_{60}^+ [17,18]).

By comparing the fragment peak intensity in the time-of-flight mass spectrum with that of the corresponding parent species, it is possible to obtain rate constants for the metastable fragmentation process[13]. Similar measurements have also been obtained by the Bowers group using laser ablation of graphite as the carbon cluster source and a double focusing, reverse geometry mass spectrometer for analysis of the metastable fragmentation[15]. A comparison of the two sets of rate constants for metastable C_2 loss is shown in Fig. 4. These data give a very clear picture of the relative stability of the different fullerenes as indicated by the variations in rate constant with size. The absolute values obtained were, however, very different, with the Bowers experiment producing fragmentation rates of about a factor of 5 higher than those from our polyimide source. This was originally attributed to very different source conditions, with much higher laser fluences required to produce carbon clusters from graphite compared with polyimide[13]. However, it has since been shown that this discrepancy is mainly due to interferences occurring in the double-focusing mass spectrometer[19,20], which led to the rate constants from the graphite experiment being overestimated by a factor of 2 to 3[21]. Even taking this into account, the fullerenes from the polyimide source are still "colder" than those from the graphite source (i.e., they still have somewhat smaller rate constants for fragmentation). This can be explained by considering that polyimide ablation also produces a large amount of small neutral molecules which, in the supersonic expansion from the polyimide surface[10] will act as a carrier gas, thus producing some cooling of the fullerenes. (The standard "Smalley" carbon cluster source using laser ablation of graphite also uses a pulsed He gas beam to produce cooling. This was not used in the Bowers experiments discussed above.) This cooling effect leads to the interesting observation, shown in Fig. 5, that if the metastable fragmentation of fullerenes with a particular average velocity is investigated as a function of the fluence of the ablating laser, the rate constant is seen to *decrease* as the laser fluence is *increased*! This is due to the increased gas pressure and thus more efficient cooling during the expansion from the polymer surface[13].

3. LASER DESORPTION AND AGGREGATION

When the Krätschmer/Huffman C_{60}/C_{70} extract became available, we started investigating the possibility of using UV laser desorption to produce intense molecular beams of these two fullerenes. We found that by using a fairly low laser fluence (about 20 mJcm^{-2}) to desorb the extract from a substrate in vacuum, we could produce positively charged beams directly without causing any substantial fragmentation of the fullerenes[22]. Figure 6(a) shows a typical time-of-flight mass spectrum taken under such conditions. However, as the laser fluence is increased some very interesting effects begin to emerge. At the intermediate laser fluence used in Fig. 6(b), the loss of neutral C_2 fragments from C_{60}^+ and C_{70}^+ is clearly seen, along with the formation of fullerenes up to C_{84}^+. At higher fluences (Fig. 6(c)), even the fullerenes beyond $n = 84$ are present in large quantities[22]. Here we are seeing strong evidence for the growth of fullerenes

Fig. 4. Comparison of rate constants for metastable loss of C_2 from hot fullerenes (a) values obtained in ref. [15] from graphite. (b) values obtained in ref.[13] from polyimide.

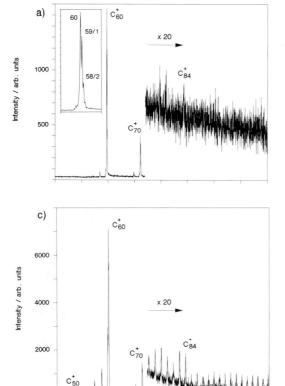

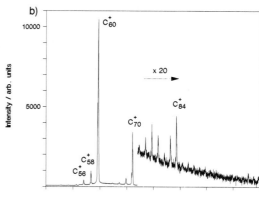

Fig. 6. Ions produced directly from desorption of C_{60}/C_{70} extract from a silicon substrate using 248 nm photons. Laser fluence (a) 20 mJcm^{-2} (b) 30 mJcm^{-2} (c) 90 mJcm^{-2}. The insert in (a) shows the isotope distribution of the C_{60}^+ peak, numbers x/y referring to $^{12}C_x{}^{13}C_{y+}$ (from ref. [22]).

due to aggregation of neutral C_2 onto C_{70}^+—the opposite of the fragmentation process discussed in section 2. Interestingly, this growth is only observed beyond C_{70}, a consequence of the considerably higher stability of C_{60}^+. The fullerenes seen between $n = 60$ and $n = 70$ are due solely to fragmentation of C_{70}^+. Evidence for growth from C_{60} has, however, been obtained in recent collision experiments between C^+ and C_{60} in which the highly unstable C_{61}^+ product was observed[23], and in recent results from our laboratory where evidence for "fusion" of C_{60} with C_{60}^+ to produce much larger fullerenes has been obtained[24]. In both these cases the relative kinetic energy of the neutral and charged species was much larger than that present between the neutral fragments and charged fullerenes in our laser desorbed beam.

4. PHOTOPHYSICS

The laser desorption source discussed above also produces intense beams of neutral C_{60}/C_{70}[22]. This allowed us to investigate the very interesting phenomenon of delayed, thermionic emission of electrons from neutral fullerenes on a timescale of microseconds[25]. The neutral beam was crossed by a second pulsed UV laser (308 nm = 4 eV), and the ions produced via photoionisation were extracted with a pulsed electric field and detected in the time-of-flight mass spectrometer. By changing the delay between the second laser and the pulsed extraction

field, ions produced *up to 15 μs after the second laser pulse had been switched off* could be detected. Figure 7 shows the detected C_{60}^+ ion intensity as a function of this delay time. The full line connecting the experimental points in Fig. 7 is a fit of the data to two exponential decays (on increasing the fluence of the second laser, a third exponential decay is needed to fit the data[26]. Later work showed that such delayed ionisation is only observed when the laser photon energy is too small to produce single photon ionisation[27]. Similar delayed ionisation effects have also been reported for the production of multiply charged "giant fullerenes" in the size range from 100 to 600 atoms[28], and for photodetachment of negatively

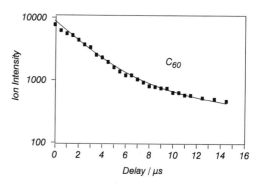

Fig. 7. Integrated C_{60}^+ peak intensity vs delay between ionising laser (20 mJcm^{-2}, 308 nm) and pulsed extraction field. The full line is a fit to two exponential decays with different lifetimes (from ref. [25]).

charged C_{60}^-[29]. In all cases it appeared reasonable to suppose that thermionic emission was responsible for the observations. When a photon is absorbed by the C_{60}, or other fullerenes, the energy is rapidly distributed among the vibrational degrees of freedom of the cluster. Consecutive absorption of many photons then leads to "superheating" of the cluster. When the cluster has sufficient internal energy, thermionic emission can occur on the timescale of the experiment, or for very hot clusters, thermionic emission accompanied by fragmentation. The experimental ionisation rate constants can be compared to a simple statistical model developed by Klots[30], which in the limit of large particles becomes the well known Richardson-Dushman equation used to describe thermionic emission from bulk metals. Excellent agreement with experiment is obtained if we assume that the neutral, laser-desorbed C_{60} has a temperature of about 3000 K (equivalent to absorbing 8 photons from the first laser) before absorbing any photons from the second laser[26]. The three different exponential decays seen in the experimental data, corresponding to three different rate constants for ionisation, can then be explained by the absorption of 2, 3, or 4 photons from the second laser, as shown by the comparison with the Klots' model (shaded area) in Fig. 8. (One can also consider a smaller initial temperature, which would then require the absorption of, for example, 4, 5, or 6 photons from the second laser. More work is in progress to obtain a more quantitative understanding of the mechanisms involved.) Further evidence for the thermal nature of the ionisation process has recently been obtained in experiments where C_{60} is "heated up" by a surface collision and is then seen to undergo delayed ionisation[31]. Many questions as to the exact nature of how and on what timescale the electronic (or collision) energy is transferred to vibrational energy (phonon bath), and how this is in turn coupled to the electron emission have still to be answered. A very recent experiment using sub-picosecond UV laser pulses to ionise

C_{60}[32] was able to show that on this short time scale the molecules are ionised directly by coherent two-photon absorption, with no observation of thermionic emission. There is not sufficient time available during the short laser pulse for the electronic energy to be converted to vibrational energy.

When using single photon ionisation the problems of energy transfer discussed above do not play such an important role. The photon excites an electron, which then leaves the molecule, taking the excess photon energy with it. For this reason, single photon ionisation of free fullerenes in molecular beams, using a source of UV light with a continuously variable wavelength (e.g., synchrotron radiation), is the most direct way of obtaining accurate ionisation potentials. Such measurements have been carried out in our group on beams of fullerenes, effusing from an oven at temperatures of approximately 500°C, using synchrotron radiation from the Berlin electron storage ring BESSY[33,34]. The ionisation potential obtained for C_{60} of 7.54 ± 0.04 eV, believed to be the adiabatic value[33], is in good agreement with results from charge transfer bracketing experiments (7.61 ± 0.11 eV)[35] and from photoelectron spectroscopy (7.61 ± 0.02 eV)[36]. However, the value obtained for C_{70} of 7.3 ± 0.2 eV was considerably lower than the charge transfer bracketing measurements (7.61 ± 0.11 eV). When the synchrotron radiation is scanned over energies above the ionisation potential, very interesting structure is observed in the photoion yield[37]. Figure 9 shows the C_{60}^+ ion yield as a function of photon energy. There is an extremely strong rise beyond the threshold towards a resonance maximum centered at around 20 eV, with a total width of about 11.5 eV (a very similar signal is obtained for C_{70}). This peak is attributed to a giant plasmon resonance which arises from collective motion of the valence electrons in the fullerenes[37]. Exactly such a resonance phenomenon was predicted theoretically by Bertsch et al.[38] using linear response theory for the determination of the optical absorption spectrum, and taking into account the electrostatic interaction of all 240 valence electrons. The

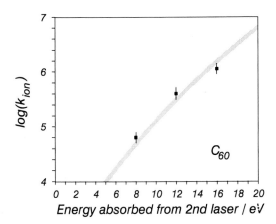

Fig. 8. Comparison of the measured ionisation rate constants with the Klots theory [30]. The shaded area gives the model prediction, taking into account the uncertainties in the ionisation potentials (from ref. [26]).

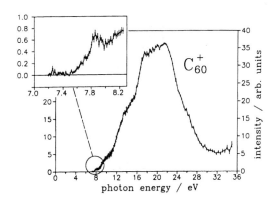

Fig. 9. C_{60}^+ photoion yield as a function of photon energy displaying excitation of the giant plasmon resonance. Inset: The threshold region on a magnified scale (from ref. [37]).

calculations predicted a resonance peak at around 22 eV, which arises from collecting the strength of the individual one-electron transitions into a single collective excitation with a total integrated oscillator strength of 95, compared to a maximum of 240 if all valence electrons are active.

5. COLLISIONS IN THE GAS PHASE

Collision experiments (hand in hand with spectroscopy) have made substantial contributions over the years in all areas of physics. In particular, collisions have played an important role in the determination of structure and in understanding the detailed interactions among nature's "building blocks." Detailed collision experiments with the fullerenes are just beginning to emerge, but already some fascinating and unexpected results have been obtained. Schwarz and co-workers were the first to show that He and Ne atoms could be captured by C_{60}^+ in collisions with energies in the laboratory frame of a few keV[39]. In a beautiful series of experiments they were then able to prove that the rare gas gets trapped inside the fullerene and is not just weakly bound to the outside[42], and could also show the successive capture of two He atoms in a double collision[43]. By measuring the collision energy dependence of the capture process, we were able to determine the threshold energy for He capture to be 6 ± 2 eV and for Ne (somewhat larger, as would be expected), 9 ± 1 eV[44]. The results for He are shown in Fig. 10, along with experimental and molecular dynamics (MD) simulations results from Mowrey et al.[45]. The agreement between the MD calculations and the experimental data is extremely good and also in good agreement with ab initio molecular orbital calculations that estimate the energetic barrier for a He atom to penetrate a $C_6H_6^+$ ring to be less than 10 eV[46].

The simulations show that the rare gas collides with a carbon atom and gets deflected inside the carbon cage, travels to the other side, gets deflected back again, and continues to "rattle around" inside the fullerene. Relative kinetic energy is converted to vibrational energy of the C_{60}^+ so that the rare gas is unable to overcome the barrier for penetration of a C_6 ring a second time, and thus gets trapped inside. At sufficiently large collision energies, enough energy is transferred to the C_{60}^+ for metastable (statistical) fragmentation to occur, thus forming even-numbered carbon clusters C_{60-2n}^+ with a He inside (written as $He@C_{60-2n}^+$), as is indeed observed in the mass spectra. This is the most likely reason for the discrepancy in the position of the maximum for the capture cross section between theory and experiment seen in Fig. 10. The timescale of the MD simulations is on the order of femtoseconds to picoseconds, whereas the experiments have a timescale on the order of microseconds. The cross section for Ne capture is about a factor of 10 smaller than that for He capture, and up until now no evidence has been obtained for capture of any other gas targets. The most dominant process is, however, not capture but fragmentation of the C_{60}^+ to give the interesting bimodal fragment distributions shown as a function of collision energy for Ne collisions in Fig. 11. The large fragments, $40 < n < 60$, could possibly be explained by metastable fragmentation of C_2 (eqn. 1) from C_{60}^+ that has been vibrationally excited in the collision, assuming that the internal energy before the collision was already on the order of 40 eV (this seems rather high, but is perhaps plausible for laser desorption sources; see section 4). The fragments around C_{15}^+, on the other hand, cannot be explained in a satisfactory manner statistically, and give evidence for the spontaneous fragmentation or "fission" of C_{60}^+ into a few large fragments[24]. It should be remembered that the experiments only detect the positively charged fragments.

The species observed in the low mass fragment distribution are those corresponding to ring structures with relatively low ionisation potentials (7.2–7.45 eV[47]), which are certainly comparable to, if not less than, the ionisation potentials of the fuller-

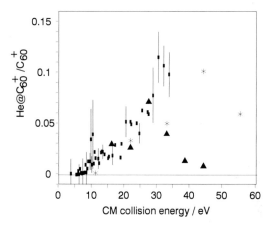

Fig. 10. Centre of mass collision energy dependence of He capture by C_{60}^+. ■: experimental results from ref. [44]; ▲: experimental results from ref. [45]; *: molecular dynamics simulations from ref. [45].

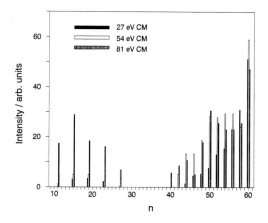

Fig. 11. Fragmentation pattern for C_{60}^+ + Ne collisions at three different collision energies. The fragment peak intensities have been multiplied by the factor 5.

enes in the range $n = 30$–50[35]. We would thus appear to detect the fission product with the lowest ionisation potential. Careful measurements of the energy thresholds for the different fragments observed in collisions between Ne^+ and C_{60}[48] carried out in a triple-sector, guided beam apparatus indicate that the fragmentation is probably initiated by the Ne passing inside the fullerene cage. One can then envisage trajectories where the Ne collides a number of times with the cage on passing through, thus "stripping off" a large fragment. More work is in progress in our laboratory to enable us to detect the neutral fragments, which should help answer the question as to whether fragmentation occurs predominantly via statistical evaporation of C_2 or via direct fragmentation into a small number of larger fragments.

6. COLLISIONS WITH SURFACES

Collisions between beams of C_{60}^+ ions and surfaces produce somewhat different results from the gas phase collisions discussed above. The first experiments with impact energies of up to 200 eV on graphite surfaces showed that the fullerenes stayed more or less structurally intact after the collision, but suffered a large loss of kinetic energy[49], with the final kinetic energy (only about 10–20 eV) being nearly independent of the impact energy[50]. On increasing the collision energy beyond this, substantial fragmentation involving the loss of C_2 units could be seen[51] in agreement with molecular dynamics simulations[52], which showed that the internal energy of the fullerene could be raised to several thousand Kelvin during the collision.

A schematic visualisation of the surface collisions is shown in Fig. 12 and combines information obtained from the experimental results[50] with molec-

ular dynamics simulations[52]. The fullerene is completely deformed on hitting the surface, with most of the initial kinetic energy being used to heat up the target. The kinetic energy of the C_{60}^+ remaining after the collision can be identified as the energy stored in the elastic deformation of the fullerene, which is set free again on recoil. By measuring the velocity components normal (v_n) and tangential (v_t) to the surface after the scattering, it could be seen that at impact energies beyond about 250 eV the normal velocity component reached a constant value[50]. When the fullerenes are slammed onto the surface a major fraction of the energy goes into heating or deforming the surface, while the deformation energy of the fullerene itself can be partially restored into motion perpendicular to the surface. The molecular dynamics simulations[52] showed that the deformation is almost complete for 250 eV, so for higher energies we could expect more energy to be transferred to the surface; but the amount of kinetic energy that can be recovered from the elastic deformation of the fullerene (and thus v_n) remains constant. The ratio of tangential velocity components after (v_t) and before (u_t) the collision was also approximately constant, independent of angle of incidence, as shown in Fig. 13. The C_{60}^+ is scattered almost specularly from the surface, with just a small shift in scattering angle towards the surface normal, which increases with increasing impact energy. Exactly such a shift in scattering angle can also be seen by bouncing a rubber ball from a hard surface, which suggests a useful analogy to explain the experimental results[60,63]. We assume that the C_{60}^+ "ball" hitting the surface can be strongly deformed by normal forces and is subject to substantial tangential frictional forces, which will tend to make it roll on the surface with an angular velocity Ω if incident with a finite tangential velocity u_t. The C_{60}^+ sphere may be deformed into an ellipse, thus changing its moment of inertia $Q = qma^2$, where m is the mass, a is the distance between the centre of the C_{60}^+ and the contact point with the surface, and $q \leq \frac{2}{3}$ (the value for a hollow sphere, i.e., the case with no

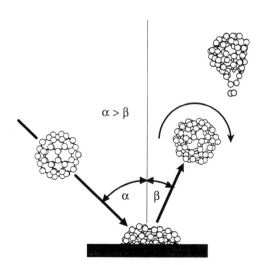

Fig. 12. Scheme for collisions between C_{60}^+ and a graphite surface, taking into consideration the main experimental results from ref. [50] and molecular dynamics simulations from ref. [52].

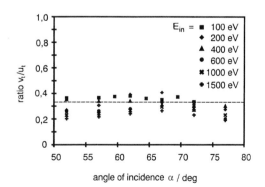

Fig. 13. Ratio of tangential velocity after (v_t) and before (u_t) collision as a function of the angle of incidence of the C_{60}^+ on the graphite surface (from ref. [53]).

deformation). Conservation of angular momentum then leads to

$$v_t = u_t - qa\Omega. \qquad (2)$$

If we make the further assumption that there is complete sticking of the contact point to the surface we can calculate the angular velocity $\Omega = u_t a$ in which case eqn (2) becomes

$$v_t = (1 - q)\, u_t. \qquad (3)$$

So, if this simple classical picture is correct, we would expect the ratio of tangential velocity components after and before the collision, v_t/u_t to be ⅓ for low collision energies where no significant deformation of the C_{60}^+ occurs ($q = ⅔$), decreasing somewhat as the deformation increases at higher impact energies. This is exactly what is seen experimentally (Fig. 13)!

7. CONCLUSIONS

Molecular beam experiments have played an important role in the fullerene story and continue to provide valuable information on the elementary properties and interactions of these wonderful molecules. They seem able to take on many guises, from bouncing rubber balls to little pieces of free-moving solid state. A well known German atomic physicist has even been heard to refer to the "C_{60} atom"! The sheer fascination and countless surprises that have been, and continue to be, thrown up by the molecular beam experiments is what drives most researchers in this field to continue to probe, in ever more detailed ways, the properties of the fullerenes. The results from their probing provide the basic building blocks needed to develop new materials based on the fullerenes and possible technological applications.

Acknowledgements—We would like to thank all the other members of the Freiburg "Buckyball" team who have contributed to the work presented here, especially Hans-Gerd Busmann and Wolfgang Kamke, and all colleagues who have kindly sent us preprints of their work. Financial support from the deutsche Forschungsgemeinschaft (through SFB 276) and the Bundesminister für Forschung und Technologie (BMFT contract numbers 13N5206, 13N5628 and 05472ABB/5) is gratefully acknowledged.

REFERENCES

1. H. W. Kroto, J. R. Heath, S. C. O'Brien, R. F. Curl, and R. E. Smalley, *Nature* **318**, 162 (1985).
2. H. W. Kroto, A. W. Allaf, and S. P. Balm, *Chem. Rev.* **91**, 1213 (1991).
3. R. F. Curl and R. E. Smalley, *Sci. Am.* (Oct. 1991).
4. G. Taubes, *Science* **253**, 1476 (1991).
5. R. F. Curl, *Carbon* **30**, xxx (1992).
6. W. Krätschmer, L. D. Lamb, K. Fostiropoulos, and D. R. Huffman, *Nature* **347**, 354 (1990).
7. R. Srinivasan and V. Mayne-Banton, *Appl. Phys. Lett.* **41**, 576 (1982).
8. R. Srinivasan, *Science* **234**, 559 (1986).
9. E. E. B. Campbell, G. Ulmer, B. Hasselberger, I. V. Hertel, *Appl. Surf. Sci.* **43**, 346, (1989).
10. G. Ulmer, B. Hasselberger, H.-G. Busmann, and E. E. B. Campbell, *Appl. Surf. Sci.* **46**, 272 (1990).
11. E. E. B. Campbell, G. Ulmer, B. Hasselberger, H.-G. Busmann, and I. V. Hertel, *J. Chem. Phys.* **93**, 6900 (1990).
12. W. R. Creasy, J. T. Brenna, *J. Chem. Phys.* **92**, 2269 (1990).
13. E. E. B. Campbell, G. Ulmer, H.-G. Busmann, I. V. Hertel, *Chem. Phys. Lett.* **175**, 505 (1990).
14. D. M. Cox, K. C. Reichmann, A. Kaldor, *J. Chem. Phys.* **88**, 1588 (1988).
15. P. P. Radi, M. T. Hsu, J. Brodbelt-Lustig, M. Rincon, and M. T. Bowers, *J. Chem. Phys.* **92**, 4817 (1990).
16. S. C. O'Brien, J. R. Heath, R. F. Curl, and R. E. Smalley, *J. Chem. Phys.* **88**, 220 (1988).
17. P. P. Radi, M. T. Hsu, M. E. Rincon, P. R. Kemper, and M. T. Bowers, *Chem. Phys. Lett.* **174**, 223 (1990).
18. C. Lifschitz, M. Iraqi, T. Peres, and J. E. Fischer, *Int. J. Mass Spec. Ion Phys.* **107**, 565 (1991).
19. D. Schröder and D. Sülzle, *J. Chem. Phys.* **94**, 6933 (1991).
20. T. Drewello, K.-D. Asmus, J. Stach, R. Herzschuh, M. Kao, and C. S. Foote, *J. Phys. Chem.* **95**, 10554 (1991).
21. M. T. Bowers, P. P. Radi, and M. T. Hsu, *J. Chem. Phys.* **94**, 6934 (1991).
22. G. Ulmer, E. E. B. Campbell, R. Kühnle, H.-G. Busmann, and I. V. Hertel, *Chem. Phys. Lett.* **182**, 114 (1991).
23. S. L. Anderson, personal communication.
24. E. E. B. Campbell, A. Hielscher, R. Ehlich, V. Schyja, and I. V. Hertel, *Nuclear physics concepts in atomic cluster physics*, In *Proc. of 88 Hereaus Sem.* (Edited by R. Schmidt, H. O. Lutz, and R. Dreizler). Springer, Berlin (1992).
25. E. E. B. Campbell, G. Ulmer and I. V. Hertel, *Phys. Rev. Lett.* **67**, 1986 (1991).
26. E. E. B. Campbell, G. Ulmer, and I. V. Hertel, *Z. Phys. D.* **24**, 81 (1992).
27. P. Wurz and K. R. Lykke, *J. Chem. Phys.* **95**, 7008 (1991).
28. S. Maruyama, M. Y. Lee, R. E. Haufler, Y. Chai, and R. E. Smalley, *Z. Phys. D.* **19**, 409 (1991).
29. L.-S. Wang, J. Conceicao, C. Jin, and R. E. Smalley, *Chem. Phys. Lett.* **182**, 5 (1991).
30. C. E. Klots, *Chem. Phys. Lett.* **186**, 73 (1991).
31. C. Yeretzian and R. L. Whetten, *Z. Phys. D.* (in press).
32. Y. Zhang, M. Späth, W. Krätschmer, and M. Stuke, *Z. Phys. D.* **23**, 195 (1992).
33. J. de Vries, H. Steger, B. Kamke, C. Menzel, B. Weisser, W. Kamke, and I. V. Hertel, *Chem. Phys. Lett.* **188**, 159 (1992).
34. H. Steger, J. de Vries, B. Kamke, W. Kamke, and T. Drewello, *Chem. Phys. Lett.* **194**, 452 (1992).
35. J. A. Zimmerman, J. R. Eyler, S. B. H. Bach, and S. W. McElvany, *J. Chem. Phys.* **94**, 556 (1991).
36. D. L. Lichtenberger, M. E. Jatcko, K. W. Nebesny, C. D. Ray, D. R. Huffman, and L. D. Lamb, *Mat. Res. Soc. Symp. Proc.* **206**, 673 (1991).
37. I. V. Hertel, H. Steger, J. de Vries, B. Weisser, C. Menzel, B. Kamke, and W. Kamke, *Phys. Rev. Lett.* **68**, 784 (1992).
38. G. F. Bertsch, A. Bulgac, D. Tomanek, and Y. Wang, *Phys. Rev. Lett.* **67**, 2690 (1991).
39. T. Weiske, D. K. Böhme, J. Hrusak, W. Krätschmer, H. Schwarz, *Angew. Chem. Int. Ed. Engl.* **30**, 884 (1991).
40. M. M. Ross and J. H. Callahan, *J. Phys. Chem.* **95**, 5720 (1991).
41. K. A. Caldwell, D. E. Giblin, C. S. Hsu, D. Cox, and M. L. Gross, *J. Am. Chem. Soc.* **113**, 8519 (1991).
42. T. Weiske, T. Wong, W. Krätschmer, J. K. Terlouw, and H. Schwarz, *Angew. Chem. Int. Ed. Engl.* **31**, 183 (1992).
43. T. Weiske and H. Schwarz, *Angew. Chem.* **104**, 639 (1992).

44. E. E. B. Campbell, R. Ehlich, A. Hielscher, J.M.A. Fra-zao, and I. V. Hertel, *Z. Phys. D.* **23**, 1 (1992).

45. R. C. Mowrey, M. M. Ross, and J. H. Callahan, *J. Phys. Chem.* (in press).

46. J. Hrusak, D. K. Böhme, T. Weiske, and H. Schwarz, *Chem. Phys. Lett.* **193**, 97 (1992).

47. S. B. H. Bach and J. R. Eyler, *J. Chem. Phys.* **92**, 358 (1990).

48. Z. Wan, J. F. Christian, and S. L. Anderson, *J. Chem. Phys.* **96**, 3344 (1992).

49. R. D. Beck, P. St. John, M. M. Alvarez, F. Diederich, and R. L. Whetten, *J. Phys. Chem.* **95**, 8402 (1991).

50. H.-G. Busmann, Th. Lill, and I. V. Hertel, *Chem. Phys. Lett.* **187**, 459 (1991).

51. H.-G. Busmann, Th. Lill, B. Reif, and I. V. Hertel, *Surf. Sci.* **272**, 146 (1992).

52. R. C. Mowrey, D. W. Brenner, B. I. Dunlap, J. W. Mintmire, and C. T. White, *J. Phys. Chem.* **95**, 7138 (1991).

53. Th. Lill, H.-G. Busmann, and I. V. Hertel, *App. Phys.* (in press).

FULLERENES AND GIANT FULLERENES: SYNTHESIS, SEPARATION, AND MASS SPECTROMETRIC CHARACTERIZATION

Deborah Holmes Parker, Kuntal Chatterjee, Peter Wurz, Keith R. Lykke, Michael J. Pellin, and Leon M. Stock

Materials Science and Chemistry Divisions, Argonne National Laboratory, 9700 S. Cass Avenue, Argonne, IL 60439, U.S.A.

John C. Hemminger

Institute for Surface and Interface Science, University of California, Irvine, CA 92727, U.S.A.

(*Accepted* 28 *April* 1992)

Abstract—We report a detailed procedure for the production of fullerenes and giant fullerenes in very high yield. Our high yields are obtained by a combination of fine control of the arc gap distance, optimal convection in the apparatus, and careful Soxhlet extraction with selected solvents. Our extraction and mass spectrometry results confirm the existence of giant fullerenes with masses in excess of 3000 amu. We find that 94% of the soot can be extracted in N-methyl-2-pyrrolidinone, indicating that a large portion of the soot has a molecular fullerene-type structure. We also present a new one-step method for the rapid separation of pure C_{60} directly from raw soot. An overview of recent results in the area of synthesis and purification of fullerenes and giant fullerenes is also presented.

Key Words—Fullerene synthesis, mass spectrometry, soot, fullerene separation.

1. INTRODUCTION

In 1984, the group at Exxon reported the observation of only even-numbered carbon cluster ions in the range C_{40} to C_{200} generated by laser ablation of graphite[1]. In the next year, the collaborative group of Kroto, Curl, Smalley, and graduate students Heath and O'Brien, proposed the structure of the truncated icosahedron to explain the exceptional stability of a carbon molecular ion consisting of 60 carbon atoms in the mass spectrum of laser ablated graphite[2]. This molecule was given the name buckminsterfullerene. Five years passed before this material was isolated in macroscopic quantities. In early 1990, Krätschmer *et al.* reported evidence for the presence of C_{60} in a sample of carbon dust prepared from vaporized graphite[3] and were able to isolate macroscopic quantities of C_{60} and C_{70}[4]. Reports of the synthesis of gram quantities of C_{60} and C_{70}[5–15] have generated a huge amount of interest in this new class of closed-caged carbon molecules. This paper discusses our recent results on improvements in the synthesis, separation, and characterization of fullerenes and presents an overview of the work of other groups in this area.

The methods used to synthesize fullerenes are of five types: evaporation of high purity carbon by using resistive heating[5,6,10]; use of an AC or DC arc discharge source[7–9,11–15]; flame production of soot

from carefully controlled combustion of benzene [16]; laser ablation of a rotating carbon disc in a furnace under flowing argon[8]; and most recently, a high frequency inductive heating method[17]. These methods produce carbon "soot" from which the fullerenes are extracted by use of appropriate solvents. Table 1 summarizes some of the experimental conditions reported in the literature for the synthesis of fullerene-containing soot by the arc discharge and resistive heating methods. The arc discharge process appears to be more efficient than the resistive heating method[7,18] since the power is dissipated in the arc rather than in heating the entire electrode[7]. Pang *et al.*[13] reported the synthesis of fullerenes from coke rods prepared from demineralized coal, and they note that DC power rather than AC power improves the percentage yield of soot for both graphite rods and rods prepared from coal. Shinohara *et al.*[15] also report that DC current is more efficient than AC current, but others report no difference in yield[19]. Koch *et al.*[12] found that lower currents gave better yields than very high AC currents. An examination of some of the conditions used for the arc synthesis (shown in Table 1) indicates that toluene- or benzene-soluble yields of about ten percent can typically be obtained for a wide range of experimental conditions. Our group uses a plasma-arc DC discharge that gives benzene soluble yields of 14% when extracted by reflux and the yield increases to 26% when Soxhlet extraction is used. This difference indicates that reflux extraction methods may be limited by solubility rather than the fullerene content of the soot. Pradeep and Rao[20] have reported a generator that produces,

Work supported by U.S. Department of Energy, BES-Materials Sciences and Chemical Sciences under Contract W-31-109-ENG-38.

29

Table 1. Summary of some fullerene synthesis methods. Yields are based on total soluble material extracted per weight of soot used except where indicated. The column labeled "Soluble yield" reports yields for which it is unclear from the reference if AC or DC current was used in the synthesis. Extraction is in benzene unless otherwise noted. Gravity feed is a special case of the contact arc method

Soluble Yield %	DC Yield %	AC Yield %	Current (A)	Voltage (V)	He Pressure (Torr)	Contact Method	Extraction Method	Ref.
	8(3†)	8 (3†)	100	50	37–75	resistive	Soxhlet	[5],[19]†
14	?	?	140–180	?	225	resistive	boiling toluene	[6]
	—	10 ± 2	100–200	10–20	100	contact arc	Soxhlet	[7]
	—	10–15	150	27	100	contact arc	?	[8]
25–40	?	?	40–60	?	200	gravity	boiling toluene	[9]
5–10	?	?	?	?	100–200	resistive	?	[10]
	—	10	200	20	150	contact arc	?	[11]
	—	7.7	55	?	100	gravity	?	[12]
	—	3	130	?	100	gravity	?	[12]
	16.2	9.3	105–110	24	250	contact arc	Soxhlet	[13]
	26	—	70	20	200	plasma arc	Soxhlet	[14]
	12		200–250	?	20	contact arc	reflux	[15]
8–12	—	—	—	–	110	high freq.	toluene	[17]

†3% is the yield of pure C_{60}; 8% is the yield of extractable material.

under certain operating conditions, pure C_{60} as measured by NMR and mass spectrometric methods. The purity is unlikely to be better than 98% since they report a red toluene soluble extract rather than the magenta color of toluene solutions of pure C_{60}[5]. A small amount of contamination of C_{60} with C_{70} leads to the red color due to the strong absorption of C_{70} in the blue region where C_{60} is transparent[6].

Separation and purification of the fullerenes is typically performed by first extracting the soot with benzene or toluene. This extract is stripped of most of the solvent in a rotary evaporator and the concentrated solution is then placed at the top of a neutral alumina column and eluted with hexane[5,6] or hexane/toluene mixtures[21] to obtain fractions containing pure C_{60} and pure C_{70}. The separation is difficult because of the low solubility of C_{60} and C_{70} in most organic solvents[6] and to the minor solubility differences among the fullerenes. Hawkins et al.[22] found good separation on a Pirkle column but poor separation by flash chromatography on silica gel and HPLC on silica gel. Cox et al.[23] report good separation by normal phase HPLC with dinitroanilinopropyl silica using a gradient from n-hexane to 50% methylene chloride. Vassallo et al.[24] have reported an improved separation of C_{60} and C_{70} by liquid chromatography using a graphite column that allows larger amounts to be purified (tens of milligrams). Repeated chromatography of the benzene soluble portion on neutral alumina has afforded the separation of the higher fullerenes C_{76}, C_{84}, C_{90}, C_{94}, and $C_{70}O$[25] and Kikuchi and co-workers[26] have recently reported the separation of milligram quantities of these species and C_{96} with high efficiency using preparative HPLC with CS_2 as eluent. Recently, Meier and Selegue have reported[27] an efficient preparative separation of C_{60} and C_{70} using gel permeation chromatography (GPC) with toluene as the mobile

phase. We have recently developed a one-step method for the extraction and purification of pure C_{60}[28] that is much faster and uses much less solvent than previously reported methods. This will be presented in further detail below.

Much of the characterization of fullerenes has relied on mass spectrometric analysis of the various extracts. For example, the first direct mass spectroscopic evidence for the presence of C_{60} in carbon soot was reported by Krätschmer et al.[4]. Meijer and Bethune[29] also reported direct evidence for the presence of C_{60} in an experiment that used laser desorption of the material followed by postionization of the desorbed neutral species. An isotope-scrambling experiment showed that the fullerenes were not formed in the analysis step but were formed in the original deposition process. Taylor et al.[5] used fast atom bombardment (FAB) mass spectrometry to further characterize the soot and extracts. Aije et al. have compared the fragmentation resulting from electron impact mass spectrometry with FAB and laser desorption mass spectrometry[6]. There are many other reports of the use of mass spectrometry in the characterization of fullerenes, and the reader is referred to excellent reviews for further details[30,31]. We also rely extensively on mass spectrometry to characterize the fullerenes[32] and details of our mass spectrometers are presented below.

There have been a number of recent reports on the "giant" fullerenes (i.e., fullerene molecules whose mass exceeds 1000 amu)[14,15,33–37]. There has been some controversey over the existence of these large carbon clusters in soot produced by the arc synthesis method. For instance, large carbon clusters including C_{60} and C_{70} and larger species have been produced and detected mass spectrometrically by the laser ablation of a number of carbonaceous species such as graphite[2], polyimide[38], ethylene-tetrafluoroethylene copolymer[39], and diamondlike carbon films[39], and coal[40]. In view of these results, it has been suggested that the giant fullerenes are produced in the laser desorption step of the analysis and not in the laboratory-produced soot. Hertel and coworkers have shown that at increasing laser fluences for the desorption step, fragmentation, and aggregation to form larger carbon clusters can occur[41]. They also suggest that laser desorption mass spectrometry alone is not sufficient evidence to prove the existence of the giant fullerenes in arc produced carbon soot[41]. In contrast to this, we will present evidence from recent work in our laboratory that confirms the existence of giant fullerenes in arc-produced soot.

2. RESULTS

2.1 Fullerene synthesis

Figure 1 shows a scale diagram of the generator used to produce the fullerene containing soot. Since our initial report[14], we added cooling coils on the outside of the generator and a pressure relief valve for increased safety. The apparatus is housed in a modified 8-in. four-way Conflat cross and pumped by a mechanical pump to a base pressure of 10^{-2} Torr. The entire apparatus is positioned in a vented fume hood for safety. After pump-down, the chamber is isolated from the pump and then backfilled with high purity helium to a pressure of 200 Torr. The production of fullerenes is carried out within the water-cooled shield region shown on the right-hand side of Fig. 1. The water cooled shield is actually a Perkin-Elmer cryoshroud (designed for ultrahigh vacuum

pumping) through which we flow cold water. This region is lined with stainless steel shims kept cool by contact with the water-cooled shield. At the end of a run, these shims can be easily removed and the soot scraped and collected using a paint brush. The large (½ in.) graphite rod ("Ultra Purity Spectroscopic Graphite Electrodes," United Carbon Products Company, Inc., Bay City, MI) is attached to a 2-in. linear feedthrough drive. A grounding strap is attached from the feedthrough shaft to the chamber to provide a low-resistance path for the high electrical current to flow to ground. This avoids high currents passing through the delicate bearings and bellows of the linear motion feedthrough. The smaller carbon rod (¼ in., National, "Special Spectroscopic Electrodes") is held in a stainless steel holder that can be fed into the chamber by means of an Ultra-Torr union. This allows almost the entire length (12 in.) of the rod to be "burned" without breaking vacuum. The holder is attached to the positive lead of the power supply. A ceramic-to-metal insulator electrically isolates the smaller carbon rod from the chamber. The negative lead of the power supply is attached to the grounded chamber. The power supply is a regulated 100 A, 50-V DC power supply.

Synthesis of the fullerenes is accomplished by first positioning the two carbon rods so that they are touching. The power supply is maintained at 25 V. Resistive heating occurs while the rods are touching and the rods are not consumed. The power supply draws full current (100 A) when the rods are touching. We quickly move the electrodes apart to ignite the arc, using the fine control provided by the linear motion feedthrough, until the plasma is burning

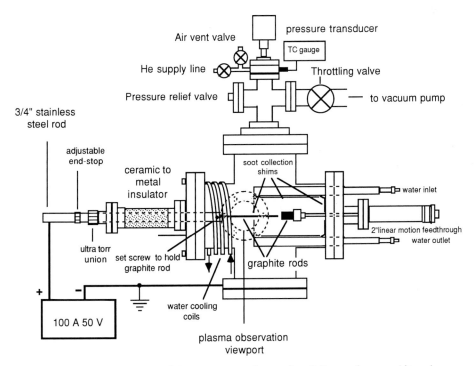

Fig. 1. Scale diagram of the apparatus used to produce fullerenes from graphite rod.

steadily. As the rod is consumed, we continually turn the linear motion feed while observing the plasma indirectly through a viewport. We adjust the gap between the electrodes to attain maximum brightness of the plasma. Under these conditions the plasma draws 60–80 A from the power supply. We do not observe the plasma directly under normal operating conditions. Instead, we observe the light emitted from the end of the collection region that is aligned directly in front of the viewport. On occasion, we have moved the electrode gap region in front of the viewport for observation (using welding goggles to observe the plasma). In this manner, we are able to determine that the gap between the rods is about 4 mm during operation. The typical power requirement under these conditions is 1–2 kW (25 V at 60–80 A). We consume the ¼ in. (positive) electrode at a rate of about 0.5 in. per minute and the larger (negative) electrode shows no detectable consumption. Occasionally, the smaller electrode is not consumed but instead transforms to look something like "popcorn." This popcorn, which is light grey in color, often breaks off and is collected with the soot, but is discarded before solvent extraction. We have not yet determined what causes the formation of this popcorn. At the end of a run, the apparatus is filled to atmospheric pressure with He. The chamber is then opened, the collection shims removed and the soot scraped from the shims and collected for further purification. Gloves and a face mask are worn during this step to avoid exposure to the very fine soot particles. With this method, tens of grams of soot per day can be produced. The addition of the cooling coils to the outside of the chamber (Fig. 1), allows the system to be operated continuously, without waiting for cooling between runs.

2.2 Mass spectrometry

We use two different mass spectrometers to characterize the soot and extracts. Our laser desorption time-of-flight (TOF) mass spectrometer has been described in more detail in a previous publication[42]. Briefly, the TOF mass spectrometer consists of a sample plug, an ion optics stack for ion acceleration, deflection plates, a field free region, and a dual microchannel plate assembly for detection of ions. The experiment measures the mass spectrum of negative ions or positive ions emitted directly from the sample in the desorption process. The mass resolution (m/Δm) of the apparatus is usually 400 under these experimental conditions. The base pressure of the system is typically 2×10^{-9} Torr. Neutral and ionized fullerene clusters are desorbed from the stainless steel substrate by a XeCl excimer laser (308 nm) or by 532 nm (266 nm) light from the doubled (quadrupled) output of a Q-switched mode-locked Nd^{3+}:YAG laser. The fluence of the desorption laser is held constant at approximately 10–100 mJ/cm², which is just above the threshold for ion production. The laser desorption at these intensities produces a minimum of fragmentation of the desorbing fullerene mole-

cules[43]. However, at slightly higher intensities, significant fragmentation occurs.

The laser desorption Fourier transform mass spectrometer (FTMS) consists of a three-region vacuum chamber with each region separated by gate valves and differential pumping apertures. The first region is a turbo-pumped rapid sample transfer region. The second region is also turbo-pumped and contains an argon ion sputter gun and gas dosing capabilities. The third region contains the optical path for an FTIR spectrometer, additional gas dosing capability, a resistive heater for vapor deposition of fullerenes, and an extension arm containing the analyzer cell for FTMS. In this region, a typical base pressure of 1×10^{-10} Torr is achieved by a cryopump. The entire apparatus is on a cart that can be moved to place the analyzer cell in the bore of a 7-T superconducting magnet. The sample is transferred between the various regions of the apparatus by a 1.5-in. polished transfer rod, and the sample can be resistively heated and cooled with liquid nitrogen. The FTMS experiments are performed with an Ionspec Omega data acquisition system. RF chirp excitation is used to accelerate the ions into cyclotron orbits inside the cell. Figure 2 shows a schematic of the FTMS cell and laser ports.

2.3 Extraction of the fullerenes

In the early stages of this work, we first determined that the fullerene synthesis was successful by stirring a portion of the soot in warm benzene for one hour. Filtration of this mixture yields a dark red solution indicating the presence of C_{60} and C_{70} and small amounts of higher fullerenes. We analyzed this solution by two procedures. First, a small amount of solution was evaporated onto a stainless steel sample holder and laser desorption TOF mass spectrometric measurements were performed on the residue. The TOF mass spectrum of this sample[14] showed primarily C_{60} and C_{70} with small amounts of C_{78} and C_{84}. Extractions were also performed with toluene and carbon tetrachloride, which gave nearly identical results. In the second procedure, we evaporated the fullerenes obtained from the same solution onto the Pd substrate using the fullerene doser in the FTMS apparatus. FTMS is more sensitive at higher molecular weights than our TOF spectrometer due to the velocity dependent detection sensitivity of the microchannel plates in the TOF detector. Figure 3 shows the FTMS spectrum indicating that small amounts of the higher fullerenes are indeed present in the benzene extract. A spectrum acquired with ejection of the ions below mass 850 is shown in the expanded portion of the spectrum. Ion ejection is accomplished with selective rf chirp excitation and does not alter the intensity of the remaining ions in the analyzer cell. This spectrum clearly shows that fullerenes up to cluster size C_{266} can be identified in the benzene extract prepared from soot formed in our fullerene generator.

We have compared the efficiency of Soxhlet extraction with the reflux technique using benzene as

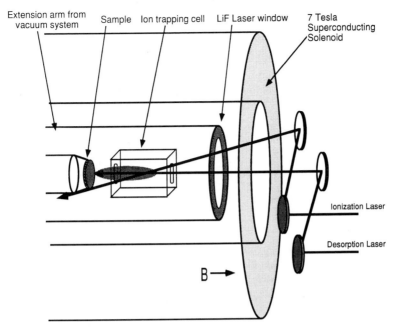

Extension arm from vacuum system · Sample · Ion trapping cell · LiF Laser window · 7 Tesla Superconducting Solenoid

Ionization Laser

Desorption Laser

B →

Laser Desorption Fourier Transform Mass Spectrometer

Fig. 2. Schematic diagram of the FTMS apparatus showing the position of the analyzer cell and laser entrance port. The ionization laser was not used in these experiments.

the extraction solvent. Both extractions were carried out for 24 h yielding strongly colored red solutions. Small amounts of the extract were saved for later analysis by mass spectrometry. The remainder of the sample was stripped of solvent in a rotary evaporator and then dried under high vacuum over a 105°C oil bath to determine the yield relative to the initial amount of soot used in the extraction. The insoluble material in the thimble was also dried over a hot oil bath under high vaccum. In all of our extractions,

we obtained a mass balance to within ~2% by measuring the weight loss of the soot compared to the weight of the dry extract. We found that the Soxhlet extraction worked much better than simple reflux, giving a 26% yield of soluble material. By contrast, the reflux extract gave a 14% yield of soluble material. Since Soxhlet extraction resulted in yields almost twice as high as with reflux, all subsequent extractions were performed by the Soxhlet method. From Table 2, it is seen that after four hours of extraction, not all

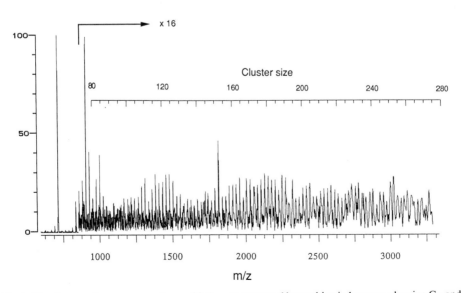

Fig. 3. Fourier transform mass spectrum of fullerenes extracted by washing in benzene, showing C_{60} and C_{70}. The inset, shown at a scale factor of $16\times$, depicts a spectrum acquired by ejecting ions with m/z less than 850, showing that fullerenes up to cluster size C_{266} are produced in our apparatus.

Table 2. Experimental results for Soxhlet extraction of soot with *n*-hexane and benzene

Extraction Solvent	Yields	He Pressure (Torr)	batch†/sample number‡	Extraction Time
n-hexane	18%	200	batch #1 sample b	4 h
n-hexane	§18.6%	200	batch #6 sample b	4 h
n-hexane	21%	200	batch #1 sample a	24 h
n-hexane	20%	200	batch #6 sample a	24 h
benzene	24%	200	batch #1	24 h
benzene	27%	200	batch #2 sample a	24 h
benzene	26%	200	batch #2 sample b	24 h

†Batch numbers refer to different sample of soot produced either by different experimenters or on different days or both.
‡Sample numbers refer to different samples of the same batch of soot.
§"Blind" analysis performed by an independent laboratory.

of the fullerenes are extracted compared to the 24-h extraction. Consequently, all extractions were performed for 24 hours unless noted otherwise with cycling times of 25 min per cycle. The cycling time and total time are important parameters to consider when comparing results between laboratories since these two parameters determine the total volume of solvent washed through the soot. A run carried out at a slower

At first, we were surprised by these high yields since they are considerably higher than yields previously reported (Table 1) with the exception of the work by Diederich *et al.*, which also reports yields in this range[25]. We spend considerable time verifying these results and a summary of several of our experimental runs using *n*-hexane and benzene extraction solvents is shown in Table 2. This table shows that the results are quite reproducible and we consistently obtain these high yields, unlike the work reported in Refs. [6,9, and 25] which gives a wide range of yields (25–35%). Our yields have been repeated by several operators and with several different Soxhlet extraction apparatus.

We were concerned that some of the yield could be attributed to insoluble micron-sized particles escaping through the cellulose Soxhlet thimble without actually becoming dissolved. If this occurred, a good mass balance would still be obtained but the yield would be artificially high. To ensure that this was not contributing to our high yields, we filtered some of the solutions through a 0.8-μm Nucleopore polycarbonate membrane filter or a 0.45-μm millipore filter. No particulates were trapped by the filter indicating that the extract is a true solution and does not contain insoluble particulates. Because benzene is a carcinogen, we now use toluene or *n*-hexane for routine extractions. The fullerenes are less soluble in toluene than benzene. Only 17% soluble material is extracted in toluene compared to 26% for benzene. Figure 4A gives the TOF mass spectrum of the material isolated from the benzene Soxhlet extractions showing that the benzene solvent dissolves primarily C_{60} and C_{70} in a ratio of approximately 3:1 with very small amounts of higher fullerenes up to mass 1200. The mass spectrum of the sample prepared by reflux was identical. However, the mass spectrum of a toluene extract is similar to the mass spectrum obtained for benzene

extraction, indicating that the solubility differences are largely independent of mass. We also found that if the extraction is performed under a nitrogen atmosphere, the yields are slightly higher (3–4%) than when extracted under ambient conditions. Also, it is important to cover the Soxhlet apparatus with foil during the extraction to keep light from the solution that will degrade the fullerenes. We refluxed a solution of pure C_{60} in toluene exposed to room light

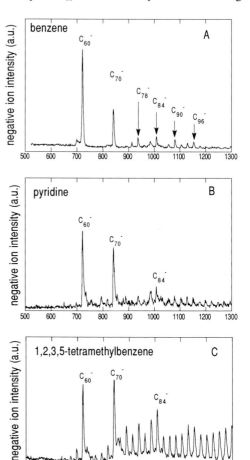

Fig. 4. Laser desorption TOF mass spectra of soot extracts resulting from sequential extractions with (A) benzene, (B) pyridine, and (C) 1,2,3,5-tetramethylbenzene. The spectra are plotted on the same mass scale for comparison.

Fullerene Separation Scheme

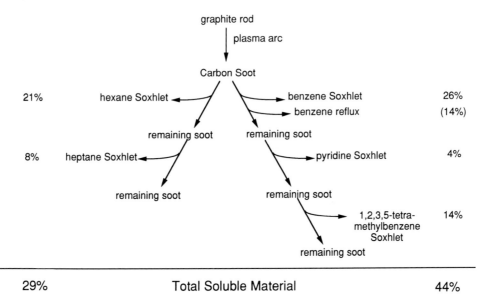

graphite rod

plasma arc

Carbon Soot

21%   hexane Soxhlet   benzene Soxhlet   26%

benzene reflux   (14%)

remaining soot   remaining soot

8%   heptane Soxhlet   pyridine Soxhlet   4%

remaining soot   remaining soot

1,2,3,5-tetra-   14%
methylbenzene
Soxhlet

remaining soot

29%   Total Soluble Material   44%

Fig. 5. Diagram of the sequential fullerene extraction scheme. The TOF mass spectra corresponding to the right-hand branch are shown in Figs. 4 and 6. The TOF mass spectra corresponding to the left-hand branch are shown in Fig. 11.

under ambient conditions. After 24 h of reflux, the solution had changed from the beautiful magenta color of pure C_{60} to an orange color, indicating decomposition of the C_{60}.

2.4 Selective molecular weight extraction by varying the extraction solvent

Following the observation of the small amounts of fullerenes up to mass 1200 in Fig. 4A, we thought that we could increase the solubility of these larger species by using higher boiling solvents. We performed the successive extraction scheme shown in the right-hand side of Fig. 5. We first attempted to isolate higher molecular weight fullerenes by extraction with a more polar solvent. The material remaining in the Soxhlet thimble was Soxhlet extracted with pyridine yielding an additional 4% yield of soluble material. The TOF mass spectrum of this material, depicted in Fig. 4B, shows C_{60} and C_{70} in a ratio of 2:1 with small amounts of the higher fullerenes. This extraction gives results very similar to the benzene extraction, and there appears to be no advantage to using pyridine over benzene. The material remaining in the thimble was then extracted with 1,2,3,5-tetramethylbenzene (TMB). This extraction yielded a dark greenish-brown solution with an additional 14% soluble material. TOF mass spectrometry (Fig. 6) shows that this solution contains significant concentrations of fullerenes up to mass 2400 (C_{200}). This TOF spectrum is also shown in Fig. 4C, where it is plotted on the same scale as the benzene and pyridine extracts for comparison. The spectrum (Fig. 6) shows only successive even-numbered peaks suggesting the presence of hollow fuller-

ene cage molecules [44]. From peak areas and abundance considerations, it appears that C_{60} and C_{70} are present in concentrations of less than 1%. Because of the decreasing channel plate detection sensitivity for higher mass species in the TOF apparatus, we suspected that there may be even larger fullerene species present in the TMB extract. Using FTMS, and carefully optimizing the excitation conditions, using low desorption laser power to avoid fragmentation, and ejecting masses below 850 from the cell, the spectrum shown in Fig. 7 was obtained. Based on the yields from the successive extractions of the soot with benzene, pyridine, and 1,2,3,5-tetramethylbenzene, the total yield of soluble material consisting of fullerenes C_{60} to C_{466} is 44($\pm$2)% for this series of extractions.

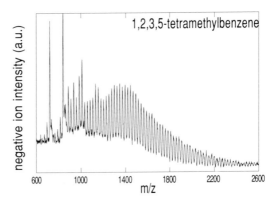

Fig. 6. Laser Desorption TOF mass spectrum of TMB extract prepared from soot previously extracted with benzene and pyridine according to the diagram shown in Fig. 5. A portion of this spectrum is also shown in Fig. 4(C).

cluster size (atoms)

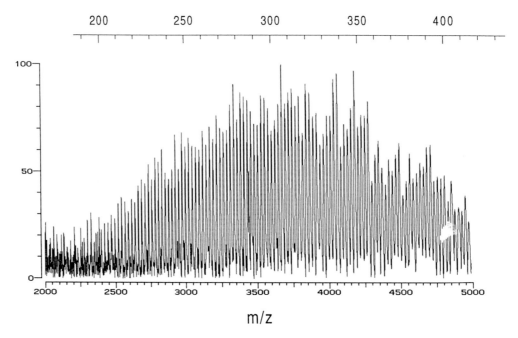

Fig. 7. Laser Desorption FTMS spectrum of the 1,2,3,5-tetramethylbenzene extract. The spectrum was acquired using low laser fluence, and ejecting masses below m/z = 850.

We were concerned that perhaps C_{60} and C_{70} were reacting with the solvent to form a polymeric species that, upon laser desorption, readily recombined to give the higher fullerenes, much as reported by Brenna and Creasy for the laser ablation of other polymers[39]. To test this, we refluxed pure C_{60} with TMB for 20 h under nitrogen to determine if polymerization was taking place. The TOF mass spectrum (Fig. 8A) of this reaction mixture shows that the solvent has added to the fullerene cage, but we see no evidence of polymerization of the solvent or fragmentation and reaggregation to form giant fullerenes. Figure 8B shows a high resolution FTMS spectrum plotted for the region around the first adduct peak (centered at m/z 854). These experiments, comparing the TMB soot extract and the TMB C_{60} reaction product, were performed under similar conditions of laser fluence. This is very good evidence that the giant fullerenes are actually present in the extract solution and are not formed by a matrix effect during the desorption step due to polymerization by the solvent. It is possible, however, that the larger fullerenes are also reacting with the solvent thus contributing to the measured yield, and we do see a small amount of the C_{60}-TMB adduct in the mass spectrum of Fig. 4C at m/z = 854. However, in our high mass resolution FTMS spectrum of the TMB extract, we do not see any evidence for the adduct reaction products of the giant fullerenes. This reaction of C_{60} with TMB is very similar to the reactions reported by Hoke *et al.*[45] in which C_{60} was shown to add five to six molecules of toluene by refluxing this solution in the presence of

$FeCl_3$. It is important to note that, in our work and in that of Hoke *et al.,* these species are not formed in the mass spectrometer. A solution of pure C_{60} in TMB (without refluxing) evaporated onto a sample holder shows only C_{60} in the mass spectrum. Interestingly, we do not require the presence of the Lewis acid $FeCl_3$ to catalyze the solvent addition to C_{60}. The reaction likely proceeds by a free radical addition mechanism. We have performed a 20-h reflux of pure C_{60} with 1,2-dimethylbenzene and 1,3,5-trimethyl benzene and we see no evidence of reaction.

We thought that it might be possible to obtain pure C_{60}, or at least greatly enriched C_{60}, by extracting the soot directly with alkanes. Two different types of experiments were performed: extraction of raw soot with different alkane solvents and successive extractions of soot residue with a series of different alkane solvents. We extracted the *raw* soot with *n*-pentane, *n*-hexane, and *n*-heptane and find that 14($\pm$2)%, 18($\pm$2)%, and 23($\pm$2)% soluble material is extracted, respectively. The mass spectra of the *n*-pentane, *n*-hexane, and *n*-heptane extracts are shown in Fig. 9, panels A–C. The mass spectra of the extracts with saturated hydrocarbons are very similar but show larger amounts of higher fullerenes for the heptane extract. The total extracted amount from raw soot increases with increasing boiling point of the solvent for *n*-pentane, *n*-hexane, and *n*-heptane (boiling points are 36°C, 69°C, and 98°C, respectively). We can show this general trend in a plot of the amount extracted versus boiling point of solvent for all of the solvents we have used to extract raw soot (Fig. 10). We cannot make a

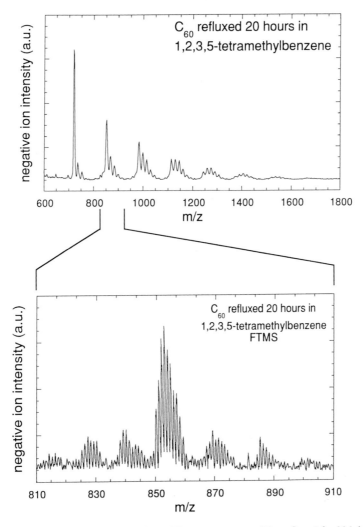

Fig. 8. The top panel shows the laser desorption TOF mass spectrum of C_{60} refluxed for 20 h in 1,2,3,5-tetramethylbenzene. The bottom panel shows a high resolution FTMS mass spectrum of the same sample, expanded about the region 810 to 910 mass units. The spectrum shows the adduct of one solvent molecule with C_{60} at mass 853. Methyl addition and methyl loss are evident in the multiplets on either side of the central peaks. Isotope peaks of carbon appear to be convoluted with hydrogen addition and loss in each of the multiplets.

van't Hoff plot as in Ref.[37] because we have used a nonequilibrium Soxhlet extraction method instead of equilibrium reflux. The van't Hoff analysis only works for processes at equilibrium.

For the successive extractions, a fresh portion of the soot was Soxhlet extracted with n-hexane for 24 h giving a pink solution. This extraction yielded 18($\pm$ 2)% soluble material. The mass spectrum shown in Fig. 11A indicates that n-hexane extracts C_{60} and C_{70} with only trace amounts of C_{76}, C_{78}, and C_{84} (<1%). The soot remaining in the Soxhlet thimble was then extracted with n-heptane, giving an orange-colored solution. This extraction yielded an additional 8% soluble material, bringing the total yield to 29($\pm$ 2)% by weight. The TOF mass spectrum of this extraction is shown in Fig. 11B. The n-heptane extract is enriched in C_{84} compared to other extracts and contains C_{60}, C_{70}, C_{76}, C_{78} and C_{84}. The ratio of fullerenes in this

extract is 2.0:1.4:0.5:1.0 for C_{60}, C_{70}, C_{78}, and C_{84}, respectively. The insoluble residue from the n-heptane extraction was then extracted with decane yielding a clear, light-yellow solution. This solution did not show any fullerenes in the TOF spectrum and may only contain trace hydrocarbons. The total yield for the hexane/heptane branch amounts of 29 $\pm$ 2%. This is lower yield than the benzene branch of the extraction, but the hexane/heptane branch does not extract fullerenes above C_{84}, whereas the TMB extracts fullerenes out to mass C_{466} and thus, the yield is expected to be lower for the hexane/heptane branch.

The solvent N-methyl-2-pyrrolidinone (NMP, bp = 202°C) was used as this is an excellent solvent for dissolving carbonaceous materials such as coal[46]. This polar aprotic solvent penetrates deep into the coal structure and swells the coal. It is also commonly used in the separation of olefins and aromatics and

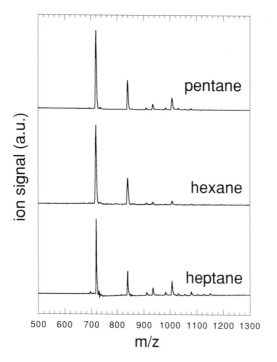

Fig. 9. Laser desorption TOF mass spectrum of pentane, hexane and heptane extracts of raw soot.

for refining oil[46]. We found that 94% of the raw soot could be desolved in this material resulting in a black solution. NMP gives high solubilities probably because of increased solvent penetration into the soot matrix, interrupting the polarization forces that hold the extractable fullerenes within the soot. The solu-

tion was evaporated onto a plug and examined by TOF MS. The result is shown in Fig. 12. Along with C_{60} and C_{70}, and giant fullerenes a significant number of smaller carbon clusters (m/z <300) were also observed in the mass spectrum.

Additionally, we have extracted giant fullerenes with 1,2,4-trichlorobenzene (TCB). We performed this extraction on soot previously extracted with toluene to deplete the soot of C_{60} and C_{70}. Figure 13 shows two LD FTMS spectra taken at different laser fluences (532-nm direct positive ion desorption). At low laser fluences, the mass distribution peaks at m/z = 2300 with fullerenes present up to m/z = 3500 and, importantly, *no C_{60} is observed*. C_{60} and C_{70} are observed for higher laser fluences along with the fragment clusters C_{68}, C_{66}, C_{58}, and C_{56} (panel B). At high laser fluence, the mass distribution peaks at a lower mass of 1500 and the high mass tail has moved down to 3000 mass units. This indicates that substantial fragmentation occurs at higher laser fluences to produce lower molecular weight fullerenes in the desorption step. At low laser fluences no C_{60} and only a trace of C_{70} is observed indicating that very little fragmentation is occurring. Thus, the spectrum in panel A likely represents the nascent distribution of giant fullerenes extracted with TCB. At worst, this spectrum is a lower limit for the mass distribution in the extract.

We have also repeated the experiment first reported by the group at Arizona[36] where toluene in a bomb reactor at high temperature and pressure was used to extract soot that had been previously extracted with toluene to deplete the soot of C_{60} and C_{70}. This result is shown in Fig. 14. We only extract about 3% of the soot in this manner. We indeed observe

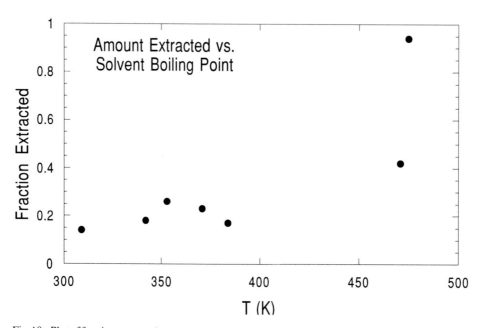

Fig. 10. Plot of fraction extracted versus solvent boiling point. The boiling points for the solvents are listed in Table 3.

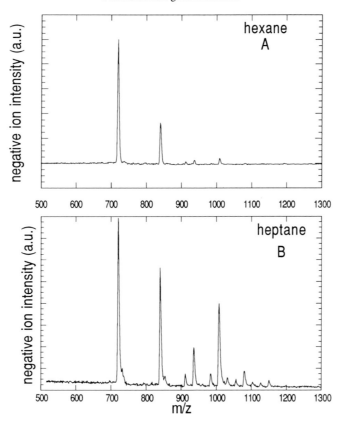

Fig. 11. Laser TOF mass spectra of extract prepared by sequential solvent extraction with (A) hexane, and (B) heptane.

higher fullerenes in this extract, but we do not observe masses as high as reported previously, probably due to difference in experimental conditions of temperature and pressure. Also, we find that once extracted from the soot, the fullerenes can be redissolved in lower boiling solvents such as hexane and methylene chloride. Further incidental evidence for the different molecular weight ranges extracted in different solvents is in the color of the solutions. These are summarized in Table 3.

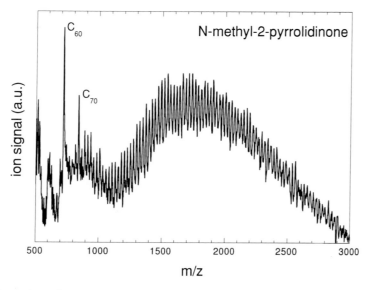

Fig. 12. Laser desorption TOF mass spectrum of raw soot extract prepared by Soxhlet extraction with N-methyl-2-pyrrolidinone.

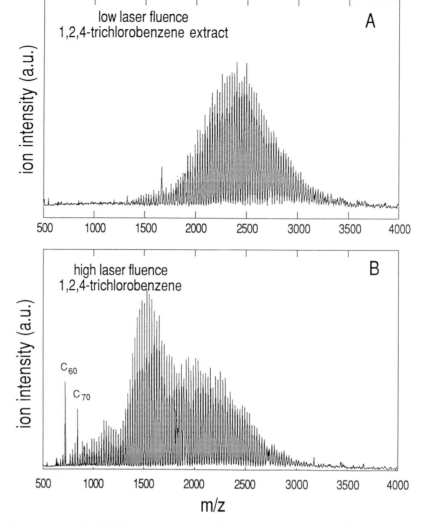

Fig. 13. Laser desorption FTMS spectra of 1,2,4-trichlorobenzene extract prepared by Soxhlet extraction of soot previously extracted in toluene to deplete the soot of C_{60} and C_{70}. Panel A depicts data collected with low laser fluence showing no C_{60} peak. Panel B shows data collected with high laser fluence, resulting in considerable fragmentation. Noise spikes have been removed from the spectrum.

2.5 *Separation of pure C_{60}*

The previously reported methods of purification of C_{60} suffer from a number of drawbacks. The liquid chromatography method on neutral alumina requires large amounts of solvent and a great deal of patience but affords excellent separation. The HPLC method is limited by the amount of material that can be injected on the column. The GPC method requires less solvent since toluene is the mobile phase, but the amount of material injected is still a limitation. Also, the HPLC and GPC methods require a considerable investment in equipment. The one-step method that we demonstrate uses only a minimum amount of solvent and runs continuously so that it does not have to be monitored constantly.

Our simple one-step method involves the combination of extraction and chromatography in a single apparatus. As discussed above, higher yields were obtained by using Soxhlet extraction instead of reflux, presumably because the soot is continually washed with hot pure solvent in the former method. We also found that 18% of the raw soot was extracted into *n*-hexane and this extract contains primarily C_{60} and C_{70}[14]. We sought to combine the efficiency of Soxhlet extraction with the selectivity of the chromatographic method to provide a one-step method

Table 3. Raw soot extract color in various solvents

Solvent	Boiling Point (K)	Color
pentane	309	pink
hexane	342	pink
heptane	371	orange
toluene	384	red
benzene	353	red
TMB	471	green-brown
NMP	475	black

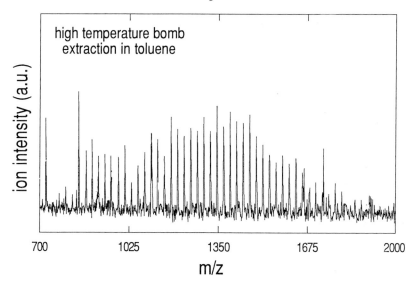

Fig. 14. Laser desorption FTMS spectra of high pressure/high temperature toluene extract prepared with soot previously extracted in toluene to deplete the soot of C_{60} and C_{70}. Noise spikes have been removed from the spectrum.

for C_{60} purification. A device to carry this out is commercially available and is known as a Kauffman chromatographic column (Ace Glass Co.). The Kauffman column is ideally suited to the purification of poorly soluble compounds[47]. In a typical experiment, aluminum oxide (200 g) was loaded in the inner column of the apparatus and raw soot (4.5 g) was added on top of the alumina. The alumina is held in place by a frit at the bottom of the inner column. Additional alumina was placed on top of the soot to immobilize it. The inner column was then placed within the outer column and tightly attached with a nylon bushing and CAPFE O-ring. The column was fitted with a condenser and was placed above a pre-weighed flask containing n-hexane (250 mL). Once reflux begins, the solvent vapors from the boiling flask rise through the space between the inner and outer columns and continue through the holes near the top of the inner column into the condenser. The condensate then falls onto the raw soot, extracts the fullerenes, and passes through the alumina column. The magenta C_{60} band is readily separated from the band containing C_{70}. Hexane and the purified C_{60} flow into the boiling flask and n-hexane is again vaporized while the purified C_{60} remains in the flask. A yield of 0.27 g (6%) of pure C_{60} was obtained by this procedure after 11 h of extraction.

The product was analyzed by laser desorption TOF mass spectrometry which showed only C_{60} with no other species detected at a signal-to-noise ratio of 100:1 (Fig. 15, panel a). Pure C_{70} cannot be obtained by this method since C_{60} is continually extracted from the soot and contaminates the C_{70} band as it moves down the column (Fig. 15, panels b and c). This one-step method for the purification of C_{60} provides several advantages. The refluxing solvent provides continuous solvent feed so that the process does not have

to be monitored constantly. Much less solvent is needed compared to the previous methods: 300 mL compared to 2–3 L for C_{60}. The new method requires one-tenth the time required in earlier methods involving sequential Soxhlet extraction, chromatographic separation, and rotary evaporation. However, not all of the C_{60} is eluted from the column. Some C_{60} continues to elute with C_{70}, contaminating the C_{70} band, and at present, the C_{70} has not been isolated in pure form with this technique. Even so, we obtain yields of 6% pure C_{60}. Pure C_{70} can be obtained by extracting the alumina with toluene to recover the remaining C_{60} and C_{70}, and using the ordinary neutral alumina liquid chromatography method on the C_{70}-enriched fraction.

3. DISCUSSION

3.1 Fullerene synthesis

The use of fine control over the gap distance is a logical development in the synthesis of fullerene soot since virtually no carbon vaporizes when the electrodes touch[18] and yields are generally lower when there is little control over the gap distance, such as under gravity-feed conditions or spring-held positions. Combining the improved yield that results from fine control over the arc gap in the generation process with efficient Soxhlet extraction generates higher yields of soluble fullerenes than are generally reported. In contradistinction to our results, Smalley's group finds that the yield does not depend on the gap spacing[8]. Another important design feature is the convection rate that determines the rate of annealing (cooling) of the carbon once it leaves the plasma. This factor is known to be important for fullerene formation from results of the laser ablation jet expansion experiments of Smalley's group that show

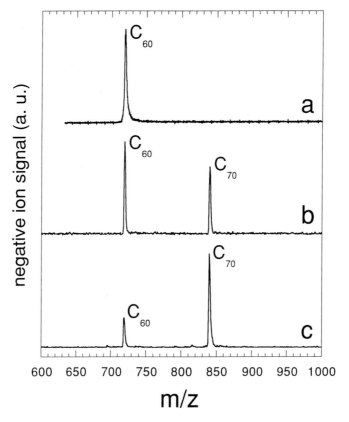

Fig. 15. Time-of-flight mass spectrum of the fullerene solutions obtained after Kauffman column chromatography. Panel a shows that pure C_{60} is obtained in the first fraction eluted from the column (5 h). Successive fractions shown in panels b (20 h) and c (additional 20 h after fraction b was removed) show the increasing enrichment of C_{70} in the solution.

dramatic differences in the carbon cluster formation depending on the nozzle conditions[2,44]. We find that, when the end shims are removed, the observed convection pattern in the chamber is markedly different and the yields of soluble materials are lower *by over a factor of two:* 8% Soxhlet extracted in toluene compared to 17% extracted in toluene with the shims in place. It is likely that careful design of a plasma arc apparatus that combines effective convection conditions with continuous graphite feed could result in a fullerene generator with even higher yields, and it may be possible to "tune" the conditions to produce fullerenes of a desired molecular weight. Even so, as the data in Table 1 indicate, fullerene generation in significant quantity can easily be performed in a wide variety of experimental setups, enabling any ordinary chemical laboratory to produce fullerenes.

3.2 Giant fullerene extraction

Our experiments[14] were the first to report the mass spectra showing the dissolution of giant fullerenes in high boiling point solvents, although Diederich and co-workers mentioned a similar experiment in trichlorobenzene in a note added in proof in their *Science* paper[25]. Since that time, other reports of

similar results have appeared in the literature. Smart and co-workers report solvation of giant fullerenes in xylenes, 1,3,5-trimethylbenzene, 1,2,4-trichlorobenzene, and 1-methylnapthalene. They raise the question, "How much of the soot is fullerenes?" From our data using TMB, it appears that nearly half of the soot consists of fullerenes. Even more of the soot may have a molecular fullerene-type structure given the tremendous amount of soot—94%—that is soluble in NMP. The helium pycnometry experiments of Ruoff *et al.*[35] support this idea. The pycnometry experiment determines an upper bound to the density of the soot that is much lower than the density of graphite. This suggests the presence of large amounts of carbon cage molecules and very little graphite. Our NMP solvation experiment suggests this as well since graphite does not dissolve in NMP. We are presently working on experiments to determine the structure and identity of the NMP soluble portion. Shinohara *et al.* have also been successful in extracting very large fullerenes from arc-produced soot using quinoline (bp 238°C). They report that fullerenes up to C_{300} are extracted with quinoline from the benzene insoluble soot. Even though they employ the highest boiling liquid used so far for the extraction of fullerenes, they do not detect fullerenes higher than 3600 mass units in their fast atom bombardment mass spectrometer, nor do they report the sharply peaked mass distributions reported by us, Lamb et al.[36], and Smart et al.[37]. In these latter three studies, the mass distributions peak about 1600—2500 mass units and taper off at about 3500 mass units.

As mentioned in the Introduction, there has been some controversy over the existence of the giant fullerenes. For example, Hertel's group states, "We are able to show that the samples prepared by Krätschmer and collaborators contain almost pure C_{60} and C_{70} molecules and that higher laser fluences can easily lead to substantial fragmentation and reaggregation forming rather large carbon clusters." Part of this statement is certainly true: Higher laser fluences can cause fragmentation and reaggregation to form large carbon clusters. This is well known from the laser ablation experiments of Smalley and collaborators[2], and from the polymer ablation experiments of Creasy and Brenna[39] and other experiments in Hertel's group[41]. However, because they see no fullerenes higher than C_{84} at lower laser powers does not mean that they are not present. It could well be that it is a detection sensitivity problem.

Let us examine the evidence for the giant fullerenes in more detail. As laser power is increased, the fragmentation of C_{60} occurs by loss of C_2 units to form C_{58} and smaller fullerenes. Fragmentation of larger fullerenes will of necessity generate C_{60} since this is an extremely stable fragmentation product. Previous experiments from our research group on pure C_{70} have shown that increasing the intensity of the positionization laser results in the creation of fragment clusters[48]. These previous experiments showed that at the highest laser fluence, the C_{60} ion signal resulting from the fragmentation of C_{70} becomes larger than the C_{70} ion signal. In our experiment with the TCB extract, we depleted the sample of C_{60} and C_{70} by Soxhlet extraction in benzene. At low laser power, we see no C_{60} and only a trace of C_{70}. As the laser power is increased, the C_{60} peak appears, as well as the C_{58} fragmentation peak, indicating that significant fragmentation is occurring. If we were using high enough laser intensity to cause the formation of significant amounts of giant fullerenes, then we should also see significant amounts of the fragmentation products C_{60} and C_{58}. Since we do not see C_{60} in our TCB extract at lower laser fluences, this is very good evidence that the giant fullerenes are indeed produced in arc-generated soot and not in the analysis step. Additional evidence fo the existence of giant fullerenes comes from the recent work of Lamb et al.[36] in which giant fullerenes are extracted in a bomb at high temperature and pressure. The species extracted in this manner were imaged with STM, showing spherical fullerenes containing up to 330 carbon atoms. Taken together, the results of our work, the He pycnometry experiments, and the STM images provide convincing evidence for the existence of giant fullerenes.

4. CONCLUSIONS

We have reported a detailed procedure for the production of fullerenes and giant fullerenes in very high yield. Our experiments indicate the fine control of the arc gap combined with proper convection in the apparatus and careful Soxhlet extraction all contribute to our high yields of fullerenes. We report several different kinds of mass spectrometric evidence for the presence of giant fullerenes extracted in high boiling solvents. Our results also indicate that the majority of the soot has a molecular fullerene-type structure and dissolves in organic solvents. We have also developed a quick one-step method for the isolation of pure C_{60} that is a factor of ten faster and uses a factor of ten less solvent than previous methods.

Acknowledgements—We thank Mukund Rangaswamy and Brad Savall for making some improvements to our fullerene generator and for making several batches of soot. One of us (K. C.) thanks Rupa Roy who had suggested, "Why don't you extract and separate in the same reaction vessel." We are grateful to Alexander J. Parker of Sherwin Williams Automotive Research Division for performing the Soxhlet extractions to verify our results and for suggesting the use of the Kauffman column. We thank Joe Gregar for constructing a large-scale Kauffman column in only one day. We acknowledge Dieter Gruen for stimulating discussions and we also acknowledge several helpful discussions with Dr. Randy Winans. We are grateful to Rod Ruoff and Stephen McElvaney for providing us with manuscripts prior to publication.

REFERENCES

1. E. A. Rohlfing, D. M. Cox, and K. Kaldor, *J. Chem. Phys.* **81**, 3322 (1984).
2. H. W. Kroto, J. R. Heath, S. C. O'Brien, R. F. Curl, and R. E. Smalley, *Nature* **318**, 162 (1985).
3. W. Krätschmer, K. Fostiropoulos, and D. R. Huffman. *Chem. Phys. Lett.* **170**, 167 (1990).
4. W. Krätschmer, L. D. Lamb, K. Fostiropoulos, and D. R. Huffman, *Nature* **347**, 354 (1990).
5. R. Taylor, J. P. Hare, A. Abdul-Sada, and H. Kroto, *J. Chem. Soc. Chem. Commun.* 1423 (1990).
6. H. Ajie, M. M. Alvarez, S. J. Anz, R. D. Beck, F. Diederich, K. Fostiropoulos, D. R. Huffman, W. Krätschmer, Y. Rubin, K. E. Schriber, D. Sensharma, and R. L. Whetten, *J. Phys. Chem.* **94**, 8630 (1990).
7. R. E. Haufler, J. Conceicao, L. P. F. Chibante, Y. Chai, N. E. Byrne, S. Flanagan, M. M. Haley, S C. O'Brien, C. Pan, Z. Xiao, W. E. Billups, M. A. Ciufolini, R. H. Hauge, J. L. Margrave, L. J. Wilson, R. F. Curl, and R. E. Smalley, *J. Phys. Chem.* **94**, 8634 (1990).
8. R. E. Haufler, Y. Chai, L. P. F. Chibante, J. Conceicao, C. Jin, L.-S Wang, S. Maruyama, and R. E. Smalley. In *Materials Research Society Symposium Proceedings.* (Edited by R. S. Averback, J. Bernholc, and D. L. Nelson), p. 627. Materials Research Society, Boston, MA (1991).
9. R. L. Whetten, M. M. Alvarez, S. J. Anz, K. E. Shriver, R. D. Beck, F. N. Diederich, Y. Rubin, R. Ettl, C. S. Foote, A. P. Darmanyan, and J. W. Arbogast. In *Materials Research Society Symposium Proceedings.* (Edited by R. S. Averback, J. Bernholc, and D. L. Nelson), p. 639. Materials Research Society, Boston, MA (1991).
10. D. M. Cox, S. Behal, K. Creegan, M. Disko, C. S. Hsu, E. Kollin, J. Millar, J. Robbins, W. Robbins, R. D. Sherwood, P. Tindall, D. Fischer, and G. Meitzner. In

Materials Research Society Symposium Proceedings. (Edited by R. S. Averback, J. Bernholc, and D. L. Nelson), p. 651. Materials Research Society, Boston, MA (1991).

11. Y. K. Bae, D. C. Lorents, R. Malhotra, C. H. Becker, D. Tse, and L. Jusinski. In *Materials Research Society Symposium Proceedings.* (Edited by R. S. Averback, J. Bernholc, and D. L. Nelson), p. 733. Materials Research Society, Boston, MA (1991).

12. A. Koch, K. C. Khemani, and F. Wudl, *J. Organic Chem.* **56**, 4543, (1991).

13. L. S. K. Pang, A. N. Vassallo, and M. A. Wilson, *Nature* **352**, 480 (1991).

14. D. H. Parker, P. Wurz, K. Chatterjee, K. R. Lykke, J. E. Hunt, M. J. Pellin, J. C. Hemminger, D. M. Gruen, and L. M. Stock, *J. Amer. Chem. Soc.* **113**, 7499 (1991).

15. H. Shinohara, H. Sato, Y. Saito, M. Takayama, A. Izuoka, and T. Sugawara, *J. Phys. Chem.* **95**, 8449 (1991).

16. J. B. Howard, J. T. McKinnon, Y. Marovsky, A. Lafleur, and M. E. Johnson, *Nature* **352**, 139 (1991).

17. G. Peters and M. Jansen, *Angew. Chem. Int. Ed. Eng.* **31**, 223 (1992).

18. J. P. Hare, H. W. Kroto, and R. Taylor, *Chem. Phys. Lett.* **177**, 394 (1991).

19. D. Walton and H. Kroto, *private communication* (1992).

20. T. Pradeep and C. N. R. Rao, *Mater. Res. Bulletin* **26**, 1101 (1991).

21. P. M. Allemand, A. Koch, F. Wudl, Y. Rubin, F. Diederich, M. M. Alvarez, S. J. Anz, and R. L. Whetten, *J. Amer. Chem. Soc.* **113**, 1050 (1991).

22. J. M. Hawkins, T. A. lewis, S. D. Loren, A. Meyer, J. R. Heath, Y. Shibato, and R. Saykally, *J. Organic Chem.* **55**, 6250 (1990).

23. D. M. Cox, S. Behal, M. Disko, S. M. Gorum, M. Greany, C. S. Hsu, E. B. Kollin, J. Millar, J. Robbins, W. Robbins, R. D. Sherwood, and P. Tindall, *J. Amer. Chem. Soc.* **113**, 2940 (1991).

24. A. M. Vassallo, A. J. Palmisano, L. S. K. Pang, and M. A. Wilson, *J. Chem. Soc., Chem. Commun.* 60–61 (1992).

25. F. Diederich, R. Ettl, Y. Rubin, R. L. Whetten, R. Beck, M. Alvarez, S. Anz, D. Sensharma, F. Wudl, K. Khemani, and A. Koch, *Science* **252**, 548 (1991).

26. K. Kikuchi, N. Nakahara, T. Wakabayashi, M. Honda, H. Matsumiya, T. Moriwaki, S. Suzuko, H. Shiromaru, K. Saito, K. Yamauchi, I. Ikemoto, and Y. Achiba, *Chem. Phys. Lett.* **188**, 177 (1992).

27. M. S. Meier and J. P. Selegue, *J. Organic Chem.* **57**, 1924 (1992).

28. K. Chatterjee, D. H. Parker, P. Wurz, K. R. Lykke, D. M. Gruen, and L. M. Stock, *J. Organic Chem.* **57**, 3253 (1992).

29. G. Meijer and D. S. Bethune, *Chem. PHys. Lett.* **175**, 1 (1990).

30. S. W. McElvany and M. M. Ross, *J. Amer. Soc. Mass Spectrometry,* in press.

31. S. W. McElvany, M. M. Ross, and J. H. Callahan, *Accounts of Chem. Res.* **25**, 162 (1992).

32. P. Wurz, K. R. Lykke, D. H. Parker, M. J. Pellin, and D. M. Gruen, *Vacuum,* **43**, 381 (1992).

33. S. Wang and P. R. Buseck, *Chem. Phys. Lett.* **182**, 1 (1991).

34. D. Ben-Amotz, R. G. Cooks, L. Dejarme, J. C. Gunderson, II, S. H. H., B. Kahr, G. L. Payne, and J. M. Wood, *Chem. Phys. Lett.* **183**, 149 (1991).

35. R. S. Ruoff, T. Thorton, and D. Smith, *Chem. Phys. Lett.* **186**, 456 (1991).

36. L. D. Lamb, D. R. Huffman, R. K. Workman, S. Howells, T. Chen, D. Sarid, and R. F. Ziolo, *Science* (1992).

37. C. Smart, B. Eldridge, W. Reuter, J. A. Zimmerman, W. R. Creasy, N. Rivera, and R. S. Ruoff, *Chem. Phys. Lett.,* in press.

38. G. Ulmer, B. Hasselberger, H. Busmann, and E. E. B. Campbell, *Appl. Surface Sci.* **46**, 272 (1990).

39. W. R. Creasy and J. T. Brenna, *J. Chem. Phys.* **92**, 2269 (1990).

40. P. F. Greenwood, M. G. Strachan, G. D. Willet, and M. A. Wilson, *Organic Mass Spectrome.* **25**, 353 (1990).

41. G. Ulmer, E. E. B. Campbell, R. Kuhnle, H. Busmann, and I. V. Hertel, *Chem. Phys. Lett.* **182**, 114 (1991).

42. J. E. Hunt, K. R. Lykke, and M. J. Pellin. In *Methods and Mechanisms for Producing Ions from Large Molecules* (Edited by K. G. Sanding and W. Ens), p. 309. Plenum Press, Minaki, Canada (1991).

43. K. R. Lykke, M. J. Pellin, P. Wurz, D. M. Gruen, J. E. Hunt and M. R. Wasielewski. In *Materials Research Society Symposium Proceedings.* (Edited by R. S. Averback, J. Bernholc, and D. L. Nelson), p. 679. Materials Research Society, Boston, MA (1990).

44. R. E. Smalley, *Accounts Chem. Res.* **25**, 98 (1992).

45. S. H. Hoke, J. Molstad, G. L. Payne, B. Kahr, D. Ben-Amotz, and R. G. Cooks, *Rapid Communi. Mass Spectrom.* **5**, 472 (1991).

46. J. Chmielowiec, P. Fischer, and C. W. Pyburn, *Fuel* **66**, 1358 (1987).

47. J. M. Kauffman and C. O. Bokman, *J. Chem. Edu.* **53**, 33 (1976).

48. P. Wurz, K. R. Lykke, M. J. Pellin, and D. M. Gruen, *J. Appl. Phys.* **70**, 6647 (1991).

FULLERENES SYNTHESIS IN COMBUSTION

Jack B. Howard, Arthur L. Lafleur, Yakov Makarovsky, Saibal Mitra,
Christopher J. Pope and Tapesh K. Yadav
Department of Chemical Engineering, Center for Environmental Health Sciences, and Energy
Laboratory, Massachusetts Institute of Technology, Cambridge, MA 02139, U.S.A.

(*Accepted* 29 *April* 1992)

Abstract—The early suggestion in fullerenes research that fullerenes might be produced in flames was soon supported by the observation of polyhedral carbon ions in flames, and in 1991 was confirmed by the recovery and identification of fullerenes C_{60} and C_{70} from benzene/oxygen flames. Recent research has determined the effects of pressure, carbon/oxygen ratio, temperature, and the type and concentration of diluent gas, on the yields of C_{60} and C_{70} in subatmospheric pressure premixed laminar flames of benzene and oxygen. Similar flames but with acetylene as fuel have also been found to produce fullerenes, but in smaller yields than with benzene fuel. The largest observed yields of $C_{60}+C_{70}$ from benzene/oxygen flames are substantial, being 20% of the soot produced and 0.5% of the carbon fed. The largest rate of production of $C_{60}+C_{70}$ was observed at a pressure of 69 Torr, a C/O ratio of 0.989 and a dilution of 25% helium. Several striking differences between fullerenes formation in flames as compared to the widely used graphite vaporization method include, in the case of flames, an ability to vary the C_{70}/C_{60} ratio from 0.26 to 8.8 (cf., 0.02 to 0.18 for graphite vaporization) by adjustment of flame conditions, and production of several isomers each of fullerenes C_{60}, C_{70}, $C_{60}O$, and $C_{70}O$. Many of the apparent isomers are thermally metastable, one C_{60} converting to the most stable form with a half-life of 1 h at 111°C. The structures of the apparent C_{60} and C_{70} isomers necessarily must include abutting five-membered rings, previously assumed to be disallowed because of their high strain energy. The chemistry of fullerenes formation in flames is in some ways similar to that of soot formation, but important differences are seen and assumed to reflect the effects of the curved, strained structures of fullerenes and their precursors.

Key Words—Fullerenes, C_{60}, C_{70}, fullerenes formation, fullerenes synthesis, flames, combustion, soot formation, polycyclic aromatic hydrocarbons.

1. BACKGROUND

Fullerenes were discovered[1] in carbon vapor produced by laser irradiation of graphite, and were later produced in macroscopic quantities[2–5] by graphite vaporization with resistive heating. The possibility that fullerenes may be formed in sooting flames was suggested[6–9] at an early stage in fullerene research. Zang *et al.*[6] considered the possible role of carbon shell structures in the formation and morphology of soot. They envisioned the growth of successive shells, separated by roughly the 0.33-nm intersheet distance in graphite, producing a soot nucleus consisting of concentric but slightly imperfect spheres, with the edge of the outermost shell giving a very rapid growth front. Kroto and McKay[7] described a carbon nucleation scheme involving quasi-icosahedral spiral shell carbon particles, and suggested it may apply to soot. Curl and Smalley[8] suggested that carbon nets in the form of spiral structures may be important to soot formation in flames. Recent reviews of these early concepts are given by Kroto *et al.*[10–12].

Evidence that fullerenes can be formed in flames was at first elusive, but progress was eventually made. All-carbon ions having charge/mass ratios similar to those reported for fullerenes in graphite vaporization were observed by Homann *et al.*[13–15] using on-line molecular beam/mass spectrometric probing of low-pressure premixed benzene-oxygen and acetylene-oxygen flames. Iijima[16] published an electron micrograph of a soot particle in which a circular fea-

ture was interpreted to be evidence of a fullerene. Malhotra and Ross[17] using field ionization mass spectrometry studied several soots from pyrolysis and combusion processes but found no peaks corresponding to C_{60} and C_{70} fullerenes.

The presence of C_{60} and C_{70} fullerenes in substantial quantities was reported from spectroscopic analysis of samples collected from low-pressure premixed benzene–oxygen flames at Massachusetts Institute of Technology (MIT)[18–19]. The presence and group behavior of high molecular weight compounds having molecular weights up to about 1000 g/mole in benzene-oxygen flames had been studied earlier at MIT using molecular beam sampling with on-line mass spectrometry[20–21] as well as probe collection of condensible material including soot with subsequent analysis by solvent extraction and other methods[22]. Intriguing but inconclusive evidence of the possible presence of fullerenes had been seen in the data from both the on-line mass spectrometer[21, 23] and exploratory fast atom bombardment mass spectroscopic analysis of collected samples[22]. The presence of fullerenes in these flames was finally established[18,19], soon after the discovery[2] and early implementation[3–5] of the process for producing macroscopic quantities of fullerenes by graphite vaporization with resistive heating. The same solvent extraction and spectroscopic techniques employed in the graphite vaporization studies were applied successfully in the study of the flame samples.

The knowledge of fullerenes formation in flames has grown significantly during the short time since this process was first established. The effects of combustion conditions on the distribution and yields of C_{60} and C_{70} fullerenes have been studied and the amount of fullerenes formed at different distances or times through the reaction zone of selected flames has been measured. The results show that flame synthesis offers not only an alternative method for large scale fullerenes production, but an ability to control the distribution of products (e.g., the C_{70}/C_{60} ratio) over a larger range than has been realized in the graphite vaporization method. Also, a range of fullerenes including metastable isomers can be produced in flames[24]. Plausible mechanisms of fullerenes formation in flames have been proposed and subjected to preliminary kinetics testing against data[25]. These advances are reviewed below.

2. EQUIPMENT AND TECHNIQUES

The combustion system used in the fullerenes synthesis reported from MIT is a premixed laminar flame stabilized on a water-cooled burner illustrated in Fig. 1. Benzene was the fuel in all the work described below, but acetylene has also been used successfully in the MIT work. Either pure oxygen or oxygen mixed with argon, helium, or nitrogen or some combination thereof, is fed with the fuel.

The burner is in a low-pressure chamber equipped with windows and feed-throughs for visual observation, optical diagnostics, electrical ignition, and monitoring and sampling probes. The chamber is exhausted into a vacuum pump not shown in Fig. 1. The burner consists of a horizontal copper plate (100-mm diameter and 12-mm thick, drilled through with 1-mm-diameter holes centered 2.5 mm apart in a tri-

angular array) upward through which the feed mixture is delivered. The flame is stabilized with a flat front uniformly displaced from the burner plate by a short distance which depends on the velocity of the gas leaving the burner and the flame speed of the mixture. Only the inner 70-mm diameter section of the burner plate is used for the experimental flame. The 15-mm wide outer section is used for an independently fed fuel-rich but nonsooting ethylene/oxygen/inert diluent flame. This annular flame shields the experimental flame, allowing it to approximate a one-dimensional core within which temperature and species concentrations vary only with distance, or residence time, from the burner surface, thereby simplifying the mathematical analysis of data.

The essential features of the burner were designed to duplicate an extensively used burner in our laboratory, so as to allow reproduction of flame conditions for which profiles of temperature and concentration of many species including radicals had already been measured[20,26]. The present burner was previously built and used in mechanistic studies of soot nucleation and growth[22] and the flames studied are of a type for which considerable data on temperature and chemical composition are available[13–15,18,20,22,26–30].

Fullerenes formation has been studied under different sets of flame conditions over the following ranges: burner chamber pressure, 12–100 Torr (1.60–13.35 kPa); atomic C/O ratio, 0.717–1.082; mol % Ar, He, N_2, or combinations thereof, 0–50; and gas velocity at the burner (298 K), 14.6–75.4 cm/s. The work has included one nonsooting flame, which does form fullerenes, produced under conditions (20 Torr, C/O = 0.72, 30% Ar, and 50 cm/s) where soot formation would be impending if the C/O ratio were increased by 5% with the other conditions not changed. Many of the flames were maintained for about 1–3 h, depending upon conditions, while a sample of condensible compounds and soot was withdrawn from the flame at a given distance from the burner using a quartz probe connected to a room-temperature filter, vacuum pump, and gas meter (Fig. 1). The probe consists of a tube tapered to an orifice at the tip. The probe is held vertically with the orifice directed upstream. The mass of the sample was primarily that of soot, except at the lowest C/O ratio, where the flame was nonsooting. Soot was also collected from the inside surface of the burner chamber after each run in which a probe sample was taken as well as after many other, exploratory, runs conducted solely for the surface deposited samples. Most of soot deposition inside the burner chamber occurs on the top surface, and hence at a large distance from the burner as compared to the probe sampling positions.

The samples of condensed compounds and soot were weighed, extracted with toluene using an ultrasonic bath at room temperature, and filtered. The toluene extracts were analyzed by high performance liquid chromatography (HPLC) using a Hewlett-Packard model 1090 HPLC equipped with a ternary

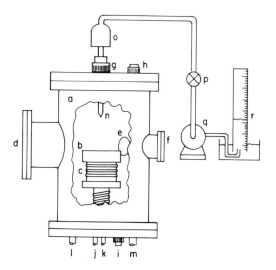

Fig. 1. Burner and associated equipment: a—low-pressure chamber; b—copper-burner plate; c, water cooling coil; d, e, and f—windows; g, h, and i—feedthroughs; j—annular-flame feed tube; k—core-flame feed tube; l and m—exhaust tubes; n—sampling probe; o—filter; p—valve; q—vacuum pump; r—gas meter.

pumping system and diode-array detector. The instrument is controlled with a model 7994 Analytical Workstation. The HPLC column is 25-cm long × 4.6-mm I.D. packed with 5-μm particles of Nucleosil (Macherey-Nagel, Duren, FRG) octadecylsilyl bonded C18 silica having 30-nm pores. A binary non-aqueous mobile phase of acetonitrile and dichloromethane was used in a gradient elution mode. The mobile phase program consists of a linear increase in dichloromethane concentration from 10 to 100% in 40 min with a holding time of 10 min at 100%. The flow rate is 1.0 mL/min. Preparative scale separations were performed with a semi-preparative Nucleosil octadecylsilyl-bonded silica analytical column 25-cm long × 1.0-cm I.D. packed with 7-μm material having 6-nm pores. Both columns were obtained from American Bioanalytical, Natick, MA.

Ultraviolet spectroscopy was conducted with ultraviolet spectroscopic detection (UVD) using a Hewlett-Packard model 8450A diode-array spectrophotometer with a 7225B plotter and 9121B disc drive. Ultra-pure glass-distilled decahydronaphthalene (decalin) was used as the solvent in order to ensure adequate dissolution and to maximize penetration into the UV. Pure C_{60} and C_{70} fullerenes were collected by HPLC and concentrated by evaporation under a stream of nitrogen. The concentrated fullerene solutions (in HPLC solvent) were exchanged into decalin by adding a measured volume to the fullerene solution and evaporating under a stream of nitrogen until the more volatile HPLC mobile phase evaporated leaving the higher-boiling decalin.

Electron impact mass spectra were obtained with a Varian-MAT model 731 mass spectrometer interfaced to a Teknivent data system. Samples of C_{60} and C_{70} fullerenes were isolated by preparative HPLC and were concentrated and evaporated to dryness in a suitable probe vessel using a vacuum centrifuge. Mass spectra were acquired as the direct injection probe was heated from 100–400°C.

The HPLC/UVD analysis of flame samples revealed the presence not only of the same C_{60} and C_{70} fullerenes as observed in graphite vaporization, but also numerous additional compounds with fullerene-like characteristics[18]. Several of these additional compounds have been analyzed using complementary techniques involving liquid chromatography coupled directly to mass spectrometry (LC/MS)[24]. The compounds were found to include thermally metastable fullerene isomers, the mass spectral analysis of which requires a more gentle ionization method than is usually used in fullerenes analysis. The method used is pneumatic nebulization coupled with atmospheric-pressure chemical ionization source in which electron transfer to benzene molecular cations is arranged to be the dominant ionization mechanism[24].

3. IDENTIFICATION OF FLAME DERIVED FULLERENES

Analysis of the toluene extract of a flame derived soot gave the electron impact mass spectrum shown in Fig. 2[18]. Comparing these data with those reported for fullerenes[1–4,31] revealed that the soot sample contained a mixture of C_{60} and C_{70} fullerenes showing molecular ions at m/e 720 and 840, respectively, and doubly charged molecular ions at m/e 360 and 420, respectively. This conclusion was confirmed[18] by Fourier transform infrared spectroscopy of the soot extract, which gave strong absorption peaks consistent with those reported for fullerenes C_{60}[2–5] and C_{70}[4].

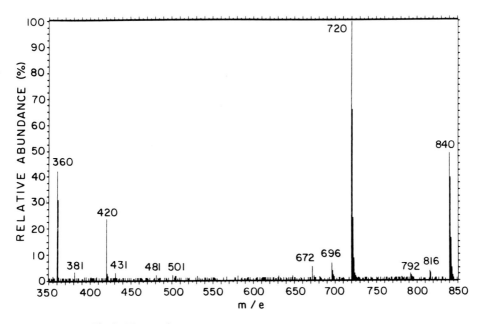

Fig. 2. Electron-impact mass spectrum of a flame soot extract[18].

Although the MS data strongly suggested that fullerenes were the major constituents of the soot, any polycyclic aromatic hydrocarbons (PAH) having molecular weights of 720 or 840 were likely to give the same mass spectra as those in Fig. 2. With improvements in chemical analysis, larger and larger PAH are being observed in combustion samples. Therefore, in order to confirm the above findings, further analysis was performed using high performance liquid chromatography with spectrophotometric diode-array detection (HPLC/DAD). This technique involves the continuous acquisition of UV spectra as peaks elute from the HPLC. The UV spectra of PAH are highly characteristic and can even permit the differentiation of isomeric PAH, a task difficult to achieve by MS[32]. We evaluated a number of HPLC separation schemes for PAH but focused primarily on those shown effective for PAH having upwards of 10 fused rings[33].

Analysis of toluene extract by HPLC/DAD gave the chromatogram shown in Fig. 3[18]. The signal consists of the broadband UV absorption in milliabsorbance units (mAU) over the 236–600-nm wavelength interval. The broadband UV response plotted in Fig. 3 is roughly proportional to mass for PAH[34]. The most striking feature of the chromatogram is the virtual absence of peaks associated with the typical PAH commonly produced in flames. The peaks labeled C_{60} and C_{70} gave UV spectra closely matching those published for C_{60} and C_{70} fullerenes, respectively[2,4]. The peaks labeled $C_{60}O$, C_{60} Isomer, C_{70} Isomer and C_{84} were identified by HPLC/UVD analysis[24] mentioned above. Two minor nonannotated peaks preceding the C_{84} peak were found to correspond to C_{76}, and evidence was found for several different $C_{60}O$ and $C_{70}O$ isomers as well as several forms of C_{90} and C_{94}[24]. Isomers of C_{60} and C_{70} fullerenes previously had not been observed experimen-

tally. These apparent isomers and many of the others were observed to be metastable[24], as discussed below. The presence of C_{60} and C_{70} isomers in flame samples has important implications discussed below.

A full-range ultraviolet-visible (UV-Vis) spectra of the C_{60} and C_{70} peaks, acquired by spectrophotometric analysis of preparative-scale HPLC fractions of the flame sample extract are illustrated in Figs. 4 and 5. The spectra are virtually identical to those reported by Ajie *et al.*[4] for fullerenes obtained from graphite vaporization. These identifications were confirmed by mass spectral analysis of the HPLC fractions. The C_{60} peak gave a mass spectrum with the reported features of C_{60} fullerene having a molecular ion base peak at m/e 720, showing no loss of hydrogen and having a doubly charged molecular ion at m/e 360[2–4,31]. Similarly, the C_{70} peak gave a mass spectrum with features closely matching those of published spectra for C_{70} fullerene showing a molecular ion base peak at m/e 840 and a doubly charged molecular ion at m/e 420[2–4,31].

4. FULLERENE YIELDS UNDER DIFFERENT FLAME CONDITIONS

The HPLC method described above, including gravimetric calibration of the C_{60} and C_{70} peaks was used to analyze toluene extracts of flame samples. The yields of fullerenes C_{60} and C_{70} and the C_{70}/C_{60} ratio found under different flame conditions are shown in Table 1. Samples 2A and 2B are from a nonsooting flame. Soot values listed in the table refer to the whole sample, consisting of the fraction soluble in toluene, which was largely fullerenes and polycyclic aromatic hydrocarbons, and the toluene insoluble material. The mass of soot collected with the probe is expressed as a fraction of the carbon in the fuel fed to the flame. The calculation of a product yield as a frac-

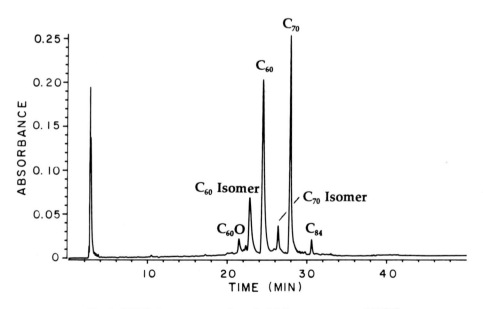

Fig. 3. HPLC chromatogram of a typical fullerene soot extract[18,24].

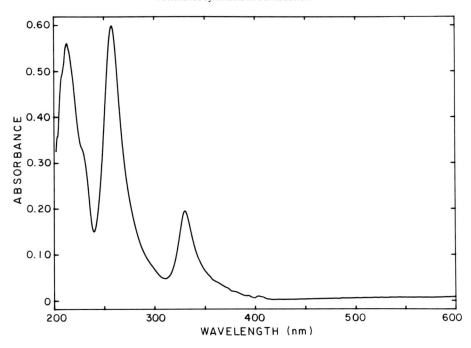

Fig. 4. UV-Vis spectrum of HPLC C_{60} fullerene peak[18].

tion of carbon fed is based on the metered volume of flame gas withdrawn with the condensed sample, the known feed rates and burner chamber pressure, and a nominal flame temperature of 1800K which is representative of previous measurements under similar conditions[21,22]. The increase in number of moles accompanying combustion under the different conditions was approximated by equilibrium calculations using STANJAN[35]. The calculation of a product yield as a fraction of carbon fed is possible only for the samples withdrawn with the probe, because the volume of flame gas associated with the collected material is not known for samples removed from the chamber surface.

The effective flame position represented by a sample withdrawn with the probe is a few probe orifice diameters upstream of the physical position of the probe tip[36]. The orifice diameter employed for the

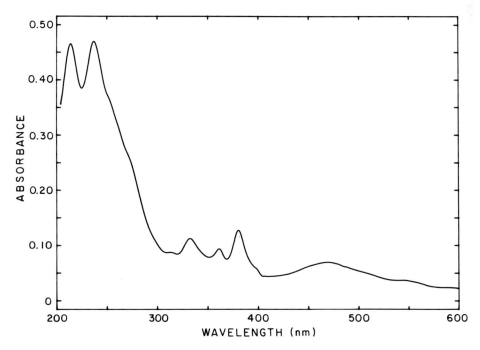

Fig. 5. UV-Vis spectrum of HPLC C_{70} fullerene peak[18].

Table 1. Experimental conditions, extent of soot formation, and fullerene yields in flames

Sample Number	P^a, Torr	C/O^b, molar	V^c, cm/s	Ar^d, %	Dist.^{e}, cm	Soot Fullerene Mass, % of soot				C_{70}/C_{60}, molar
						% of C^f	C_{60}	C_{70}	$C_{60}+C_{70}$	
1A	12	0.960	75.4	0	3.69	0.75	0.73	1.1	1.8	1.2
1B	12	0.960	75.4	0	CS^g	—	0.0063	0.0065	0.013	0.83
2A	20	0.717	50	30	1.425	0	—	—	—	0.74
2B	20	0.717	50	30	CS	—	—	—	—	0.86
3A	20	0.784	50.3	30	1.68	0.23	0.24	0.44	0.68	1.6
3B	20	0.784	50.3	30	CS	—	—	—	—	1.4
4A	20	0.960	50.3	10	2.11	—	2.2	3.2	5.4	1.3
4B	20	0.960	50.3	10	CS	—	0.15	0.39	0.54	2.2
5A	20	0.996	49.1	10	8	3.6	3.4	3.5	6.9	0.88
5B	20	0.996	49.1	10	CS	—	2.6	4.6	7.3	1.5
6A	40	0.960	25.1	10	0.8	3.0	0	0	0	0/0
6B	40	0.960	25.1	10	CS	—	0.76	1.6	2.4	1.8
7A	40	0.960	25.1	10	1.175	—	0	0	0	0/0
8A	40	0.960	25.1	10	1.5	1.8	1.6	3.1	4.7	1.6
8B	40	0.960	25.1	10	CS	—	1.0	1.8	2.8	1.5
9A	40	0.960	25.1	10	3.0	1.5	3.8	5.4	9.2	1.2
9B	40	0.960	25.1	10	CS	—	1.8	2.9	4.7	1.7
10A	40	1.072	23.4	11	2.15	7.3	0.023	0.0072	0.030	0.26
10B	40	1.072	23.4	11	CS	—	0.0022	0.0070	0.0092	2.7
11A	100	0.960	14.6	38	1.45	8.4	0.0089	0.024	0.033	2.4
11B	100	0.960	14.6	38	CS	—	0.0017	0.0062	0.0079	3.1
12A	100	0.996	14.4	39	1.21	12	0.0038	0.025	0.029	5.7
12B	100	0.996	14.4	39	CS	—	0.0016	0.0010	0.0026	0.56
13A	100	0.996	37.2	39	1.5	2.4	1.2	1.7	2.9	1.2
13B	100	0.996	37.2	39	CS	—	3.5	4.3	7.8	1.1

^{a}Pressure in combustion chamber.
^{b}Carbon to oxygen ratio in flame.
^{c}Gas velocity at burner surface, at pressure P and 25°C.
^{d}Molar percentage of argon in feed gas.
^{e}Perpendicular distance from burner surface to orifice at tip of sampling probe.
^{f}Total carbon fed.
^{g}Denotes samples collected from combustion chamber surface.

data in Table 1 was 2 mm except in the cases of samples 6A, 8A, and 9A, where it was 0.7 mm, and sample 7A, where it was 1.0 mm. The smaller orifice diameters were used to achieve a finer resolution of distance from the burner. If the upstream displacement is assumed to be two to three orifice diameters, sample 7A would be equivalent to a sample taken with a 0.7-mm diameter probe orifice located approximately 1.1 cm from the burner surface. Accordingly, data from sample 7A may be combined with those from samples 6A, 8A, and 9A in the study of concentration profiles. With similar adjustments for the upstream displacement, samples 2A and 3A are the equivalent of samples taken with a 0.7-mm diameter probe orifice at distances from the burner surface of 1.10 and 1.36 cm, respectively. Based on previous measurements in these two flames[20,21,26], the last two distances mark the positions of the peak concentration of the sum of all species of molecular weight 700 g/mol or larger.

The data shown in Table 1 provided the basis for the first reported characteristics of fullerenes formation in flames[18]. This information has since been supplemented by many other experimental runs at different sets of conditions of pressure, C/O ratio, gas velocity at the burner, and diluent identity and concentration. The conditions were selected with emphasis on larger yields of fullerenes C_{60} and C_{70}, the objective being to achieve an approximate multivariable optimization. For expediency, the flame probe was not employed and samples were collected only from the burner chamber surface. A separate set of supplementary data was obtained using the probe to collect samples from a given flame at different distances from the burner. These data provide more information on the effect of distance or residence time in the flame on fullerenes formation than can be seen in Table 1. All the data are summarized below.

The mass of $C_{60}+C_{70}$ produced under the different sooting flame conditions is in the range 0.0026–20% of the soot mass, compared to 1–14% from graphite vaporization[3–5,37], using similar solvents and extraction procedures in both cases. The $C_{60}+C_{70}$ yield, expressed as a percentage of fuel carbon, can only be computed directly in cases where the flame samples were taken with the probe, as described above. The yields thus obtained range from 2×10^{-4}% for the nonsooting flame (Table 1, sample 2A) to 0.26% at a pressure of 20 Torr, a C/O ratio of 0.996, 10% Ar, and a gas velocity at the burner of 49.1 cm/s. However, larger yields can be estimated as follows from the recent experiments in which flame samples were collected only from the burner chamber surface. Data from the experiments in which both probe and sur-

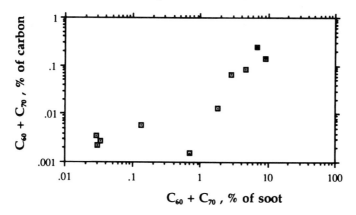

Fig. 6. Yield of $C_{60}+C_{70}$ fullerenes as a percentage of carbon fed and as a percentage of soot, from probe
samples (Table 1).

face samples were collected (Table 1) reveal an approximate relationship between the amounts of soot and fullerenes depositing on the chamber surface and the amounts actually produced in the flame. As can be seen in Fig. 6, the data from the probe samples of Table 1 reveal considerable correlation between the $C_{60}+C_{70}$ as a percentage of the carbon fed and the $C_{60}+C_{70}$ yield as a percentage of the soot produced. Considering the trend seen in these data in the region of higher yields (Fig. 6), the highest $C_{60}+C_{70}$ yield of 20% of the soot, which was deduced from surface samples collected in the multivariable optimization experiments, would appear to correspond to approximately 0.5% of the carbon fed. The same conclusion can be reached from Fig. 7 where both the probe sample data from Table 1 and the surface sample data from the optimization experiments are shown to-

gether. Although the data shown on this basis are very scattered, owing in part to the wide ranges of experimental conditions represented, the $C_{60}+C_{70}$ yields as a percentage of soot tend to be distributed around a peak at a soot yield of about 1% of the carbon in the case of the surface samples, and at about 2–3% of the carbon in the case of the probe samples. The implication is that the largest $C_{60}+C_{70}$ yield in the optimization experiments is about 0.4–0.6% of the carbon, essentially the same as the 0.5% deduced above. Thus the flame synthesis can convert a kilogram of benzene to over 4 g of $C_{60}+C_{70}$. Given the ability to scale up combustion processes in flow reactors, as for example in carbon black production, flame synthesis of fullerenes would appear to offer potential for large-scale production.

The highest production rate of fullerenes ($C_{60}+$

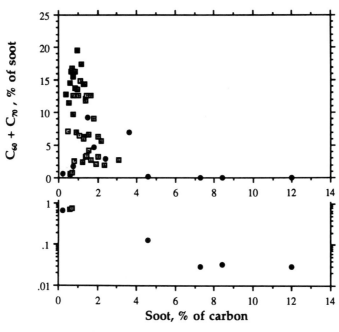

Fig. 7. Yields of $C_{60}+C_{70}$ fullerenes and yields of soot, from probe samples (●) and chamber surface deposits (□).

$C_{60}+C_{70}$ is shown in Fig. 14(a) for a flame at 40 Torr with a C/O ratio of 0.99 and a helium concentration of 25%. As invoked above, the yield of $C_{60}+C_{70}$ in a given flame is seen to increase with increasing burner gas velocity. At higher velocities the flame front is stabilized farther from the burner and a smaller fraction of the heat released by combustion is lost by conduction into the water-cooled burner plate, thus giving a hotter flame. The C_{70}/C_{60} ratio [Fig. 14(b)] for this flame exhibits a maximum as the velocity is increased. The trend of increasing C_{70}/C_{60} with increasing velocity appears to be an exception to the almost general rule that the C_{70}/C_{60} ratio is larger the smaller the yield.

The effect of pressure on the $C_{60}+C_{70}$ yield and production rate is shown in Fig. 15 for several flames operated under the same set of conditions (C/O = 1.0, velocity 50 = cm/s, and 0 to 25% helium) except for pressure. The measurements were performed by collecting the material deposited on the burner chamber surface. This method underestimates the true production rate of soot and fullerenes by a factor of about 2.5 as described above. Data obtained with a sampling probe in a 20 Torr flame under these conditions (Table 1, sample 5A) give a production rate 3.6 times larger than the point at 20 Torr in Fig. 15(b), but the smaller factor of about 2.5 is probably appropriate at the larger pressures. As can be seen in the figure, both the fullerenes yield and production rate under these conditions appear to exhibit a peak

at about 40 Torr. However, other data at higher pressures and lower velocities but otherwise identical conditions to those represented here, together with consideration of the effect of velocity, would suggest that the apparent peaks may be data scatter. More work is required to establish the behavior at higher pressures.

The effect of diluent concentration on $C_{60}+C_{70}$ yields is shown in Fig. 16(a) for two flames, one with high yields (37.5 Torr, C/O = 2.5, 40 cm/s and helium diluent) and the other with low yields (20 Torr, C/O = 2.6, 50 cm/s and helium diluent). The effect is more pronounced for the low yield flame. The yield goes through a maxima as helium concentration is varied from 10 to 50%. For the high-yield flame, the increased helium concentration marginally reduces the fullerene yield. As can be seen in Fig. 16(b), the C_{70}/C_{60} ratio increases significantly as the helium concentration is increased in the low-yield flame, but changes little in the high-yield flame.

Helium, argon and nitrogen have been studied as diluents as indicated in Fig. 17. Helium consistently gives higher yields than the other two. The highest $C_{60}+C_{70}$ yields as percentage of soot achieved with each gas are 19.6% with helium, 14.5% with argon, and 13.8% with nitrogen. Yields as high as 16.8% have been produced with no diluent. The yields presented in Fig. 17(a) are for two different flames, one with high yields and the other with low yields. The effect of diluent is prominent for low-yield flames. For

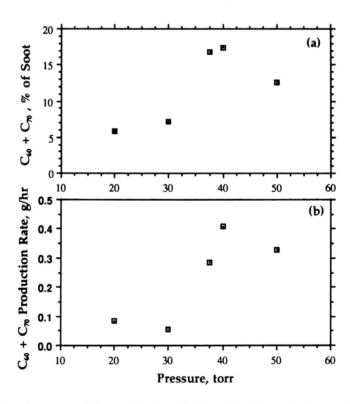

Fig. 15.   Effect of pressure on fullerenes $C_{60}+C_{70}$ yield (a) and $C_{60}+C_{70}$ production rate (b) for flames at
C/O = 1.0, 50 cm/s and 0–25% helium, from chamber surface deposits.

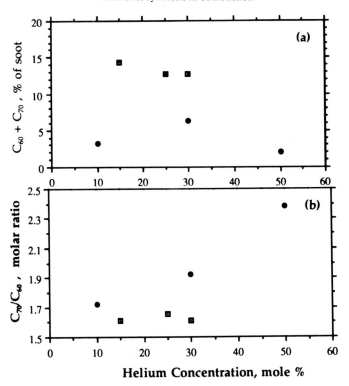

Fig. 16. Effect of diluent helium concentration on fullerenes $C_{60}+C_{70}$ yield (a) and C_{70}/C_{60} ratio (b) for two sets of flame conditions: ($\square$) 37.5 Torr, C/O = 1.00 and 40 cm/s; ($\bullet$) 20 Torr, C/O = 1.04 and 50 cm/s (from chamber surface deposits).

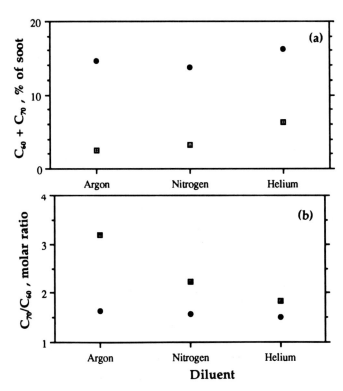

Fig. 17. Effect of diluent type on fullerenes $C_{60}+C_{70}$ yield (a) and C_{70}/C_{60} ratio (b) for two sets of flame conditions: ($\bullet$) 20 Torr, C/O = 1.04, 50 cm/s and 30% diluent; ($\square$) 37.5 Torr, C/O = 0.96, 50 cm/s and 25% diluent (from chamber surface deposits).

example, replacement of argon with helium in the low-yield flame more than doubles the yield of C_{60} + C_{70}. In high-yield flames, a change from argon to helium only changes the yield by about 10%. This general observation is further supported by the achievement, mentioned above, of large yields without any diluent. Thus, the presence of diluent is not essential for flames favoring fullerene production.

The change in C_{70}/C_{60} ratio with change in diluent shows a pattern similar to that of yield. As can be seen in Fig. 17(b), changing the diluent from argon to helium reduces the C_{70}/C_{60} molar ratio. This effect is more pronounced for low-yield flames than for high-yield flames.

These data indicate that the presence of diluent is not critical except under conditions only moderately favorable to fullerenes formation. The relative effectiveness of the three diluents studied in the promotion of larger yields of C_{60} and C_{70} is in the order of their diffusivities. Whether the effect has to do with thermal conductivity and the temperature profile, and/or molecular diffusion and species concentration profiles, both in the fullerenes formation zone of the flame, requires more study.

5. CONTRASTS WITH GRAPHITE VAPORIZATION

Comparison of fullerenes synthesis by combustion with the widely used technique of carbon-rod vaporization reveals several differences. Although the two methods achieve comparable yields of fullerenes C_{60} and C_{70} as a percentage of the soot produced, the 12.2% yield reported above from flame synthesis with the conditions adjusted to give the largest observed production rate is apparently considerably larger than the yield achieved when the carbon-rod vaporization conditions are similarly adjusted. This relative attribute of flame synthesis is especially noteworthy since the data for that method were obtained with an improvised system (i.e., a burner and associated equipment designed for other purposes and in no way modified for the fullerenes research). With the main part of the system being a continuous-flow gas-phase reactor, the application of familiar optimization and control techniques should lead to even better performance. Also, this type of reactor can be scaled up as needed.

A striking contrast between the two production methods is the much larger C_{70}/C_{60} ratio obtained from combustion, and the ability to set this ratio to different values over more than a 30-fold range by the adjustment of flame conditions. The ability to promote the yield of certain products apparently extends, at least to some degree, to larger fullerenes. Exploratory extraction of some of the flame samples with more potent solvents such as chlorobenzene, trichlorobenzene, and tetramethylbenzene followed by liquid chromatography shows evidence of larger fullerenes, in amounts varying with flame conditions. Whether flame generated fullerenes extend to or beyond the largest sizes produced in graphite vaporization[38–41] remains to be determined.

The production of metastable fullerene isomers in flames, mentioned above, is another major contrast with the graphite vaporization method. Several apparent isomers of C_{60}, C_{70}, $C_{60}O$, and $C_{70}O$ were detected in toluene extract of flame-derived soot, but evidence for only two $C_{70}O$ compounds was observed in an extract of soot produced by graphite vaporization[24]. The flame-derived fullerenes also include isomers of C_{76}, C_{84}, C_{90}, and C_{94}, and many hydrogen-containing complexes including $C_{60}H_2$, $C_{60}H_4$, $C_{70}H_2$, $C_{60}CH_4$, which may be $[C_{60}(CH_2)(H_2)]$ or $[C_{60}(H)(CH_3)]$, as well as tentatively identified $C_{60}(CH_4)_2$ and $C_{60}(C_7H_6)$[24,42].

The observation of isomers of C_{60} and C_{70} fullerenes in flame-derived soot extract would be the first production and collection of fullerenes having adjacent 5-rings in their structure. The separation of 5-rings by at least one 6-ring so as to avoid the strain energy added by adjacent 5-rings has been a widely accepted constraint in the study of fullerene structures[43]. This constraint must be violated by C_{60} and C_{70} isomers since the separation of 5-rings rule can be satisfied by only one C_{60} fullerene and one C_{70} fullerene, which are the stable flame-derived C_{60} and C_{70} fullerenes[18]. The constraint must also be violated by the apparently C_{50} fullerene, seen as a prominent peak in mass specrtra from molecular beam sampling of carbon vapor[1,6,44] and flame[13–15] systems, but not yet reported from analyses of macroscopic samples. A number of different C_{60} isomers have been considered theoretically in structural or stability analyses[43,45–50]. Experimental study of the structure of the flame-synthesized C_{60} isomers is now underway.

Metastability of the apparent C_{60} isomers is revealed by the data[24] presented in Fig. 18. At the temperature of boiling toluene (111°C), the isomer is converted with a half-life of about 1 h. As the reaction proceeds, the amount of the isomer consumed is essentially equal to the amount of the stable C_{60} fullerene produced, thus indicating that the isomer is converted to the stable structure. The lower stabilities of the experimentally observed C_{60} isomers relative to that of the stable C_{60} fullerene is qualitatively consistent with the presence of adjacent five-membered rings in the isomeric structures. The possibility of intramolecular rearrangements from less symmetrical C_{60} isomers to the icosahedral C_{60} has been predicted theoretically[49]. Whether the data in Fig. 18 reflect only isomerization in the presence of the solvent, or transformation assisted by $C_{60}O$ and/or $C_{70}O$ (the latter being present but not shown in the figure) requires more study.

The higher reactivity of the metastable isomers relative to the more stable form, or forms in the case of higher fullerenes, presumably extends to chemical reactions, such as complexing and cross linking, as well as physical transformations. This higher reactivity may be advantageous in some approaches to fullerene applications.

Another contrast between fullerenes synthesis in

CHEMICAL REACTIVITY AND FUNCTIONALIZATION OF C₆₀ AND C₇₀ FULLERENES

GEORGE A. OLAH, IMRE BUCSI, ROBERT ANISZFELD and G. K. SURYA PRAKASH

Loker Hydrocarbon Research Institute and Department of Chemistry, University of Southern California, University Park Los Angeles, CA 90089-1661, U.S.A.

(*Accepted* 28 *April* 1992)

Abstract—This account discusses the chemical reactivity and various functionalizations of C_{60} and C_{70} fullerenes with emphasis on results derived from the authors' laboratory.

Key Words—Fullerenes, chemical reactivity, oxidation, reduction halogenation, free radicals.

1. INTRODUCTION

In 1985 it was discovered that vaporization of graphite by laser irradiation produces remarkably stable molecular allotropic forms of carbon C_{60} cluster (Fig. 1) and to a lesser extent a stable C_{70} cluster (Fig. 2) as was shown by mass spectrometry[2]. Kroto *et al.*[2] proposed the structure for the sixty carbon cluster to be a truncated icosahedron composed of 32 faces of which 12 are pentagonal and 20 are hexagonal, a structure analogous to a soccerball and reminiscent of the geodesic domes of Buckminster Fuller. Thus, C_{60} is commonly referred to as "buckminsterfullerene."

Initial experimental support for the spherical structure of C_{60} included mass spectrometric studies using lanthanum impregnated graphite which showed intense $C_{60}La$ peaks[3]. A multitude of theoretical studies[4] also indicated the truncated icosahedral structure (Fig. 1) for C_{60} to be a stable closed shell system and aromatic. In early 1990 Krätschmer *et al.* obtained spectroscopic evidence for C_{60} in smoke from thermal evaporation of graphite rods[5a] by observing four weak but distinct IR bands which were consistent with theory[4h–k]. Later Krätschmer *et al.* reported isolation and characterization of pure C_{60} and C_{70} using a chemical extraction technique [5b]. At the same time Kroto and co-workers[6] and Johnson *et al.*[7] have prepared a mixture of C_{60} and C_{70} using similar techniques. The fullerenes were separated by Kroto[6] via column chromatography on neutral alumina and their structure characterized by ¹³C NMR spectroscopy. The ¹³C NMR spectrum of C_{60} in benzene consists of a single line at 142.7 ppm confirming the icosahedral structure. The ¹³C NMR spectrum of C_{70} in benzene consists of five lines (150.7, 148.1, 147.4, 145.4, and 130.9 ppm in 1:2:1:2:1 ratio, respectively) confirming a highly symmetrical egg-shaped structure[8] (Fig. 2) (C_{5h} symmetry). The structure of C_{70} has been recently confirmed by two-dimensional INADEQUATE NMR spectroscopy[9]. The C–C bond lengths in C_{60} has been determined by solid state ¹³C NMR spectroscopy[10]. Furthermore, X-ray crystal structures of several C_{60} derivatives have also been determined[11–13].

C_{60} and C_{70} is now being routinely prepared in gram quantities using bench-top reactors[14]. Authors themselves have built a nine rod reactor using gravity feed technique allowing continuous burning of graphite electrodes without the need of opening the reactor after evaporation of each rod[15]. In one day in a single operation with this reactor 25–30 g of soot can be produced containing 8–10% of fullerenes. Separation of C_{60} and C_{70} and related higher fullerenes from fullerene mixture is also relatively well established[6,11,16]. In an elegant work Diederich *et al.*[16b,c] have isolated and characterized two isomers of C_{78} and one distinct chiral C_{76} allotrope. Other higher fullerenes have also been isolated[16d].

Most of the chemical reactivity and functionalization studies have been centered on C_{60} because of its relative abundance in the fullerene extracts compared to other higher fullerenes. In this account we will discuss various functionalization chemistry that has been carried out with C_{60} and related fullerenes. Some pertinent chemical reactivity studies will also be considered. The emphasis is placed on results of our own laboratory without diminishing the contributions made by others in this field, notably by Wudl and his co-workers. Gas phase[17a], metal ion encapsulation[17b] and doped fullerene chemistry[17c] is considered outside the scope of our discussion.

C_{60} is a fully delocalized closed shell system showing "ambiguous" aromatic character[18]. The relatively high electronegativity of C_{60} has been proposed to be due to pyracyclene character of certain five-membered ring bonds (fulvalene-like character), see Fig. 3.

2. REDUCTION STUDIES

Theory predicts an extremely high electron affinity (facile reduction) for both C_{60} and C_{70} fullerenes[4a,k]. Initial experimental support for the ease of reduction of C_{60} was the formation of $C_{60}H_{36}$ via a Birch reduction (Li in ammonia) of C_{60} by Smalley

Fig. 1. Buckminsterfullerene, C_{60}.

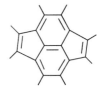

Fig. 3. Fulvalene-like structure of C_{60}.

and co-workers[19b]. They successfully dehydrogenated $C_{60}H_{36}$ back to C_{60} by treatment with DDQ reagent and have also carried out cyclic voltammetry studies which indicated that C_{60} undergoes reversible two electron reduction. Wudl, Diederich, and co-workers[19a] carried out cyclic voltammetry studies on chromatographically pure samples of C_{60} and C_{70}, which showed that each fullerene undergoes reversible three electron reduction (down to -1.5 V vs. Ag/AgCl electrode). More recently, even reversible four, five, and six electron reduction of C_{60} have been carried out[20,21].

C_{60} and C_{70} fullerene mixtures (in approximate 85:15 ratio) undergo reduction with Li metal (reduction potential of $Li^0 \sim -3.0$ V in THF-d_8) under ultrasound irradiation[22a]. The C_{60} and C_{70} fullerenes are only slightly soluble in THF. However, the reduced fullerenes are highly soluble and gave a deep red-brown solution after sonication. The ^{13}C NMR spectrum at room temperature shows a single resonance at δ ^{13}C 156.7 ppm for the reduced C_{60}. The deshielding of 14 ppm per carbon atom is remarkable because generally carbanionic carbons are shielded compared to their neutral precursors. Such deshielding in the case of C_{60} polyanion may be rationalized by populating the anti-bonding LUMO[22c] (paratropic deshielding). The ^{13}C NMR spectrum which indicates the presence of reduced C_{70} was obtained at $-80°C$ to improve the signal-to-noise ratio. Five resonances were observed at δ ^{13}C 158.3, 152.3, 149.6, 137.9, and 133.7 in a 1:2:1:2:1 ratio, respectively, showing a slight overall deshielding compared to neutral C_{70} by about 0.9 ppm per carbon atom. Reduction of chromatographically purified (alumina and hexane/toluene as eluant) sample of C_{60} confirmed the spectral assignments of reduced C_{60}. Further, the ^{7}Li NMR spectrum of the reduced C_{60}/C_{70} solution at $-80°C$ showed a fairly sharp resonance at $+1.6$ ppm (vs. 1M LiCl in THF). The ^{7}Li NMR spectrum at room temperature was extremely broad indicating a

solvent separated ion pair/contact-ion pair equilibrium in the temperature range studied[23].

The polyanion obtained contains an even number of electrons judging from the sharp ^{13}C NMR signals indicating a diamagnetic species. At early stages of sonication, cloudy green-colored solutions were obtained and no ^{13}C NMR signals for the solution could be detected. This solution was ESR active and showed a strong signal at the g value close to that of a free electron. Since previous cyclic voltammetry study[19a,b,20,21] indicates reversible four or higher electron reduction for each fullerene, it seems likely under the conditions used for the reduction C_{60} and C_{70} that each has accepted four or more electrons. Theoretical calculations[4a,b,d,e,f,g,m] indicate a triply degenerate LUMO for C_{60} making it probable that a hexaanion of C_{60} is most likely generated. A similar situation exists for C_{70} in which the LUMO and doubly degenerate LUMO^{+1} are closely spaced[4d]. Attempts to determine the exact number of electrons added to C_{60} and C_{70} fullerenes by quenching the polyanions with D_2O were unsuccessful. The isolated mixture was shown by field ionization mass spectrometry (FIMS) to be a mixture of C_{60} and C_{70}. Presumably, the polyanions either transfer electrons to D_2O or the deuteriated product mixture undergoes rapid oxidation to regenerate the more stable starting fullerenes in both cases. Similar observations have been made by Volhardt *et al.*[24] in the case of [3] phenylene dianion.

There is a flurry of activity[25] on partially reduced fullerene C_{60} obtained by metals such as potassium, cesium, thallium, rubidium etc. in the area of molecular superconductivity (onset of Tc at temperatures up to 40°K). Radical anion salt of C_{60} has also been isolated[26]. Furthermore, tetra(dimethylamino)-ethylene (TDAE) forms a redox complex with C_{60} which behaves as a soft ferromagnet at 16°K[27]. However, these interesting results are outside the scope of the present discussion.

3. OXIDATION OF C_{60}

Contrary to reduction experiments, oxidation of C_{60} was found to be difficult and electrochemical oxidations are irreversible[28]. In the mass spectrum of C_{60}, a peak at exact half mass (360 amu) was observed indicating formation of $C_{60}2^+$ under electron impact in the gas phase[29]. However, attempts to generate the diamagnetic dication of C_{60} in solution have been unsuccessful[22a]. Attempted oxidation of C_{60} in

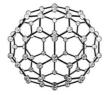

Fig. 2. Fullerene C_{70}.

large excess of SbF_5/SO_2CIF gave a dark green-colored solution that showed extremely broad ^{13}C NMR spectra at all temperatures employed ($-80°C$ to room temperature). Similar spectra was obtained using SbF_5 and Cl_2 as an oxidant in SO_2CIF solution. It appears that radical cations have been generated and no diamagnetic di(or poly)cations were formed. This is not surprising since electrochemical studies[19] and FT–ICR experiments[30] indicate a high oxidation potential for C_{60}. It is also in accord with theoretical studies[4] that predict the oxidation potential for C_{60} to be comparable to that of naphthalene, a molecule not oxidized to a stable dication under superacidic conditions[31].

C_{60} also dissolves in oleum to give a dark green-colored solution. An electron spin resonance spectrum of the solution indicated formation of a radical cation[32]. Miller and co-workers[33a] have used Magic Acid ($FSO_3H:SbF_5$) to oxidize C_{60} to its radical cation. The oxidized C_{60} has been trapped with nucleophiles to obtain nucleophilic addition products. Two, four, and six groups add symmetrically to the C_{60} radical cation. Methanol and butanol and even benzene (vide infra) was found to add to C_{60} radical cation. Reaction of $C_{60}^{\cdot+}$ with 1,6-hexanediol produced 1,6-difulleroxyhexane. Christe recently[33b] reported preliminary results of the oxidation of C_{60} by XeF_2 to the fullerene radical cation $C_{60}^{\cdot+}$.

4. ALKYLATION AND RELATED REACTIONS OF FULLERENE, C_{60}

Alkylation of C_{60} polyanion with excess methyl iodide yielded a light brown solid that field ionization mass spectrometry (FIMS) indicated to be a mixture of polymethylated fullerenes[22a]. ^1H and ^{13}C NMR studies in benzene-d_6 indicate methyl absorptions at δ^1H 0.06 and $\delta^{13}C$ 1.0, respectively. The FIMS analysis clearly shows a range of methylated products from 1 all the way to 24 methyls.

$$C_{60}^{6-}6Li^+ \xrightarrow{\text{xs } CH_3I} C_{60}(CH_3)_n.$$
$$n \leq 24$$

There is a preponderance of products with an even number of methyl groups (with six and eight methyl groups predominating). The nominal masses of the products with an odd number of methyl groups corresponds to the addition of a methyl group(s) and hydrogen atom(s). However, the exact mechanism of the observed alkylation is not yet clear, but possibly involves electron transfer to methylated fullerenes during quenching. The polymethylation of these fullerene polyanions represented the first functionalization of C_{60} and C_{70} via alkylative C–C bond formation[22a]. Attempts to obtain selective methylation has, however, been unsuccessful. Reaction of the polyanion with trimethylsilyl chloride did not result in a well-defined trimethylsilylated product.

Wudl and co-workers[34] have shown that C_{60} reacts with organometallic reagents such as alkyl/aryl lithiums and Grignard reagents. C_{60} stoichiometrically reacts with $LiB(Et)_3H$ to form the salt $C_{60}HLi \cdot 9H_2O$ which is indefinitely stable and does not reject with methyl iodide or methanol. In contrast, organometallic adducts of C_{60} react with methyl iodide to give mixed alkyl/methyl and aryl/methyl adducts of C_{60}.

$$C_{60} + 60\ PhMgBr \xrightarrow[\text{MeI}]{\text{THF,RT}} C_{60}Ph_{10}Me_{10},$$

$$C_{60} + \text{t-BuLi} \xrightarrow[\text{MeI}]{\text{THF,RT}} C_{60}\text{t-BuMe},$$

$$C_{60} + \text{t-BuMgBr} \xrightarrow[\text{MeI}]{\text{THR,RT}} C_{60}tBu_{10}Me_{10}.$$

C_{60} also reacts with trimethylsilyl chloride in the presence of magnesium metal in THF under sonication[22a] (Barbier conditions) to give di- and tetra(trimethylsilyl) C_{60} adducts as determined by FIMS analysis[22a].

Negative ion ionization of C_{60} in the mass spectrometer in the presence of methane gives methylene adducts[34]. The maximum number of carbons attached to C_{60} would seem to be 13 based on mass spectrometric analysis.

5. REACTIONS WITH NEUTRAL BASES

Neutral nucleophiles such as amines were found to add to C_{60}. Wudl and co-workers[34] have added up to 12 propylamine molecules. The diethylamine, propylamine, and morpholine adducts are soluble in dilute hydrochloric acid. The dodecylamine adduct is insoluble. The mechanism of addition appears to occur in a stepwise fashion with electron transfer preceding covalent bond formation. The intermediate radical anions have been identified by ESR. The morpholine adduct $H_6C_{60}(\text{morph})_6$ has been identified by elemental analysis and ^1H NMR spectroscopy[35]. Many of these amine adducts are water soluble. Triethyl phosphite also adds to C_{60} to produce an Arbutzov-type product as determined by IR spectroscopy[36]. Up to six phosphonate groups are added across C_{60} as determined by FAB mass spectrometry.

6. CYCLOADDITION REACTIONS OF C_{60}

The double bonds of C_{60} show good dienophilic reactivity towards dipolarophiles and dienes. Wudl *et al.* have, for example, obtained evidence for addition of p-nitrophenyl azide to C_{60}[34]. Diazomethane, substituted diazomethanes and sulfonium ylides also add smoothly[37a]. The diazomethane adducts upon nitrogen loss give methylene annulated fulleroids. The intermediately formed cyclopropane adducts undergo electrocyclic ring opening to give methan-

Fig. 4. Diphenylfulleroid.

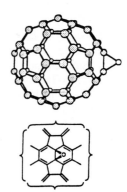

Fig. 5. Monoepoxide of C_{60}.

oannulene structures (e.g., in the case of diphenyldi-azomethane diphenylfulleroid, Fig. 4).

In the case of diphenyldiazomethane, Wudl *et al.* were able to isolate C_{61} to C_{66} polyphenyl fulleroids[38]. Photochemically generated disubstituted silylenes also add to C_{60}[37b].

C_{60} also reacts with dienes such as anthracene, furan, and cyclopentadiene to give Diels-Alder cycloadducts[34].

$$C_{60} + \text{xs Anthracene} \xrightarrow[120°C]{ODCB} C_{60}(C_{14}H_{14})_x$$

$$C_{60} + \text{xs Cyclopentadiene} \xrightarrow{RT} C_{60}(CpH)_x$$

West and co-workers[37b] have carried out reactions of C_{60} with disubstituted silenes to obtain (2 + 2) adducts.

7. EPOXIDATION AND OXYGENATION

Reactions of C_{60} generally give complex mixture of products. Thus far, only a few well-defined monofunctionalized products have been isolated, Diederich *et al.* have isolated $C_{70}O$ as a minor component of the fullerene mixture generated by resistive heating of graphite[16d]. Wood *et al.*[39] have detected $C_{60}O$ in a similar mixture by mass spectrometry and Kalsbeck and Thorp reported[40] generating $C_{60}O_n$ ($n = 1-4$) by electrochemical oxidation of C_{60}. More recently, Cox, Smith *et al.*[41] have obtained well-characterized monooxide $C_{60}O$ as the sole isolable product (7% yield) upon photochemical oxidation of C_{60} in benzene. ^{13}C NMR and other spectroscopic data seem to support the formation of monoepoxide (Fig. 5) and not the 1,6-oxo[10]annulene (Fig. 6). The isolated $C_{60}O$ is efficiently reduced back to C_{60} (in 91% yield) during chromatography on neutral alumina. The widespread use of alumina for purification of the fullerenes may explain why $C_{60}O$ has not been isolated previously. Facile preparation of monoepoxide opens new doors for monofunctionalization of C_{60}. See Fig. 4.

8. HALOGENATION

The first halogenation carried out on C_{60} was fluorination. Smith *et al.*[42] have fluorinated C_{60} with elemental fluorine and obtained a mixture of poly-

fluorinated products. The major component of the mixture was $C_{60}F_{36}$ analogous to the Birch reduction product $C_{60}H_{36}$[19b]. Subsequently, Holloway and co-workers[43a] have claimed the formation of $C_{60}F_{60}$, when 10 mg of C_{60} was exposed to fluorine gas over a period of 12 days. The product shows single broad fluorine absorption in the ^{19}F NMR spectrum. However, further studies indicate that the cage structure may have been lost during the fluorination reaction and some kind of C_nF_n product resulted (fluorographites of such compositions are known)[43b].

Polychlorination as well as polybromination of C_{60} has also been achieved[44]. Treatment of C_{60}/C_{70} mixture or pure C_{60} with neat chlorine or with chlorine in the presence of catalysts such as $AlCl_3$ or $AlBr_3$ in chloro organic solvents at various temperatures did not furnish any chloro products. However, in a hot glass tube under a slow stream of chlorine gas, both the C_{60}/C_{70} mixture and pure C_{60} react readily. At 250°C, C_{60}/C_{70} were chlorinated (chlorine flow rate 10 mL/min) and maximum uptake of chlorine was achieved within five hours. The weight increase of the yellowish brown product indicated that on average 24 chlorine atoms (confirmed by elemental analysis) were added.

$$C_{60} \xrightarrow{Cl_2(g)} C_{60}Cl_n$$

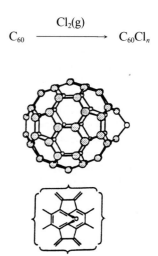

Fig. 6. 1,6-oxo[10]annulene structure.

Zahurak, R. Tycko, G. Dabbagh, and F. A. Thiel, *Nature* (London) **350**, 320 (1991); (b) R. C. Haddon, A. F. Hebard, M. J. Rosseinsky, D. W. Murphy, S. H. Galrum, T. T. M. Palstra, A. P. Ramirez, S. J. Duclos, R. M. Flemming, T. Seigrist, and R. Tycko, In *Fullerenes, ACS Symposium Series 481* (Edited by G. S. Hammond and V. J. Kuck), Chap. 5. American Chemical Society, Washington D.C. (1992). Also see R. C. Haddon, *Acc. Chem. Res.* **25**, 127 (1992); (c) K. Holczer, O. Klein, S-M. Haung, R. B. Kaner, K-J. Fu, R. L. Whetten, and F. Diederich, *Science* **252**, 1154 (1991); (d) K. Tanigaki, T. W. Ebbesen, S. Saito, J. Mizuki, J. S. Tsai, Y. Kubo, and S. Kuroshima, *Nature* (London) **352**, 222 (1991).

26. A. Penicaud, J. Hsu, C. A. Reed, A. Koch, K. C. Khemani, P-M. Allemand, and F. Wudl, *J. Am. Chem. Soc.* **113**, 6698 (1991).

27. P-M. Allemand, K. C. Khemani, A. Koch, F. Wudl, K. Holczer, S. Donovan, G. Grüner, and D. Thomson, *Science* **253**, 301 (1991).

28. See Refs. [19–21].

29. See Ref. [16a].

30. J. Zimmermann, J. R. Eyler, S. B. H. Bach, and S. W. MacElvany, *J. Chem. Phys.* **94**, 3556 (1991). Also see S. W. MacElvany, M. M. Ross and J. H. Callahan, *Acc. Chem. Res.* **25**, 162 (1992).

31. G. A. Olah and D. A. Forsyth, *J. Am. Chem. Soc.* **98**, 4086 (1976).

32. G. A. Olah, I. Bucsi, G. K. S. Prakash, and R. Aniszfeld, unpublished results.

33. (a) G. P. Miller *et al.* unpublished results from Exxon Laboratory, see *Chem. & Eng. News,* Dec. 16 (1991); (b) K. O. Christe and W. W. Wilson, Paper presented at the Fluorine Division, 203rd American Chemical Society Meeting, San Francisco, CA (1992), Paper No. 66.

34. F. Wudl, A. Hirsh, K. C. Khemani, T. Suzuki, P-M Allemand, A. Koch, H. Eckert, G. Srdanov, and H. M. Webb, In *Fullerenes, ACS Symposium Series 481* (Edited by G. S. Hammond and V. J. Kuck), Chap. 11. American Chemical Society, Washington D.C. (1992). Also see F. Wudl, *Acc. Chem. Res.* **25**, 157 (1992).

35. A. Hirsh, Q. Li, and F. Wudl, *Angew. Chem. Int. Ed. Engl.* **30**, 1309 (1991).

36. A. Hirsh and F. Wudl, unpublished observation, see Ref. [34].

37. (a) T. Suzuki and F. Wudl, unpublished results, see Ref. [34]; (b) K. Oka, M. Miller, B. Dreczewski, and R. West, Presented at XXV Silicon Symposium, Los Angeles, CA (1992), Poster No., 97P.

38. T. Suzuki, Q. Li, O. Almarsson, K. C. Khemani, and F. Wudl, *Science* **254**, 1186 (1991).

39. J. M. Wood, B. Kahr, S. H. Hoke, II, L. Dejarme, R. G. Cooks, and D. Ben-Amotz, *J. Am. Chem. Soc.* **113**, 5907 (1991).

40. W. A. Kalsbeck and H. H. Thorp, *Electroanal. Chem.* **314**, 363 (1991).

41. K. M. Creegan, J. L. Robbins, W. K. Robbins, J. M. Millar, R. D. Sherwood, P. J. Tindall, D. M. Cox, A. B. Smith, III, J. P. McCauley, Jr., D. R. Jones, and R. T. Ghallagher, *J. Am. Chem. Soc.* **114**, 1103 (1992).

42. H. Selig, C. Lifshitz, T. Peres, J. E. Fischer, A. R. McGhie, W. J. Romanow, J. P. McCauley, Jr., and A. B. Smith, III, *J. Am. Chem. Soc.* **113**, 5475 (1991).

43. (a) J. H. Holloway, E. G. Hope, R. Taylor, G. J. Langley, A. G. Avent, J. H. Dennis, J. P. Hare, H. W. Kroto, and D. R. M. Walton, *J. Chem. Soc., Chem. Commun.* 966 (1991); (b) A. B. Smith, III, personal communication.

44. G. A. Olah, I. Bucsi, C. Lambert, R. Aniszfeld, N. J. Trivedi, D. K. Sensharma, and G. K. S. Prakash, *J. Am. Chem. Soc.* **113**, 9385 (1991).

45. F. N. Tebbe, J. Y. Becker, D. B. Chase, L. E. Firment, E. R. Holler, B. S. Malone, P. J. Krusic, and E. Wasserman, *J. Am. Chem. Soc.,* in press.

46. P. R. Birkett, P. B. Hitchcock, H. W. Kroto, R. Taylor, and D. R. Walton, *Nature* (London), submitted for publication.

47. G. A. Olah, I. Bucsi, C. Lambert, R. Aniszfeld, N. J. Trivedi, D. K. Sensharman, and G. K. S. Prakash, *J. Am. Chem. Soc.* **113**, 9387 (1991).

48. S. H. Hooke, II, J. Molstad, G. L. Payne, B. Kahr, D. Ben-Amotz, and R. G. Cooks, *Rap. Commun. Mass. Spec.* **5**, 472 (1991).

49. P. J. Krusic, E. Wasserman, P. N. Keizer, J. R. Morton, and K. F. Preston, *Science* **254**, 1183 (1991).

50. C. N. McEwen, R. G. McKay, and B. S. Larsen, *J. Am. Chem. Soc.,* **114**, 4412 (1992).

51. D. A. Loy and R. A. Assink, *J. Am. Chem. Soc.* **114**, 3977 (1992).

52. J. W. Hawkins, T. A. Lewis, S. D. Loren, A. Meyer, J. R. Heath, Y. Shibato, and R. J. Saykelly, *J. Org. Chem.* **55**, 6259 (1990). Also see J. W. Howkins, *Acc. Chem. Res.* **25**, 150 (1992).

53. P. J. Fagan, J. C. Calabrese, and B. Malone, In *Fullerenes, ACS Symposium Series 481* (Edited by G. S. Hammond and V. J. Kuck), Chap. 12. American Chemical Society, Washington D.C. (1992). Also see P. J. Fagan, J. C. Calabrase, and B. Malone, *Acc. Chem. Res.* **25**, 134 (1992).

tually not known at the time we started our work. At that time, many of the published reports tended to treat C_{60} as a relatively chemically inert aromatic molecule. We had had considerable experience reacting the organometallic fragment $[C_p*Ru(CH_3CN)_3]^+$ $(C_p* = \eta^5 - C_5(CH_3)_5)$ with relatively electron-rich aromatic molecules[70]. Reaction of this reagent with planar aromatic molecules containing six-membered rings has led without exception thus far to loss of all three acetonitrile molecules. The resulting product is a sandwich complex with ruthenium bound to six carbons of the aromatic. We asked if this same chemistry might occur for C_{60} with ruthenium attaching to the six-membered rings. However, upon reaction, the complex $\{[C_p*Ru(CH_3CN)_2]_3C_{60}^{3+}(X^-)_3$ $(X = O_3SCF_3)$ was isolated in which two acetonitrile molecules are retained on each ruthenium atom (eqn 1)[19]:

$$3C_p*Ru(CH_3CN)_3^+X^- + C_{60} \rightarrow$$
$$\{[C_p*Ru(CH_3CN)_2]_3C_{60}\}^{3+}(X^-)_3. \quad (1)$$

We concluded that each ruthenium was simply bound to one carbon-carbon double bond of the C_{60} cluster (i. e., to an edge of the cluster and not to a hexagonal face). This reactivity is more typical of electron-poor alkenes, which react to give the alkene complexes $\{[C_p*Ru(CH_3CN)_2](\eta^2 - CH_2 = CH - E)\}$ $^+X^-$ where E is an electron-withdrawing group. Although we could not obtain an x-ray structure of this molecule, we were able to conclude that the reactivity of C_{60} parallels that of relatively electron-poor alkenes. Overall, each of the 30 carbon-carbon double bonds on C_{60} can be regarded as being strained, having a high electron affinity, and being only partially conjugated to the surrounding alkene units. The double bonds in C_{60} do not react as those in relatively electron-rich planar aromatic molecules or alkenes. This is in accord with the well established bond-length alternation in C_{60} six-membered rings[71,72], and with the high electron affinity of the sphere[73].

The above conclusions led us to investigate the interaction of zero-valent, electron-rich, platinum reagents with C_{60}[19]. These compounds are well known to react with electron-poor and/or strained alkenes[74–76]. This enabled us to prepare the first structurally well defined derivative with a metal atom bound directly to the carbons of C_{60}, namely $[Ph_3P]_2Pt(v^2 - C_{60})$ (Ph = phenyl, eqn 2)[19]:

Ph₃P:
　　＼　　CH₂　　　　　　　　　−CH₂ = CH₂　Ph₃P:
　　　　Pt ←　‖　+ C₆₀　 toluene 　　　＼
　　╱　　CH₂　　　　　　　　　　　　　　Pt ← C₆₀.
Ph₃P:　　　　　　　　25 deg, 2 h 　Ph₃P:╱

$$(2)$$

The structure of this compound was determined by x-ray crystallography and is shown in Fig. 1. There are two types of carbon-carbon bonds in C_{60}, those at the six-six ring junctions (1.39 Å), and those at the

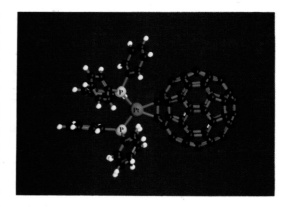

Fig. 1. X-ray structure of $(Ph_3P)_2Pt(\eta^2 - C_{60})$. Carbon atoms are grey, phosphorus atoms pink, and hydrogen atoms white. Double bonds are in red and single bonds in cyan.

six-five ring junctions (1.45 Å). The platinum binds at a six-six ring junction of C_{60}, pulling the two carbons to which it is attached out from the sphere. This same point of binding is observed for (4-*t*-butyl-pyridine)$_2$OsO$_4$C$_{60}$[13,14] and $(Ph_3P)_2(CO)ClIr(\eta^2 - C_{60})$[22]. Presumably because the six-six ring junctions are shorter, these contain the most double-bond character and are the most likely to exhibit such reactivity. We have now extended this chemistry to prepare the related Ni, Pd, and Pt derivatives, $(Et_3P)_2M$ $(\eta^2 - C_{60})$ (Et = C_2H_5), from reaction of the complexes $M(Et_3P)_4$ (M = Ni, Pd, Pt) with C_{60}[16]. Raman spectroscopy was essential for establishing their structure (vide infra)[77]. Other groups have subsequently prepared low-valent metal complexes of fullerenes, including $(R_3P)_2(CO)ClIr(\eta^2 - C_{60})$ [22], $(Ph_3P)_2(CO)ClIr(\eta^2 - C_{70})$[23], and $(\eta^5 - indenyl)$ $(CO)Ir(\eta^2 - C_{60})$[24].

We wondered how many metal fragments might be attached to the C_{60} core. This led to the reaction of C_{60} with an excess of the reagents $M(Et_3P)_4$ (M = Ni, Pd, Pt) and production in high yield (ca. 85%) of the novel hexasubstituted metal derivatives shown in eqn 3[16,18,78]:

$$6 (Et_3P)_4M + C_{60} \xrightarrow{-12\ Et_3P} [(Et_3P)_2M]_6C_{60}$$
$$M = Ni, Pd, Pt. \quad (3)$$

These metal complexes were shown by NMR[18] and Raman spectroscopy[77] to be a single structural isomer. We subsequently determined the platinum and palladium structures by X-ray crystallography (Fig. 2)[16,18,78]. Ignoring the ethyl groups and the fact that the P_2M planes twist slightly with respect to the MC_2 planes, the molecules have T_h point group symmetry with an octahedral array of metal atoms about the C_{60} core. As with the mono-substituted complexes, each of the metals binds at six-six ring junctions. These complexes represent the first multiply substituted buckminsterfullerene derivatives with a well defined structure. Later, the structure of $C_{60}Br_{24}$ discussed below was also established[25].

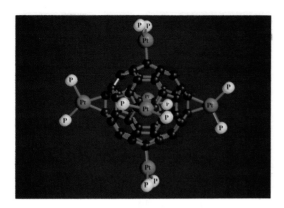

Fig. 2. X-ray structure of $[(Et_3P)_2Pt]_6C_{60}$. Carbon atoms are grey, phosphorus atoms pink, and hydrogen atoms white. Double bonds are in red and single bonds in cyan. For clarity, ethyl groups on phosphine ligands are excluded.

Having characterized the monosubstituted and hexasubstituted derivatives, we wanted to determine how the final T_h symmetry structure of the $[(Et_3P)_2M]_6C_{60}$ complexes arises. By NMR spectroscopy, we were able to determine that the disubstituted complex $[(Et_3P)_2Pt]_2C_{60}$ exists as a mixture of three isomers[16,78]. The trisubstituted and tetrasubstituted metal derivatives also exist as complex mixtures of isomers. It was clear in these complexes that the relative positions of the platinum atoms with respect to one another on the C_{60} surface did not correspond to the final orientations seen in the hexasubstituted metal derivatives. We concluded that in order to achieve the octahedral array of metal atoms in the final T_h symmetry structure, the platinum atoms must either dissociate or wander across the C_{60} surface during complex formation. Since addition of a trapping agent such as diphenylacetylene to a benzene solution of $[(Et_3P)_2Pt]_6C_{60}$ yields primarily $[(Et_3P)_2Pt]_5C_{60}$ and $(Et_3P)_2Pt(\eta^2-diphenylacetylene)$, we concluded that the $(Et_3P)_2Pt$ fragments equilibrium on and off the C_{60} sphere[16,78]. This provides a mechanism to lead to the final T_h symmetry structure. Examination of a space-filling model reveals why only six $(Et_3P)_2Pt$ groups attach to the C_{60} core: Sterically it is impossible to fit more of these units onto the sphere[16]. The octahedral arrangement allows these six units to be as far away from one another as possible. Thus it is a combination of chemical reversibility, steric factors, and the electronic preference of these metals to bind to the six-six ring junctions, which leads to this highly symmetrical arrangement.

We have investigated the electrochemical properties of the metal derivatives in detail[79].* Basically, attachment of a metal fragment to the C_{60} core lowers the electron affinity of the cluster. Indeed, there is a

linear relationship between the number of metals on the cluster, and the electron affinity of the sphere. For example, addition of each $(Et_3P)_2Pt$ group to C_{60} makes the molecule harder to reduce by 0.36 V. It may be that the reactivity of C_{60} can be tuned by attachment of metals, which may serve as removable protecting groups. Interestingly, the complexes $(Et_3P)_2M(\eta^2-C_{60})$ (M = Ni, Pd, Pt) all have the same reduction potential. (These reduction events are C_{60}-centered, as has also been found for $(\eta^5-indenyl)IrCO(\eta^2 - C_{60})[80]$.) Because of strong metal d-orbital back bonding with a $(Et_3P)_2M$ fragment, the metal-bound double bond is electronically detached from the π system of the remaining 29. This π system determines in part the electron affinity of the sphere. The metal also inductively adds electron density to the C_{60} σ bond system, further decreasing the electron affinity of the sphere. This would contribute to the relatively large 0.36 V shift of the reduction potential upon binding the metal fragment. The 0.36 V shift on attachment of a metal fragment explains why we can selectively form the monosubstituted derivatives in high yield. The tendency of a second metal fragment to add is significantly reduced. Even if some $[(Et_3P)_2Pt]_2C_{60}$ or more highly substituted derivatives do form in the reaction mixture, there is a sufficient driving force for these to disproportionate with free C_{60} in solution to form the monosubstituted derivative. (This will occur as long as the chemistry is reversible, and may perhaps be mediated by single electron transfer steps.) We believe this effect will be general to all C_{60} chemistry. The magnitude of the shift of the electron affinity of the cluster with a particular reagent will to some extent control the selectivity one can achieve in addition reactions.

In summary, the metal chemistry of C_{60} has already revealed several aspects of fullerene reactivity, and is expected to be extensive.

3. HALOGENATION

In spite of its high electrochemical oxidation potential[29], C_{60} easily undergoes oxidative halogenation with elemental fluorine, chlorine, and bromine. Often the products are mixtures of polyhalogenated compounds, which probably consist of several compositions and their isomers. The first structurally characterized halogenated derivative of C_{60} was $C_{60}Br_{24}$, isolated as $C_{60}Br_{24}\cdot(Br_2)_n$ ($n \approx 1 - 2$), with molecular bromine in the crystal lattice[25].† Subsequently, $C_{60}Br_6$ and $C_{60}Br_8$ have been characterized[26]. The structure of $C_{60}Br_{24}$ is a new addition pattern on the C_{60} surface and suggests a structural type for derivatives with a large number of addends. We begin this section with a discussion of $C_{60}Br_{24}$, and then review published information on chlorinated and fluorinated derivatives.

*Work done in collaboration with Prof. Dennis Evans and Susan Lerke of the Chemistry Dept., University of Delaware.

†Unpublished information on the bromination of C_{60} and formation of $C_{60}Br_{24}$ is mentioned in ref. [47].

Chemical and electrochemical reductions of "$C_{60}Cl_{12}$" to C_{60} have been reported[37]. The isolated yield of C_{60} (80%) obtained by reduction with PPh_3 suggests that the integrity of the C_{60} framework is largely retained during chlorination at low temperatures. Cyclic voltammetry on "$C_{60}Cl_{12}$" in benzonitrile showed a multielectron irreversible wave near -0.2 V (vs Ag/AgCl) and reversible reduction waves characteristic of C_{60}.* The ease with which "$C_{60}Cl_{12}$" is electrochemically reduced is characteristic of unusual compounds such as CCl_3NO_2 with carbon attached to highly electron-withdrawing groups. It will be of interest to see if the chemistry of chlorinated C_{60} compounds can be related to their facile electrochemical reduction.

Early reports suggest there will be a rich chemistry of halogenated C_{60} derivatives. Under basic conditions, "$C_{60}Cl_{24}$" in methanol is converted to mixed compounds which were assigned the compositions $C_{60}(OMe)_n$ based on mass spectral data[33]. Evidence for substitutions of fluoride in $C_{60}F_n$ by methoxide and a number of other nucleophilic species has been presented[47,82]. A mechanism of nucleophilic substitution of halogenated C_{60} derivatives by addition-elimination was considered [10,47]. Under Friedel-Crafts conditions, "$C_{60}Cl_{24}$" in benzene or toluene is converted to arylated derivatives "$C_{60}(Ar)_n$"[33]. Phenylated products have been reported from combinations of benzene and C_{60} in the presence of Br_2 and $FeCl_3$[48,49].

4. VIBRATIONAL SPECTROSCOPY

The high degree of symmetry of C_{60} makes it an extremely interesting system for vibrational spectroscopy. The sparse spectra seen in the infrared and Raman arise from the restrictive selection rules present in I_h symmetry. Clearly, if the molecular symmetry were to be reduced, many more vibrational modes would be observed. We have used this relationship between symmetry and spectral density to probe the extent of chemical reactions involving C_{60} for both organometallic systems[77] and halogenated products[25].

As noted in several studies [83–86], there are four allowed infrared modes and ten allowed Raman modes for C_{60}. The infrared intensities are relatively weak, due to dipole moment vectors of very small magnitude associated with these vibrations. The Raman spectrum is quite strong due to the high polarizability of the molecular framework. Since we hope to monitor changes in the number of observed vibrational modes associated with the C_{60} moiety, Raman is the logical spectroscopic probe. Additional vibrational modes due to the chemically attached ligands should be relatively weak compared to the parent C_{60} Raman modes. This hypothesis was confirmed by the Raman spectra of $(Ph_3P)_2Pt(\eta^2-C_{60})$

[77]. The Raman spectrum of this material exhibited a large number (>40) of bands and only four were due to the triphenylphosphine ligand. The spectra observed were independent of the metal (Pt, Pd, Ni) and relatively unaffected by the ligand, either triphenylphosphine or triethʊlylphosphine. The large increase in the number of observed Raman bands was due entirely to the reduction in symmetry. For $[(Et_3P)_2Pt]_6C_{60}$ the point group is T_h (ignoring ethyl groups)[18], which is lower symmetry than C_{60} (I_h) but higher than C_{2v} (ignoring phenyl groups) found for the monosubstituted complex $(Ph_3P)_2Pt(\eta^2-C_{60})$. In agreement with the hypothesis of lowered symmetry producing more complex Raman spectra, the hexasubstituted material shows a band pattern of intermediate complexity between the parent C_{60} and the monosubstituted C_{60} complex. Spectra are shown for all three species in Fig. 4.

The situation for halogenation reactions is somewhat different. The number of vibrational modes associated with the carbon-halogen bonds is much smaller, and these are not as strong infrared absorbers as the phosphine phenyl groups. Therefore, IR spectra might be more useful. The product of a chlorination reaction shows a broad feature near 900 cm^{-1} in the IR spectrum, characteristic of C—Cl stretching modes[37]. There are no indications of residual C_{60}, and the Raman spectrum shows only a broad background fluorescence. These observations are consistent with the assumption of multiple species or various isomers of chlorinated C_{60}. The vibrational modes for the various isomers are superimposed, yielding a broad, featureless infrared spectrum. It was

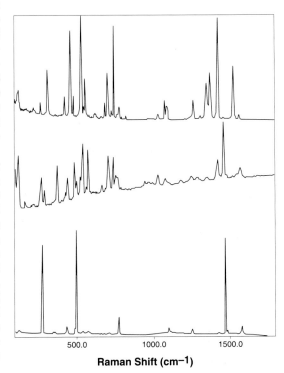

Fig. 4. Raman spectrum of C_{60} (bottom), $(Et_3P)_2Pd(\eta^2-C_{60})$ (middle), and $[(Et_3P)_2Pd]_6C_{60}$ (top).

---

*Electrochemical studies were done in collaboration with Prof. James Y. Becker.

observed that heating the sample while taking infra-red spectra resulted in absorption bands due to the parent C_{60} molecule. The chlorination reaction is thermally reversible, and this back reaction can be observed in the IR spectrum, establishing that the C_{60} framework remains intact.

It was expected that similar results would be obtained with the product of bromination reactions. However, the infrared and Raman spectra of the bromination product $C_{60}Br_{24}$ were quite different from those obtained in the chlorination study[25]. The spectra showed relatively few bands and the linewidths were quite narrow. The spectral features were characteristic of a single species of well defined structure. A close comparison of the IR and Raman showed that there were no modes common to both spectra. This is indicative of a center of symmetry for the molecule. Pending x-ray structure determination, a tentative structure having 24 bromines distributed on the C_{60} sphere in T_h symmetry was postulated [25]. This structure contained isolated $C=C$ double bonds. Symmetry analysis predicted five Raman active $C=C$ stretches and three IR active $C=C$ stretches. Figure 5 shows the Raman and IR spectra in the 1600 cm^{-1} region, and this data is in good agreement with theory. The x-ray structure determination confirmed the structure proposed with the aid of vibrational spectroscopic data.

Although it is clear that precise structure determination requires x-ray data, vibrational spectroscopy of C_{60} reaction products can quickly and easily establish the extent of reaction and provide, in some cases, clear insight into possible structures.

5. ESR STUDIES OF ALKYL RADICAL ADDITION*

Considering that C_{60} behaves as an electron deficient polyolefin having 30 weakly conjugated double bonds, it is not surprising that the molecule has a high reactivity towards free radicals [42,43]. Halogenation may involve addition of chlorine and bromine atoms to some extent[25,81]. A variety of organic free radicals, generated either photochemically or thermally by established free radical reactions, add very rapidly and in substantial numbers to C_{60} to afford mixtures of diamagnetic and radical derivatives, warranting characterization of this molecule as a free radical sponge[87]. For example, mass spectrometry shows that benzyl and methyl radicals react to form R_nC_{60} with $n = 1$–15 for R = benzyl and $n = 1$–34 for R = methyl. In this section we outline the information obtained by ESR for the radical adducts resulting when one, three, and five radicals add to C_{60}.

Particularly revealing was the addition of t-butyl radicals to C_{60} [44,45]. The t-butyl radical adduct t-BuC$_{60}\cdot$ can be generated in a variety of ways, including the photolysis directly in the ESR cavity of a saturated benzene solution of C_{60} (ca. 0.002 M) containing one equivalent of di-t-butylmercury. After complete decomposition of the radical precursor, the light can be shuttered and the spectrum (Fig. 6A) is invariant at constant temperature. At 70°C it consists of ten narrow lines of binomial relative intensities appropriate for a hyperfine interaction of the unpaired electron with the nine equivalent protons of the t-butyl group (0.17 Gauss). The satellite structure due to ^{13}C nuclei in natural abundance visible at high gain on each side of the main spectrum could be simplified by replacing the t-butyl group with its perdeutero analog (Fig. 6B). Since the deuteron splitting is ca. ⅙ the proton splitting, each ten-line multiplet collapses into an unresolvable single line. Ten pairs of ^{13}C satellites can now be clearly resolved. Consideration of their relative intensities and splittings leads to the conclusion that the unpaired electron is mostly confined to the two fused, six-membered rings on the C_{60} surface having the t-butyl substituent at one of the points of fusion. Each half of the resulting radical structure of C_s symmetry can be viewed as a cyclohexadienyl radical and can be described by the canonic resonance forms shown in Fig. 7. Thus, the largest ^{13}C splitting (17.80 Gauss) belongs to C1, and the next largest (13.11 Gauss) to the tertiary carbon of the t-butyl group, as confirmed by isotopic labelling. The remaining important resonance forms place the unpaired electron in 2p orbitals on C3, C3′ and C5, C5′ to which we assign the next two splittings (9.33 and 8.84 Gauss). Such an assignment is consistent with the observation that the latter satellites are twice as intense as those belonging to the previous

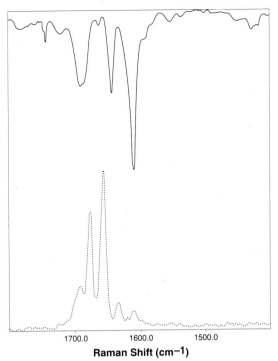

Fig. 5. Infrared (top) and Raman (bottom) spectra of $C_{60}Br_{24}$.

Raman Shift (cm^{-1})

1700.0   1600.0   1500.0

*In collaboration with Drs. J. R. Morton and K. F. Preston of the National Research Council of Canada.

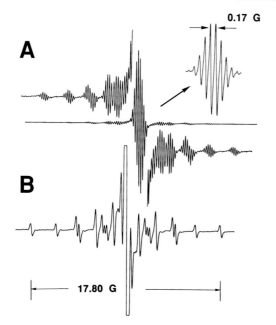

0.17 G

A

B

17.80 G

Fig. 6. ESR spectra at 80°C in benzene: (A) *t*-butyl-C_{60}· showing (in the wings at high gain) the hyperfine satellites due to ^{13}C in natural abundance and (insert) the 0.17 G hyperfine splitting due to the 9 methyl protons. (B) $(CD_3)_3C$—C_{60}· at high gain to show the ^{13}C satellites.

carbons that lie in the plane of symmetry. By analogy with the cyclohexadienyl radical[88,89], commensurate hyperfine interactions of negative sign due to spin polarization effects are also expected for the carbons adjacent to those bearing the unpaired electron in the principal resonance forms (C2, C2′, C4, C4′, and C6). Only very small hyperfine splittings are left for the more remote carbon atoms, ruling out appreciable delocalization of the unpaired electron over the surface of the C_{60} framework.

Well resolved ESR spectra were obtained for a variety of RC_{60}· monoalkyl radical adducts ($R =$

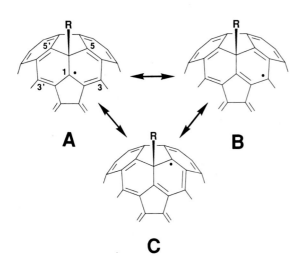

Fig. 7. The five major canonical resonance forms of R—C_{60}· radicals (B and C are doubly degenerate due to the plane of symmetry passing through C1, C6, and R).

CH_3, CH_3CH_2, $CH_3CH_2CH_2$, $(CH_3)_2CH$, 1-adamantyl, CCl_3, benzyl, etc.) using the photochemical decomposition of the corresponding mercurials and other methods (e.g., the photolysis of alkyl bromides in the presence of C_{60}). In all cases the proton hyperfine structures were those expected for the entering radicals (Table 1). The isopropyl radical adduct *i*-PrC_{60}·, for example, displayed a doublet of septets at 160°C appropriate for a unique proton (0.48 Gauss) and six equivalent protons (0.15 Gauss). Unusually high temperatures, necessitating the use of *t*-butylbenzene as a high boiling solvent, were needed to obtain good spectra when R was a radical of small size. The reason for this is considered next.

A remarkable feature of the spectra of these adducts was the strong temperature dependence of their intensity. For *t*-BuC_{60}·, for example, the spectrum grew dramatically as the temperature was raised from 30 to 130°C, and decreased again as it was lowered, disappearing at ~10°C. For the isopropyl radical adduct, the temperature dependence of the intensity was shifted to 130–200°C. This behavior is expected for the thermal dissociation of a dimer into two paramagnetic monomers and with a dimer bond strength that depends on the size of the entering radical[46] (eqn 5):

$$RC_{60}\!\!-\!\!C_{60}R \rightleftharpoons 2\ RC_{60}\cdot. \qquad (5)$$

Table 1. Hyperfine interactions (Gauss) of RC_{60}· radicals[a]

R	hfi/H	hfi/^{13}C	Temp./°K
CH_3	3H < 0.08	1C = 16.64	383
CH_3CH_2[b]	2H = 0.28		473
	3H = 0.13		
$CH_3{}^{13}CH_2$[b]	2H = 0.28	1C = 15.5	473
	3H = 0.13		
$(CH_3)_2CH$[b]	1H = 0.48		433
	6H = 0.15		
$C_6H_5CH_2$[c]	2H = 0.42		353
	2H = 0.19		
$C_6H_5{}^{13}CH_2$[c,d]	2H = 0.42	1C = 14.9	353
	2H = 0.19		
$C_6D_5CH_2$[d]	2H = 0.42		353
$(CH_3)_3C$[e]	9H = 0.17		373
$(CD_3)_3C$[d,e]	9D = 0.03	1C = 17.80	373
		1C = 13.11	
		2C = 9.33	
		2C = 8.84	
		3C = 5.63	
		2C = 4.45	
		2C = 4.02	
		2C = 3.61	
		4C = 2.43	
		8C = 0.89	
$(CH_3)_2(^{13}CH_3)C$[d,e]	9H = 0.17	1C = 0.40	373
$(CH_3)_3{}^{13}C$[d,e]	9H = 0.17	1C = 13.11	373
1-adamantyl[e]	3H = 0.26		373
	6H = 0.046		

[a]All *g*-factors are in the range 2.0022–2.0023.
[b]Solvent: *tert*-butylbenzene.
[c]Solvent: toluene.
[d]Isotopically enriched compound was used.
[e]Solvent: benzene.

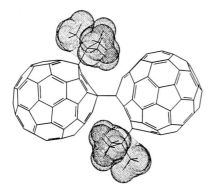

Fig. 8. Molecular model of the dimer [t-butyl-C_{60}]$_2$.

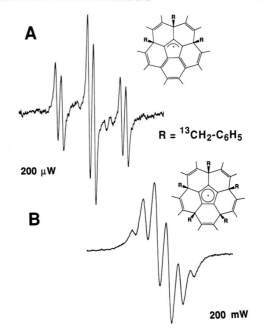

$R = {}^{13}CH_2\text{-}C_6H_5$

200 μW

200 mW

Fig. 9. ESR spectra obtained at 50°C by photolysis of a solution of C_{60}, 99% [α-^{13}C]toluene, and di-t-butylperoxide at (A) 200 μW and (B) 200 μW of incident microwave power ($R = {}^{13}CH_2C_6H_5$).

Plots of ln ($T \times$ Intensity) against $1000/T$ gave straight lines for various $RC_{60}\cdot$ radicals. On the assumption of a small degree of dissociation, the slope of the graphs is proportional to the enthalpy change for the dissociation, and therefore to the bond strength uniting the two halves of the dumbbell-shaped dimers. The following values were obtained: i-PrC$_{60}\cdot$, 35.5 kcal/mol; t-BuC$_{60}\cdot$, 22.1 kcal/mol; 1-adamantyl-$C_{60}\cdot$, 21.6 kcal/mol; $CCl_3C_{60}\cdot$, 17.1 kcal/mol. The dependence of the dimer bond strength on the size of R strongly suggests that the dimer bonding occurs at a carbon atom in the immediate vicinity of the substituent R and is consistent with the localization of the unpaired electron, as shown in Fig. 7. Molecular mechanics calculations* show that dimerization at C1, C5, or C5′ is impossible in most cases, but that it is quite feasible at C3 or C3′, even for the t-butyl radical adduct, provided the two t-butyl groups stay out of each other's way. The optimized structure of the t-butyl dimer in Fig. 8 clearly shows the anticipated puckering of the spherical surface of each C_{60} unit at the nearly tetrahedral carbons and the formation of boat-shaped surface 1,4-cyclohexadienyl rings.

Prolonged irradiation and large excesses of radical precursors in the above experiments gave rise to broad, featureless ESR spectra that were often quite persistent. Although the details have not yet been deciphered completely, we attribute these absorptions to the superposition of various radical species, including species of higher spin multiplicity[42], resulting from multiple radical addition to C_{60} and to the RC_{60}-$C_{60}R$ dimers. Some insight concerning multiple radical addition came from the addition of benzyl radicals labelled with ^{13}C at the benzylic position[43]. The latter were generated by $in situ$ photolysis of saturated solutions of C_{60} in labelled toluene containing $\sim$5% di-t-butyl peroxide[90]. Photochemically generated t-butoxy radicals readily abstract a benzylic hydrogen atom, and the resulting benzyl radicals add to C_{60}. Two remarkably persistent radical species were formed with widely different power saturation behav-

ior. The spectrum of each species could be recorded unencumbered by that of the other simply by changing the microwave power level. The spectrum at low microwave powers (Fig. 9A) is appropriate for an allylic radical formed by the addition of three benzylic radicals to adjacent double bonds radiating from a five-membered ring on the C_{60} surface (Fig. 10). It displays a relatively strong hyperfine interaction with two equivalent ^{13}C nuclei (triplet of 9.70 Gauss) and a much weaker interaction with a unique ^{13}C nucleus (doublets of 1.75 Gauss). The two equivalent ^{13}C nuclei are optimally disposed to interact by hyperconjugation with the unpaired electron at each terminus

Scheme 1

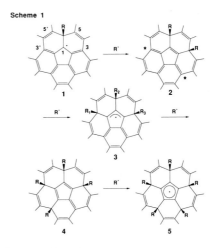

Fig. 10. Formation of $R_3C_{60}\cdot$ and $R_5C_{60}\cdot$ radicals ($R = CH_2C_6H_5$).

*Insight II Molecular Modelling Software, Biosym Technologies Inc., San Diego, CA 92121.

of the allylic structure, as shown below. The single, weakly interacting ^{13}C nucleus is similarly positioned with respect to the central allylic carbon, where a smaller, negative spin density is expected[91].

At very high microwave powers a binomial sextet spectrum emerged (Fig. 9B) appropriate for five equivalent ^{13}C nuclei interacting with the unpaired electron. This is assigned to a cyclopentadienyl radical structure on the surface of C_{60} resulting from the addition of five labeled benzyl radicals, as shown in Fig. 10. The ^{13}C hyperfine interaction (3.56 Gauss) is again due to hyperconjugation. It is smaller than in the allylic radical, however, since only $\frac{1}{5}$ of the unpaired electron density resides on each carbon atom of the cyclopentadienyl ring. The widely different power saturation behavior of the two radicals can now be understood, since the $R_5C_{60}\cdot$ radical is orbitally degenerate due to the five-fold axes of symmetry. Such radicals have efficient spin relaxation caused by strong spin-orbit coupling, and do not power saturate as readily as ordinary carbon-centered radicals. Computer molecular modeling suggests that the extraordinary stability of these two radicals derives from the steric protection afforded by three or five benzyl groups sheltering the radical sites[43]. It also explains the particular stability of the mono-*t*-butyl radical adduct. The size of the *t*-butyl group is such that another *t*-butyl radical can no longer add at C3 or C3′ (Fig. 10), precluding in this case the formation of C_{60} allyl and cyclopentadienyl radicals.

A major unifying view for the formation of covalent derivatives of C_{60} followed from the Scheme in Fig. 10. The direct EPR detection of the trisubstituted allylic-like radical (Fig. 10A) and the pentasubstituted cyclopentadienyl-like radical (Fig. 10B) suggested that multiple "1,4"-additions to diene units were involved in the reactions (eqn 6).

The structure of the heavily brominated $C_{60}Br_{24}$ may be regarded as arising from twelve "1,4"-additions chosen to minimize the Br to Br nonbonded interactions, which, as mentioned previously, led to the proposal of a structure that was later confirmed from x-ray diffraction. More recent evidence has provided further support for this scheme. Of particular note is the structure of $C_{60}Br_6$[26] which may be viewed as the cyclopentadienyl-like radical (Fig. 10B) pairing with a bromine atom (eqn 7):

Two such substitution patterns well-separated on the C_{60} sphere might rationalize formulations such as $C_{60}Cl_{12}$[37].

6. THEORETICAL STUDIES

There has been a wide variety of theoretical studies of C_{60}, some of which predate the initial mass spectrometric discovery of C_{60}. Our initial interest in theoretical modeling of C_{60} was in predicting its nonlinear optical properties[92] as part of a general series of theoretical studies searching for molecules with large nonlinearities[93,94]. Because of the size of C_{60} and the number of molecules we were studying with full geometry optimization, we employed semi-empirical molecular orbital methods based on the MNDO Hamiltonian[95–99]. Our geometry optimized at this level is consistent with the experimental results showing sets of long bonds, which we call single bonds, and short bonds, which we call double bonds. A simple geometrical model for C_{60} is thus sets of connected five-membered radialene rings that are curved into a sphere. Our studies on the value of γ for C_{60} predicted a value much smaller than the value obtained by extrapolating the macroscopic experimental values[100–103] to a value for a single molecule at zero frequency. The semi-empirical results of a small value for γ at zero frequency have been confirmed by local density functional calculations[104].

After this initial foray into C_{60} chemistry, we began to study the addition chemistry of C_{60} as an aid in interpreting the experimental work underway at Du Pont. Again, we initially used the semi-empirical molecular orbital method based on the MNDO Hamiltonian. For the first study, we used the PM-3 parameterization of the Hamiltonian[95–99]. Our first study of how molecules add to C_{60} was based on adding H_2 external to the cage[105]. There are three ways to add an X_2 molecule such as H_2 to C_{60}: (a) Addition directly to a double bond without changing the remainder of the bonding of the cage. We denote this "1,2 addition." (b) Conjugate addition of the two atoms which forces changes in the canonical bonding structure of C_{60} in order to maintain a closed-shell Kekulé structure. An example of this is addition to the two carbons of a single bond. (c) Hydrogenolysis leading to breaking of a C—C single bond.

We initially studied the 23 possible closed-shell singlet structures generated by exo-addition of types 1 and 2 (i. e., the cage is not disrupted on addition, although the polyene structure may be). The lowest energy structure was found to be the one where addition is to a double bond. If the first atom to which

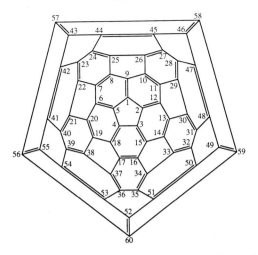

Fig. 11. Numbering system and valence bond diagram for C_{60}.

the H is added is labelled atom 1, then this results in the second atom being added at C9 (see Fig. 11 for C_{60} numbering). The next lowest energy structure is the one where addition occurs "1,4" to a cyclohexatriene, leading to formation of the 1,7-isomer. This isomer is 3.9 kcal/mol higher in energy than the 1,9-isomer generated by "1,2-addition." The next highest energy isomers are predicted to be 15.5 (1,23-isomer) and 18.4 (1,2-isomer) kcal/mol higher in energy than the 1,9-isomer. These are the only isomers that can be formed exothermically from the addition of H_2 to C_{60}. The energy of the $C_{60} + H_2$ asymptote is calculated to be 798.3 kcal/mol, with the energy of the most stable isomer calculated to be 776.1 kcal/mol at the PM-3 level.

If one examines the optimized geometries after addition, one structural feature becomes quite clear. Except for the 1,9-isomer generated by "1,2 addition" to a cyclohexatriene, all of the structures have at least one double bond placed into a five-membered ring. The number of double bonds placed into five-membered rings corresponds to the minimum number of double bonds that must migrate to obtain a Kekulé structure. This holds except for the largest number of double-bond migrations (eight), which yields a zwitterion structure, and for a delocalized polyene structure resulting from seven-bond migrations. Plots of the energy versus the number of migrated double bonds show that each migrated double bond adds 8.5 kcal/mol to the energy. We also calculated the electronic structure of a number of triplet species. These isomers have a calculated ΔH_f of ~ 815 kcal/mol. Thus, if the number of migrated double bonds is less than five, the ground state of $1,n\text{-}C_{60}H_2$ will be a closed shell singlet, and if it is greater than or equal to five it will have an open-shell ground state.

Clearly there are only two low-energy isomers of $C_{60}H_2$, and it is worth discussing their relative energies. The 1,9-isomer is the most stable, and the bond-

ing of the polyene framework of C_{60} has not been changed by the addition, other than removal of a π bond. There are eclipsing interactions between the two hydrogens bonded to the two sp^3 centers that one can estimate as being 3–5 kcal/mol destabilizing [106, 107]. The 1,7 isomer introduces a double bond into a five-membered ring, which should cost 8.5 kcal/mol, but does not have any eclipsing H,H interactions. Thus the 1,9-isomer should be about 4–6 kcal/mol more stable than the 1,7-isomer, in agreement with our calculated energy difference of 4 kcal/mol.

The presence of steric interactions in the 1,9-isomer suggests that if the steric bulk of the substituents becomes large, then the relative energies of the 1,9- and 1,7-isomers could change. For example, the rotation barrier in ethane is about 3 kcal/mol, but the rotation barriers for eclipsing two halogens in the 1,2-dihaloethanes for chlorine, bromine, and iodine are 9–10 kcal/mol[108]. Thus the larger steric effects could lead to the 1,7-isomer being lower in energy. In order to test this, we performed calculations on the 1,7- and 1,9-isomers for $C_{60}X_2(X = H, F, Cl, Br, I)$. The AM1 parameterization of the MNDO Hamiltonian worked somewhat better than the PM-3 parameterization in treating nonbonded interactions, so we used the former in our studies of halogenated C_{60}s. Careful studies of the barriers to rotation in the 1,2-dihaloethanes and of non bonded interactions showed that the semi-empirical molecular orbital methods do not have large enough repulsive nonbonded interactions and, consequently, the steric effects will be underestimated[108,117]. We thus used the local-density functional (LDF) method[109–116] to calculate the energy differences between the two isomers based on the optimized MNDO/AM1 geometries[81]. The energy difference between the 1,7- and 1,9-isomers for $X = H$ is 6.8 kcal/mol at the LDF level, favoring the 1,9-isomer. The energy difference between the isomers is essentially the same for $X = F$, and decreases to 2.5 kcal/mol for $X = Cl$. For $X = Br$, the energy difference is 0.5 kcal/mol favoring the 1,7-isomer, and the 1,7-isomer becomes even more stable, 9 kcal/mol, for $X = I$. Thus, as the steric size of the substituent increases, the 1,7-isomer formed by "1,4-addition" to a cyclohexatriene becomes favored energetically. This is consistent with the formation of $T_h - C_{60}Br_{24}$, which is formed by multiple "1,4-addition" processes. This result is also consistent with the structures of the open-shell isomers derived from ESR measurements on adding various hydrocarbon radicals to C_{60}.

We used a combined experimental/theoretical approach to estimate the energy of adding X_2 to C_{60}. This reaction is predicted to be weakly to moderately exothermic for addition of H_2 and Cl_2, and strongly exothermic for addition of F_2. For Br_2, the reaction is predicted to be weakly endothermic and for I_2, strongly endothermic. Thus fluorination should lead to both "1,2-" and "1,4-addition" products. Since fluorinations are strongly exothermic, there is a low probability of equilibration of the various sites, and the products from multiple additions are expected to be mixtures of kinetic products. In other words, a wide range of products is expected. Hydrogenation and chlorination are mildly exothermic, and again mixtures of isomers containing both "1,2-" and "1,4-addition" products could be expected, depending on reaction conditions. Chlorination differs from hydrogenation in that steric effects will become more important as larger numbers of chlorines are added, and under certain conditions "1,4-addition" could dominate. It is important to note here some qualitative results. As more atoms are added "1,2" the energy increment becomes smaller for H_2 and F_2 addition [117]. For "1,4-addition," adding more chlorines or bromines is more energetic than adding the first X_2 molecule, up to 12 molecules being added. Thus one should not multiply the energy of the first addition by 12 to get the energy of adding 12 X_2 molecules. For bromine, the additions will produce mixtures of "1,2-" and "1,4-addition" products. As more substituents are added, steric effects become more important and semi-empirical calculations indeed show that the energy for adding additional bromines "1,4" becomes more exothermic, whereas for adding "1,2" it becomes less exothermic. The formation of $C_{60}Br_{24}$ of high symmetry can only occur if there is an equilibrium adding and removing Br_2 from $C_{60}(Br_2)_n$. If the first addition is "1,4," there are no obvious energetic constraints on how the next "1,4-addition" will occur. Thus, in order to reach the highly symmetrical structure of $C_{60}Br_{24}$, many equilibrations with Br_2 must occur. In this way, the least sterically hindered structure can be attained. Such equilibrations are likely for bromine, and are observed for addition of six $(Et_3P)_2Pt$ groups that attain a T_h symmetry structure by equilibration on and off the C_{60} sphere (vide supra). The observed T_h structure for $C_{60}Br_{24}$ is the least sterically hindered structure. If one takes the observed structure for $C_{60}Br_{24}$ and moves two appropriate Br atoms, three "1,2-" or "1,6- eclipsing" steric interactions occur, but only one double bond is placed in a five-membered ring; this should lead to a structure with significantly higher energy. If one Br atom is moved, a diradical structure is found. The further addition of two Br atoms to $C_{60}Br_{24}$ (if they can penetrate the van der Waals cloud of the bromines bonded to C_{60} in order to react), will lead to five "1,2-" or "1,6-steric" interactions, clearly an unfavorable process. In order to add more atoms, it is likely that C—C bond rupture would occur, but this has not yet been definitively observed in C_{60} chemistry.

We also modeled the addition of alkyl groups to form the radicals observed in the ESR experiments [118]. Our initial studies used H, F, and Cl as models of the alkyl group. The groups were added to the ends of double bonds originating from one five-membered ring. The calculations were done with the semi-empirical MNDO/PM-3 method in the RHF and ROHF framework. Spin populations are in reasonable agreement with the ESR results, as are the valence bond diagrams derived from the populations and the geometry parameters. The major energetic result is the difference between the energies of the two isomers derived by adding three groups. The structure with three adjacent groups all added "1,4" (A) is lower in energy by 8.5 kcal/mol compared to the energy of the structure (B) where the groups are added at non-adjacent positions.

A   B

Acknowledgements—We would like to thank Brian Malone for synthesizing the initial batches of C_{60} soot, and Ed Holler for chromatographic purification. We thank Ronald J. Davis, Stephen Hill, and Robert Young for invaluable technical assistance. The efforts of Bruce Parkinson, as well as Susan Lerke and Prof. Dennis Evans of the University of Delaware and James Y. Becker of the Ben-Gurion University of the Negev on the electrochemistry of C_{60} derivatives, have been exceptionally helpful. We thank Joe Lazar and Charles McEwen for mass spectroscopy; Frederick Davidson and Creston G. Campbell, Jr., for NMR studies; James E. Krywko (Dupont-Merck Co.) for molecular modelling studies; and Norman Herron and Ying Wang for valuable discussions and collaborations.

REFERENCES

1. Special issue on fullerenes, *Acc. Chem. Res.* **25**, 98–175 (1992).
2. R. F. Curl and R. E. Smalley, *Science* **242**, 1017 (1988).
3. H. Kroto, *Pure Appl. Chem.* **62**, 407 (1990).
4. R. F. Curl and R. E. Smalley, *Sci. Am.* **265**, 54 (1991).
5. F. Diederich and R. L. Whetten, *Angew. Chem., Int. Ed. Eng.* **30**, 678 (1991).
6. J. S. Miller, *Adv. Mater.* **3**, 262 (1991).
7. H. W. Kroto, A. W. Allaf, and S. P. Balm, *Chem. Rev.* **91**, 1213 (1991).
8. F. J. Stoddart, *Angew, Chem.* **103**, 71 (1991).
9. H. Kroto, *Science* **242**, 1139 (1988).
10. H. Schwarz, *Angew. Chem. Int. Ed. Engl.* **31**, 293 (1992).
11. *Fullerenes* (Edited by George S. Hammond and Valerie J. Kuck) (ACS Symp. Ser., Vol. 481, pp. 1–195), American Chemical Society, Washington, D.C. (1992).
12. W. Krätschmer, L. D. Lamb, K. Fostiropoulos, and D. R. Huffman, *Nature* **347**, 354 (1990).
13. J. M. Hawkins *Acc. Chem. Res.* **25**, 150 (1992).
14. J. M. Hawkins, A. Meyer, T. A. Lewis, S. Loren, and F. J. Hollander, *Science* **252**, 312 (1991).
15. J. M. Hawkins, T. A. Lewis, S. D. Loren, A. Meyer, J. R. Heath, Y. Shibato, and R. J. Saykally, *J. Org. Chem.* **55**, 6250 (1990).
16. P. J. Fagan, J. C. Calabrese, and B. Malone, *Acc. Chem. Res.* **25**, 134 (1992).

17. P. J. Fagan, J. C. Calabrese, and B. Malone, In *Fullerenes* (Edited by George S. Hammond and Valerie J. Kuck) (ACS Symp. Ser. 481, pp. 177–186), American Chemical Society, Washington, D.C. (1992).

18. P. J. Fagan, J. C. Calabrese, and B. Malone, *J. Am. Chem. Soc.* 113, 9408 (1991).

19. P. J. Fagan, J. C. Calabrese, and B. Malone, *Science* 252, 1160 (1991).

20. K. M. Creegan, J. L. Robbins, W. K. Robbins, J. M. Millar, R. D. Sherwood, P. J. Tindall, D. M. Cox, A. B. Smith III, J. P. McCauley, Jr., D. R. Jones, and R. T. Gallagher, *J. Am. Chem. Soc.* 114, 1103 (1992).

21. Y. Elemes, S. K. Silverman, C. Sheu, M. Kao, C. S. Foote, M. M. Alvarez, and R. L. Whetten, *Angew. Chem., Int. Ed. Eng.* 31, 351 (1992).

22. A. L. Balch, V. J. Catalano, and J. W. Lee, *Inorg. Chem.* 30, 3980 (1991). A. L. Balch, V. J. Catalano, J. W. Lee, and M. M. Olmstead, *J. Am. Chem. Soc.* 114, 5455 (1992).

23. A. L. Balch, V. J. Catalano, J. W. Lee, M. M. Olmstead, and S. R. Parkin, *J. Am. Chem. Soc.* 113, 8953 (1991).

24. R. S. Koefod, M. F. Hudgens, and J. R. Shapley, *J. Am. Chem. Soc.* 113, 8957 (1991).

25. F. N. Tebbe, R. L. Harlow, D. B. Chase, D. L. Thorn, G. C. Campbell, Jr., J. C. Calabrese, N. Herron, R. J. Young, Jr., and E. Wasserman, *Science* 256, 822 (1992).

26. P. R. Birkett, P. B. Hitchcock, H. W. Kroto, R. Taylor, and D. R. M. Walton, *Nature* 357, 479 (1992).

27. F. Wudl, *Acc. Chem. Res.* 25, 157 (1992).

28. T. Suzuki, Q. Li, K. C. Khemani, F. Wudl, and Ö. Almarsson, *Science* 254, 1186 (1991).

29. R. E. Haufler, J. Conceicao, L. P. F. Chibante, Y. Chai, N. E. Byrne, S. Flanagan, M. M. Haley, S. C. O'Brien, C. Pan, Z. Xiao, W. E. Billups, M. A. Ciufolini, R. H. Hauge, J. L. Margrave, L. J. Wilson, R. F. Curl, and R. E. Smalley, *J. Phys. Chem.* 94, 8634 (1990).

30. F. Wudl, A. Hirsch, K. C. Khemani, T. Suzuki, P.-M. Allemand, A. Koch, H. Eckert, G. Srdanov, and H. M. Webb, In *Fullerenes* (Edited by George S. Hammond and Valerie J. Kuck) (ACS Symp. Ser. 481, pp. 161–175), American Chemical Society, Washington, D.C. (1992).

31. J. W. Bausch, G. K. S. Prakash, G. A. Olah, D. S. Tse, D. C. Lorents, Y. K. Bae, and R. Malhotra, *J. Am. Chem. Soc.* 113, 3205 (1991).

32. G. A. Olah, I. Bucsi, C. Lambert, R. Aniszfeld, N. J. Trivedi, D. K. Sensharma, and G. K. S. Prakash, *J. Am. Chem. Soc.* 113, 9387 (1991).

33. G. A. Olah, I. Bucsi, C. Lambert, R. Aniszfeld, N. J. Trivedi, D. K. Sensharma, and G. K. S. Prakash, *J. Am. Chem. Soc.* 113, 9385 (1991).

34. A. Hirsch, Q. Li, and F. Wudl, *Angew. Chem., Int. Ed. Engl.* 30, 1309 (1991).

35. H. Nagashima, A. Nakaoka, Y. Saito, M. Kato, T. Kawanishi, and K. Itoh, *J. Chem. Soc., Chem. Comm.*, 377 (1992).

36. R. Taylor, J. P. Parsons, A. G. Avent, S. P. Rannard, T. J. Dennis, J. P. Hare, H. W. Kroto, and D. R. M. Walton, *Nature* 351, 277 (1991).

37. F. N. Tebbe, J. Y. Becker, D. B. Chase, L. E. Firment, E. R. Holler, B. S. Malone, P. J. Krusic, and E. Wasserman, *J. Am. Chem. Soc.* 113, 9900 (1991).

38. J. H. Holloway, E. G. Hope, R. Taylor, G. J. Langley, A. G. Avent, T. J. Dennis, J. P. Hare, H. W. Kroto, and D. R. M. Walton, *J. Chem. Soc., Chem. Commun.*, 966 (1991).

39. H. Selig, C. Lifshitz, T. Peres, J. E. Fischer, A. R. McGhie, W. J. Ramanow, J. P. McCauley, Jr., and A. B. Smith, III, *J. Am. Chem. Soc.* 113, 5475 (1991).

40. D. A. Loy and R. A. Assink, *J. Am. Chem. Soc.* 114, 3977 (1992).

41. A. Pénicaud, J. Hsu, and C. Reed, *J. Am. Chem. Soc.* 113, 6698 (1991).

42. P. J. Krusic, E. Wasserman, B. A. Parkinson, B. Malone, E. R. Holler, P. N. Keizer, J. R. Morton, and K. F. Preston, *J. Amer. Chem. Soc.* 113, 6274 (1991).

43. P. J. Krusic, E. Wasserman, P. N. Keizer, J. R. Morton, and K. F. Preston, *Science* 254, 1183 (1991).

44. J. R. Morton, K. F. Preston, P. J. Krusic, S. A. Hill, and E. Wasserman, *J. Phys. Chem.* 96, 3576 (1992).

45. J. R. Morton, K. F. Preston, P. J. Krusic, and E. Wasserman, *J. Chem. Soc., Perkin Trans.* (in press).

46. J. R. Morton, K. F. Preston, P. J. Krusic, S. A. Hill, and E. Wasserman, *J. Amer. Chem. Soc.* 114, 5454 (1992).

47. R. Taylor, J. H. Holloway, E. G. Hope, A. G. Avent, G. J. Langley, T. J. Dennis, J. P. Hare, H. W. Kroto, and D. R. M. Walton, *J. Chem. Soc., Chem. Comm.*, 665 (1992).

48. R. Taylor, G. J. Langley, M. F. Meidine, J. P. Parsons, A. K. Abdul-Sada, T. J. Dennis, J. P. Hare, H. W. Kroto, and D. R. M. Walton, *J. Chem. Soc., Chem. Comm.*, 667 (1992).

49. S. H. Hoke, II, J. Molstad, G. L. Payne, B. Kahr, B.-A. Dor, and R. G. Cooks, *Rapid Commun. Mass Spectrom.* 5, 472 (1992).

50. R. Seshadri, A. Govindaraj, R. Nagarajan, T. Pradeep, and C. N. R. Rao, *Tetrahedron Lett.* 33, 2069 (1992).

51. J. R. Heath, S. C. O'Brien, Q. Zhang, Y. Liu, R. F. Curl, H. W. Kroto, F. K. Tittel, and R. E. Smalley, *J. Am. Chem. Soc.* 107, 7779 (1985).

52. D. M. Cox, D. J. Trevor, K. C. Reichmann, and A. Kaldor, *J. Am. Chem. Soc.* 108, 2457 (1986).

53. Y. Chai, T. Guo, C. Jin, R. E. Haufler, L. P. F. Chibante, J. Fure, L. Wang, J. M. Alford, and R. E. Smalley, *J. Phys. Chem.* 95, 7564 (1991).

54. F. D. Weiss, J. L. Elkind, S. C. O'Brien, R. F. Curl, and R. E. Smalley, *J. Am. Chem. Soc.* 110, 4464 (1988).

55. R. D. Johnson, M. S. de Vries, J. R. Salem, D. S. Bethune, and C. S. Yannoni, *Nature* 355, 239 (1992).

56. M. M. Alvarez, E. G. Gillan, K. Holczer, R. B. Kaner, K. S. Min, and R. L. Whetten, *J. Phys. Chem.* 95, 10561 (1991).

57. T. Guo, C. Jin, and R. E. Smalley, *J. Phys. Chem.* 95, 4948 (1991).

58. R. E. Smalley, In Fullerenes (Edited by George S. Hammond and Valerie J. Kuck) (ACS Symp. Ser. 481, pp. 141–159), American Chemical Society, Washington, D.C. (1992).

59. B. C. Guo, K. P. Kerns, and A. W. Castleman, Jr., *Science* 255, 1411 (1992). B. C. Guo, S. Wei, J. Purnell, S. Buzza, and A. W. Castleman, Jr., *Science* 256, 515 (1992).

60. L. M. Roth, Y. Huang, J. T. Schwedler, C. J. Cassady, D. Ben-Amotz, B. Kahr, and B. S. Freiser, *J. Am. Chem. Soc.* 113, 6298 (1991). Y. Huang, B. S. Freiser, *J. Am. Chem. Soc.* 113, 9418 (1991).

61. R. L. Garrell, T. M. Herne, C. A. Szafranski, F. Diederich, F. Ettl, and R. L. Whetten, *J. Amer. Chem. Soc.* 113, 6302 (1991).

62. R. J. Wilson, G. Meijer, D. S. Bethune, R. D. Johnson, D. D. Chambliss, M. S. de Vries, H. E. Hunziker, and H. R. Wendt, *Nature* 348, 621 (1990).

63. J. L. Wragg, J. E. Chamberlain, H. W. White, W. Krätschmer, and D. R. Huffman, *Nature* 348, 623 (1990).

64. Y. Zhang, G. Edens, and M. J. Weaver, *J. Am. Chem. Soc.* 113, 9395 (1991).

65. C. Jehoulet, A. J. Bard, and F. Wudl, *J. Am. Chem. Soc.* 113, 5456 (1991).

66. T. Chen, S. Howells, M. Gallagher, L. Yi, D. Sarid, D. L. Lichtenberger, K. W. Nebesny, and C. D. Ray, *J. Vac. Sci. Technol., B* 9, 2461 (1991).

67. Y. Zhang, X. Gao, and M. J. Weaver, *J. Phys. Chem.* 96, 510 (1992).

68. T. R. Ohno, Y. Chem, S. E. Harvey, G. H. Kroll, J. H. Weaver, R. E. Haufler, and R. E. Smalley, *Phys. Rev. B: Condens. Matter* **44**, 13747 (1991).
69. R. C. Haddon, *Acc. Chem. Res.* **25**, 127 (1992).
70. P. J. Fagan, M. D. Ward, and J. C. Calabrese, *J. Am. Chem. Soc.* **111**, 1698 (1989).
71. W. I. F. David, R. M. Ibberson, J. C. Matthewman, K. Prassides, T. J. S. Dennis, J. P. Hare, H. W. Kroto, R. Taylor, and D. R. M. Walton, *Nature* **353**, 147 (1991).
72. K. Hedberg, L. Hedberg, D. S. Bethune, C. A. Brown, H. C. Dorn, R. D. Johnson, and M. de Vries, *Science* **254**, 410 (1991).
73. R. F. Curl and R. E. Smalley, *Science* **242**, 1017 (1988).
74. S. D. Ittel and J. A. Ibers, *Adv. Organomet. Chem.* **14**, 33 (1976).
75. A. Kumar, J. D. Lichtenhan, S. C. Critchlow, B. E. Eichinger, and W. T. Borden, *J. Am. Chem. Soc.* **112**, 5633 (1990).
76. K. Morokuma and W. T. Borden, *J. Am. Chem. Soc.* **113**, 1912 (1991).
77. B. Chase and P. J. Fagan, *J. Am. Chem. Soc.* **114**, 2252 (1992).
78. P. J. Fagan, submitted for publication, and unpublished results.
79. S. A. Lerke, D. Evans, P. J. Fagan, and B. Parkinson, *J. Am. Chem. Soc.* **114**, (in press).
80. R. S. Koefod, C. Xu, W. Lu, J. R. Shapley, M. G. Hill, and K. R. Mann, *J. Phys. Chem.* **96**, 2928 (1992).
81. D. A. Dixon, N. Matsuzawa, T. Fukunaga, and F. N. Tebbe, *J. Phys. Chem.*, **92**, 6107 (1992).
82. R. Taylor, A. G. Avent, T. J. Dennis, J. P. Hare, H. W. Kroto, D.R.M. Walton, J. H. Holloway, E. G. Hope, and G. J. Langley, *Nature* **355**, 27 (1992).
83. D. S. Bethune, G. Meijer, W. C. Tang, H. J. Rosen, W. Golden, H. Seki, C. Brown, and M. S. de Vries, *Chem. Phys Lett.* **179**, 181 (1991).
84. J. Hare, J. Dennis, H. Kroto, R. Taylor, A. Allaf, S. Balm, and D. Walton, *J. Chem. Soc., Chem. Comm.*, 412 (1991).
85. D. M. Cox, S. Behal, M. Disko, S. Gorun, M. Greaney, C. Hsu, E. Kollin, J. Millar, J. Robbins, W. Robbins, R. Sherwood, and P. Tindal, *J. Am. Chem. Soc.* **113**, 2940 (1991).
86. B. Chase, N. Herron, and E. Holler, *J. Phys. Chem.* **96**, 4262 (1992).
87. C. N. McEwen, R. G. McKay, and B. S. Larsen, *J. Am. Chem. Soc.* **114**, 4412 (1992).
88. R. W. Fessenden and R. H. Schuler, *J. Chem. Phys.* **38**, 773 (1963).
89. D. M. Chipman, *J. Chem. Phys.* **71**, 761 (1979).
90. P. J. Krusic and J. K. Kochi, *J. Amer. Chem. Soc.* **90**, 7155 (1968).
91. J. E. Wertz and J. R. Bolton, In *Electron spin resonance; Elementary theory and practical applications,* Chapter 6. McGraw-Hill, New York (1972).
92. N. Matsuzawa and D. A. Dixon, *J. Phys. Chem.* **96**, 6241 (1992).
93. N. Matsuzawa and D. A. Dixon, *Int. J. Quantum Chem.* **96**, 6232 (1992).
94. N. Matsuzawa and D. A. Dixon, *J. Phys. Chem.* (in press).

95. M. J. S. Dewar and W. Thiel, *J. Am. Chem. Soc.* **99**, 4899 (1977).
96. M. J. S. Dewar, E. G. Zoebisch, E. F. Healy, and J. J. P. Stewart, *J. Am. Chem. Soc.* **107**, 3902 (1985).
97. J. J. P. Stewart, *J. Comp. Chem.* **10**, 209; 221 (1989).
98. J. J. P. Stewart, QCPE Program 455, 1983; version 6.00 (1991).
99. J. J. P. Stewart, *J. Comp.-Aided Mol. Design* **4**, 1 (1990).
100. W. J. Blau, H. J. Byrne, D. J. Cardin, T. J. Dennis, J. P. Hare, H. W. Kroto, R. Taylor, and D. R. M. Walton, *Phys. Rev. Lett.* **67**, 1423 (1991).
101. Z. H. Kafafi, J. R. Lindle, R. G. S. Pong, F. J. Bartoli, L. J. Lingg, and J. Milliken, *Chem. Phys. Lett.* **188**, 492 (1992).
102. H. Hoshi, N. Nakamura, Y. Maruyama, T. Nakagawa, S. Suzuki, H. Shiromaru, and Y. Achiba, *Jpn. J. Appl. Phys.* **30**, L1397 (1991).
103. Y. Wang and L.-T. Cheng, *J. Phys. Chem.* **96**, 1530 (1992).
104. N. Matsuzawa and D. A. Dixon, *J. Phys. Chem.* **96**, 6872 (1992).
105. D. A. Dixon, N. Matsuzawa, and T. Fukunaga, *J. Phys. Chem.*, (in press).
106. E. L. Eliel, In *Stereochemistry of carbon compounds.* McGraw-Hill, New York (1962).
107. P. W. Payne and L. C. Allen, In *Applications of electronic structure theory* (Edited by H. F. Schaefer III) Chapter 2, p. 29. Plenum Press, New York, (1977).
108. N. Matsuzawa and D. A. Dixon, J. Phys. Chem. (submitted for publication).
109. R. G. Parr and W. Yang, In *Density functional theory of atoms and molecules,* Oxford University Press, New York (1989).
110. D. R. Salahub, In *Ab initio methods in quantum chemistry-II* (Edited by K. P. Lawley), p. 447. Wiley & Sons, New York (1987).
111. E. Wimmer, A. J. Freeman, C.-L. Fu, P.-L. Cao, S.-H. Chou, and B. Delley, In *Supercomputer research in chemistry and chemical engineering* (Edited by K. F. Jensen and D. G. Truhlar), *ACS Symp. Ser.,* **353** p. 49. American Chemical Society, Washington, D.C. (1987).
112. R. O. Jones and O. Gunnarsson, *Rev. Mod. Phys.* **61**, 689 (1989).
113. D. A. Dixon, J. Andzelm, G. Fitzgerald, E. Wimmer, and B. Delley, In *Science and engineering on supercomputers* (Edited by E. J. Pitcher), p. 285. Computational Mechanics Publications, Southampton, England (1990).
114. D. A. Dixon, J. Andzelm, G. Fitzgerald, E. Wimmer, and P. Jasien, In *Density functional methods in chemistry* (Edited by J. K. Labanowski and J. W. Andzelm) Chapter 3, p. 33. Springer-Verlag, New York (1991).
115. B. Delley, *J. Chem Phys.* **92**, 508 (1990).
116. B. Delley, In *Density functional methods in chemistry* (Edited by J. K. Labanowski and J. W. Andzelm), Chapter 11, p. 101. Springer-Verlag, New York, (1991).
117. N. Matsuzawa, T. Fukunaga, and D. A. Dixon, *J. Phys. Chem* (in press).
118. N. Matsuzawa, D. A. Dixon, and P. Krusic, *J. Phys. Chem.,* (in press).

ELECTRONIC, MAGNETIC, AND STRUCTURAL PROPERTIES OF AMORPHOUS CARBONS AND THE DISCOVERY OF THE FULLERENES

Lawrence J. Dunne,[1,2] Anthony D. Clark,[1] Martin F. Chaplin,[1] and
Harsha Katbamna[1]
[1]South Bank University, London SE1 OAA, England
[2]School of Molecular Sciences, University of Sussex, Falmer, Brighton, BN1 9QJ, England

(*Accepted* 28 *April* 1992)

Abstract—Although amorphous carbons which may contain heteroatoms have been subjected to extensive study over many years, their structure and properties are not well understood. The discovery of the closed fullerene cages and related structures, and their possible occurrence in amorphous carbons offers a new approach to the understanding of the chemical and physical properties of amorphous carbons. Most significantly electron micrographs of amorphous carbons and carbon blacks clearly depict the nonplanar nature of the carbon networks and three types of curved structures which may contain pentagonal rings are discussed. The nonplanarity of such networks may lead to voids in the structure which can trap free radicals (probably of the odd-alternate π-type), and this allows us to offer an explanation for the electrical conductivity and the magnetic properties of such materials. We also consider the effect of heteroatoms (N, S, and O) in amorphous carbon structures.

Key Words—Fullerenes, electron microscopy, carbon blacks, paramagnetism, electrical properties.

1. INTRODUCTION

An old but outstanding problem in the physics and chemistry of the organic solid state is the relationship between the structure and properties, both electric and magnetic, of amorphous carbons prepared by thermal decomposition of organic materials in an inert atmosphere. In this paper, we examine the impact which the discovery of the fullerene cage structures, containing five-membered carbon rings, has on this field and we will review a few old but selected problems in the light of these outstanding discoveries which are described in this special issue of *Carbon* devoted to the fullerenes. The possible occurrence of nonplanar carbon networks containing pentagonal rings gives a new insight into the understanding of amorphous carbon structures.

Our particular interest in the work on the fullerenes arose out of our detailed work on the structure and properties of amorphous carbons prepared from the thermal decomposition of sulphanilic acid in the solid state[1]. While our experience with carbon structure is largely derived from these studies, we will consider the problem of solid carbon structure and properties more generally, and hence we will also use the published results of other workers.

Amorphous carbons (which we loosely define to embrace carbon blacks and other graphitised materials) may be prepared in a variety of ways of which thermal degradation of organic materials in an inert atmosphere is the most common. They are not a form of pure carbon but usually contain a few percent of other atoms from the parent organic material, although carbon dominates. The structure of such ma-

terials is an outstanding problem in the physics and chemistry of the organic solid state. A comprehensive review of the physics and chemistry of carbons has been given by Jenkins and Kawamura[2]. These materials are collectively known as paracrystalline carbons and their structure was believed to consist predominately of layers of carbon atoms[3]. We have previously proposed[1] the existence in amorphous carbons of nonplanar structures containing pentagonal rings which are related to the newly discovered carbon clusters found by Kroto *et al.*[4–7], and hence we discuss our data on sulphanilic acid carbons (SACs) in this context. Many of the ideas which Kroto and co-workers[5–7] have put forward to describe soot particles are also applicable to amorphous carbons where the former are formed in the gas phase while the latter in the solid or liquid state. It is very significant indeed that C_{60} and C_{70} were both isolated from a form of pure solid carbon[8,9] although amorphous carbons also contain heteroatoms. Our task then is to ask what particular features of amorphous carbon structure and properties are attributable to structures related to the fullerenes.

2. ELECTRON MICROSCOPY STUDIES OF AMORPHOUS CARBONS

Considerable work has been undertaken on the electron microscopy of amorphous carbons but to our knowledge the existence of nonplanar carbon networks containing pentagonal rings or pentagonal disclinations was not invoked to explain electron micrographs of chars, before the discovery of the fuller-

ene and related structures. Carbon "spheres" have
been observed on many occasions[3,10–12] in electron microscopy studies of carbon. A comprehensive
review by Oberlin[13], of high resolution transmission electron microscopy studies of carbonaceous
materials has recently appeared, and this shows many
examples of curved structures, arising from a wide
range of carbons. There are three main curved structures discussed which are an onionlike one consisting
of concentric spherical layers, a rolled cylindrical
structure and crumpled lamellae which we liken to
sheets of lasagne. The interpretation of these structures in this review[13], however, makes no mention
of the possibility of fullerene type structures, but it
does show that curved structures are a general feature
of the morphology of carbons.

Figure 1 shows a transmission (TEM) mode electron micrograph of a sulphanilic acid carbon (SAC).
There is little apparent variation in the morphology
of these carbons over the HTT range 700–1100K and
hence this micrograph (which refers to 1100K) may
be regarded as typical for these materials. Structures
with a high degree of curvature are visible, such as the
ringlike structures with a diameter range of up to
6000Å. Notice also the microscopic voids in the bulk
of the structure. Since the walls are several hundred
Angström units thick, they must contain several hundred layers. The TEM plates suggest that these carbon
clusters are not empty spheroidal bodies but are filled.
Scanning electron microscopy gives the impression
that the materials are formed of crystalline platelets.
However, both X-ray diffraction and transmission
electron diffraction patterns from the materials were
totally diffuse, indicating the absence of long-range
crystalline order in the sulphanilic acid carbons.
However, although we must not forget SACs contain
nitrogen and sulphur, as well as carbon, hydrogen

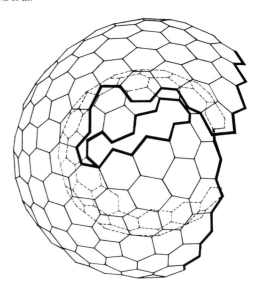

Fig. 2. A spheroidal carbon network modified from Zhang
et al.[7].

and oxygen atoms, the review of Oberlin[13] in particular shows these curved features are typical of all
carbons.

3. STRUCTURE OF AMORPHOUS CARBONS IN RELATION TO FULLERENES

The principles relating to the formation of spheroidal carbon shells may be extended[7,6] to amorphous carbon particle formation. A possible model of
a forming carbon particle (Fig. 2) is one in which carbon networks incorporate pentagonal rings as they
grow so as to generate spheroidal structures. These
pentagons generate a curvature which brings the network back on itself and closes up where possible. It is
unlikely that any such carbon network in the growth
environment appropiate to amorphous carbon formation will succeed in closing perfectly before the
growth of a new layer has commenced. This can easily lead to the formation of layered spheroidal structures which are imperfect. Such a mechanism can
give rise to the spherical structures considered above,
which contain voids or cavities in the structure, not

Fig. 1. Transmission electron micrograph of an amorphous
carbon prepared by the thermal decomposition of sulphanilic acid in a nitrogen atmosphere at 1100K. Note the ringlike structures on the left of the picture and the microscopic
voids in the bulk of the structure.

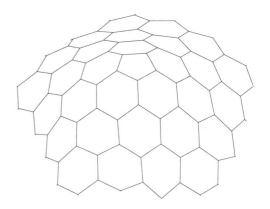

Fig. 3. A bowl-like carbon network containing a single pentagonal ring.

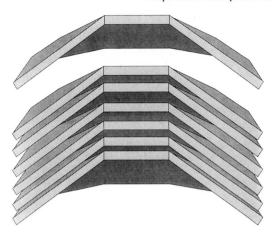

Fig. 4. A section through a stack of "bowls" giving rise to structures with an acute angle as seen in electron micrographs of some solid carbons (e.g., Oberlin[13], Fig. 25).

necessarily at the centre, due to imperfect formation of the inner shells of the carbon clusters.

However, it is clear that allowing pentagons into a carbon network allows almost any topography to be constructed and it seems likely that this is occurring in amorphous carbons. Figure 3 shows a bowl-like structure where the curvature arises from a single five-membered carbon ring at its base. Forming a stack of such bowls (Fig. 4) gives a kinked structure which also resembles structures seen in electron microscopy of carbon blacks (Oberlin[13], Fig. 25).

The many degrees of freedom introduced into a growing carbon network can give rise to the crumpled lamellae or lasagne structure as shown in Fig. 5 and also as noted by Oberlin[13], Fig. 62.

In the case of amorphous carbons which contain heteroatoms, we must consider, in general terms, how the above structures are modified as very few amorphous carbons are free of heteroatoms. X-ray photoelectron spectroscopy (XPS) of SACs[14] show only two environments for the nitrogen atoms, and these are in the form of tertiary amine and heterocyclic nitrogen. The same technique suggests all the sulphur is heterocyclic, although the sulphur content of our SACs is small (about 0.06 moles/100g), and varies little over the HTT range. We interpret this as the presence of a few thiophene-type rings on the edge of the structures. The presence of oxygen in the carbons is of wider interest however, as the oxygen containing functional groups can be used for enzyme binding, and this has been demonstrated in the use of these carbons as electrode materials in a glucose biosensor by Roulston et al.[15]. We have used XPS, IR, and base neutralisation studies to determine the nature of the oxygen containing groups in these carbons. Four functional groups are found, being in decreasing order of acidity, carboxylic, phenolic, lactonic, and quinolic[15]. The proportion of oxygen accounted for by the sum of these four functional groups ranges between 15% and 45% of the total oxygen present. Clearly therefore these form a significant proportion of the oxygen which is either on the surface of these carbon particles or is readily accessible to reagents.

Treatment of SAC prepared at a HTT of 600K with concentrated sulphuric acid yields the tetraaza-dibenzanthracene fragment below (Fig. 6); the XPS data fits closely with this structure and suggests most of the nitrogen is in units of this type. The XPS data

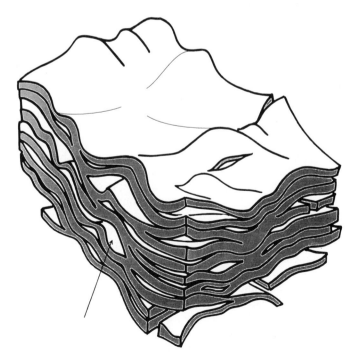

Fig. 5. Crumpled lamallae or lasagne structure; note the arrow which depicts voids in the structure which may trap planar odd-alternate π-radicals. See also Oberlin[13], Fig. 62.

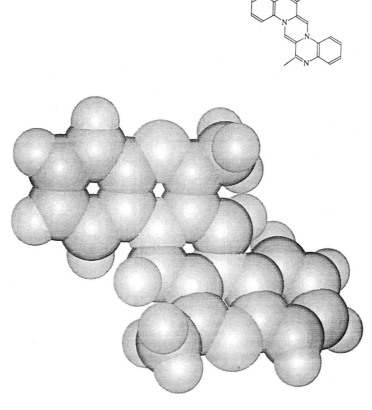

Fig. 6. Three-dimensional structure of tetraazadibenzanthracene ring system.

of SACs prepared at higher HTTs is similar so although the fragment has not been isolated from these SACs, it seems probable this structural unit is still present at higher HTTs. Consideration of the carbon to nitrogen mole ratios (our complete analytical data on the SACs can be found in Ref.[1]), shows that the nitrogen content and hence, the amount of tetraazadibenzanthracene units decreases as the HTT increases. For the sample with a HTT of 700K, all the carbon content is required for the tetraazadibenzantracene structure so there cannot be extensive regions of carbon only condensed aromatic structures in this SAC.

As the tetraazadibenzantracene ring system is nonplanar as shown in Fig. 6, SACs, and possibly other nitrogen containing carbons, may possess this additional source of curvature in their structures.

4. PARAMAGNETISM IN CARBONS

As soon as the process of carbonisation in an inert atmosphere begins, paramagnetic spin centres can be observed in the carbon by electron spin resonance (ESR) spectroscopy. The onset of significant electrical conductivity is also generally related to the appearance of spin centres in the chars. The room-temperature resistivity of chars falls rapidly as the HTT increases. In amorphous carbons generally, the g-factor of the ESR absorption remains close to, though slightly greater than, the free-spin value throughout

the HTT range and the ESR absorption signals of the outgassed samples have linewidths between 0.1 and 1 mT. From a histogram of g-factors collected for a wide range of carbons, there is a large peak between g = 2.0030 and 2.0034; this corresponds to the values expected for odd-alternate π-radicals and suggests these as the source of the paramagnetism in chars (J. Pradissitto, unpublished results). (An alternate hydrocarbon is a conjugated molecule such that the atoms can be divided into two sets, starred and unstarred so that no two members of the same set are neighbours and odd-alternate hydrocarbons in addition have an odd number of atoms.) Molecular oxygen when admitted to an outgassed sample broadens the ESR absorption by one or two orders of magnitude, and this process is usually reversible, suggesting that the spin centres are accessible to paramagnetic broadening reagents but are unreactive. Chars prepared at low HTTs generally obey Curie's law below room temperature and have paramagnetic susceptibilities corresponding to spin concentrations of typically 1×10^{20} spins g^{-1}[16].

These phenomena are still not well understood. It has generally been assumed that the initial appearance of the ESR absorption signal is associated with the free-spin centres formed as a consequence of the loss of hydrogen during carbonisation and stabilised by aromatic ring systems. Ingram[17] suggested as an explanation of free-spin centres in chars that the unpaired spins remaining after removal of the edge groups by homolytic bond fission become stabilised in the aromatic ring systems as π electrons. The range of HTT where the spin concentration falls is typically accompanied by a very large rise in the electrical conductivity of the material as noted by Ingram. This effect is shown in Fig. 7 for amorphous carbons prepared by the thermal decomposition of sulphanilic acid in the HTT range 700–1100K.

Ingram[17] suggested that the rise in the electrical conductivity with falling spin concentration would follow simply if at higher HTT aromatic free radical units in the char bond together at the edges. The spins on such units will be paired giving up a loss of paramagnetism and the increased extent of delocalisation in the newly extended aromatic network will give an enhanced electrical conductivity.

We note, from the definition above, that an alternate system cannot contain pentagonal rings in conjugation with the main conjugated system. If, therefore the free radical centres in amorphous carbons are of the odd-alternate type, which are likely to be the most stable, they are unlikely to be associated with a nonplanar carbon network where the nonplanarity arises from pentagonal rings, in which all the atoms are bonded.

The spin concentration in our samples varies between 1 spin/170 C atoms for a HTT of 900K to 1 spin/500 C atoms for 700 or 1100K; this behaviour is typical of amorphous carbons. Clearly therefore, even the smallest structures revealed by EM contain a large number of spins. However voids in the lasagne or "crumpled lamellae" structure of carbons as shown in Fig. 5, or at the centre of spherical structures, do contain sufficient space to trap an effectively planar odd-alternant π-radical such as the triphenylmethyl radical. This introduces localised electronic states at the Fermi level and such a process in a fullerene-type cage can account for the electrical conductivity behaviour discussed below. Such radicals would be inaccessible to radical scavenging attack but could strongly exchange (in ESR experiments) with each other and be broadened by molecular oxygen as is observed experimentally. Reactive material in the centre of such a cluster can evidently be trapped.

5. ELECTRICAL PROPERTIES

Amorphous carbons are, broadly speaking, semiconducting materials since their electrical resistance falls with increasing ambient temperature, a behav-

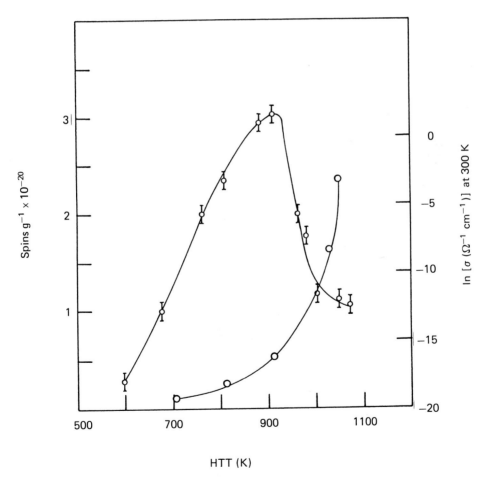

Fig. 7. Dependence of free spin concentration/heat treatment temperature of sulphanilic acid carbon (Room temperature conductivities are plotted for comparison). This behaviour is typical of amorphous carbons.

iour that is associated with activated electrical conduction.

In amorphous materials generally, following Mott and Davis[18], several mechanisms of electronic conduction can be distinguished. Provided that the temperature is sufficiently high the variation of conductivity with temperature will not be exponential but many consist of several distinct exponential regions. The ideas behind the work of Mott and Davis[18] have been applied to chars by Delhaes and Carmona[19] and Jenkins and Kawamura[2], and have been applied by us to sulphanilic acid carbons[1].

The major influence of structural disorder on the electronic properties of solids is the formation of localised states at the top of the valence band and the bottom of the conduction band. As emphasised particularly by Mott and Davis[18] and Mott[20,21], the electrical properties of amorphous materials require models for their interpretation different for those of crystalline materials. In centain energy ranges, electronic states in amorphous solids may be localised whereas, in other energy regions they may be extended. Localised states tend to have much smaller electrical mobilities than extended states and may appear at band extremities in an amorphous phase or they may exist, at some energy, a narrow band of localised states of high density particularly associated with unsatisfied valencies, and these may give rise to "kinks" in the plots of log(conductivity) versus $1/T$ for sulphanilic carbons as shown in Fig. 8.

From a detailed analysis of our conductivity data for SACs, we[1] have concluded that the mechanism for conduction is due to carriers excited into localised states at the band edges and to tunneling of carriers between localised states at the Fermi energy, and these two processes produce the discontinuity shown.

Although a detailed theoretical analysis of the electronic properties of an amorphous carbon is pro-

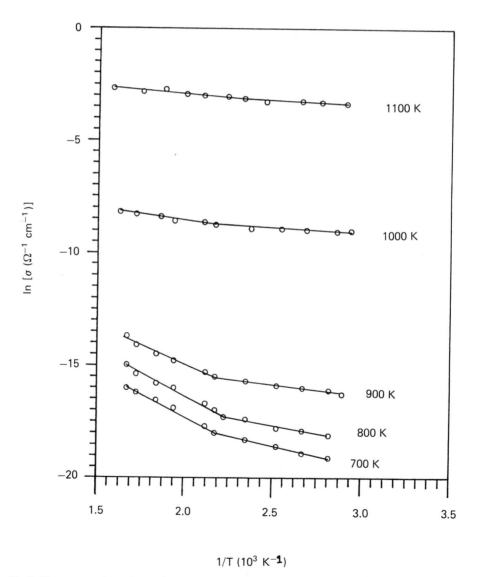

Fig. 8. Temperature dependence of the DC electrical conductivity of compressed and outgassed sulphanilic acid carbon pellets for the heat treatment temperature range 700–1100K.

hibitively difficult, at the qualitative level some comments can be made. Following Mott and Davis[18], the conductivity data shown in Fig. 8 can be understood if a narrow band of localised states exists near the centre of the gap ($E = 0$) and these pin the Fermi energy over a wide temperature range. Following Coulson and Rushbrook[22], an odd-alternant free radical will have a nonbonding orbital at $E = 0$ and if their concentration is sufficiently high, this should pin the Fermi level. We have suggested above that odd alternate π-radicals trapped inside fullerene-type voids can account for the paramagnetism; it also appears they can account in part for the electrical properties. Indeed, Lewis and Singer[16] have given convincing arguments that the free spins in chars are associated with odd-alternant radicals.

CONCLUSIONS

Although amorphous carbons, prepared by the pyrolysis of organic materials in an inert atmosphere, and carbon blacks prepared by incomplete combustion of organic materials have been studied for many years, their structure and properties are not well understood. The discovery of the fullerene cages and related structures to which this special issue of *Carbon* is devoted, and their possible occurrence in amorphous carbons gives a totally new and revealing insight into the chemical and physical properties of amorphous carbons.

Most significantly, as pointed out by Kroto[5] in the late 1980s, electron micrographs of solid carbons clearly suggest the existence of nonplanar carbon networks. Indeed C_{60} and C_{70} were isolated from solid carbon[8] and the confirmation of its structure does therefore prove that such carbons contain nonplanar carbon networks.

The nonplanarity of such networks may lead to voids in the structure which can trap free radicals which as we argue above are most likely to be of the odd-alternate π-type, but if this is so they cannot contain pentagonal rings in conjugation. This situation may have far reaching consequences for the electric and magnetic properties of these materials.

Heteroatoms can considerably complicate the situation as we have discussed above. Many problems remain to be solved, but a new chapter in solid state science has opened.

REFERENCES

1. S. A. Roulston, L. J. Dunne, A. D. Clark, and M. F. Chaplin, *Phil. Mag. B* **62**, 243 (1990).
2. G. M. Jenkins and K. Kawamura, *Polymeric Carbons-Carbon Fibre, Glass and Char.* Cambridge University Press City, Cambridge (1976).
3. J. B. Donnet, *Carbon* **20**, 266 (1982).
4. H. W. Kroto, J. R. Heath, S. C. O'Brien, R. F. Curl, and R. E. Smalley, *Nature* **318**, 162 (1985).
5. H. W. Kroto, *Science* **242**, 1139 (1988).
6. H. W. Kroto and K. G. McKay, *Nature* **331**, 328 (1988).
7. Q. L. Zhang, S. C. O'Brien, J. R. Heath, Y. Liu, H. W. Kroto, and R. E. Smalley, *J. Phys. Chem.* **90**, 525 (1986).
8. W. Krätschmer, D. L. Lowell, K. Fortiropoulos, and D. R. Huffmann, *Nautre* **347**, 354 (1990).
9. R. Taylor, J. P. Hare, A. K. Abdul-Sada, and H. W. Kroto, *J. Chem. Soc. Chem. Comm.* 1423 (1990).
10. S. Iijima, *J. Cryst. Growth* **50**, 675 (1980).
11. M. Inagaki, K. Kuroda, N. Inoue, and M. Sakai, *Carbon* **22**, 617 (1984).
12. H. Marsh, F. Dachille, F. Melvin, and P. L. Walker, *Carbon* **9**, 159 (1971).
13. A. Oberlin, *Chem. Phys. Carbon* **22**, 1 (1990).
14. L. J. Dunne and H. Harker, *Phil. Mag.* **30**, 1313 (1974).
15. S. A. Roulston, M. F. Chaplin, L. J. Dunne, and A. D. Clark, *Biosensors and Bioelectronics* **6**, 325 (1991).
16. I. C. Lewis and L. S. Singer, *Chem. Phys. Carbon* **17**, 1 (1981).
17. D. J. E. Ingram, *Free Radicals as Studied by Electron Spin Resonance.* Butterworths, London (1958).
18. N. F. Mott and E. A. Davis, *Electronic Processes in Non-Crystalline Materials.* Clarendon, Oxford (1971).
19. P. Delhaes and F. Carmona, *Chem. Phys. Carbon* **17**, 89 (1989).
20. N. F. Mott (Ed.), *Amorphous and Liquid Semiconductors.* North Holland, Amsterdam (1970).
21. N. F. Mott, *Contemp. Phys.* **18**, 225 (1977).
22. C. A. Coulson and G. S. Rushbrook, *Proc. Camb. Phil. Soc.* **36**, 193 (1940).

ISOMERISATIONS OF THE FULLERENES

Patrick W. Fowler,[1] David E. Manolopoulos,[2] and Robert P. Ryan[1,*]

[1]Department of Chemistry, University of Exeter, Exeter EX4 4QD, England
[2]Department of Chemistry, University of Nottingham, Nottingham NG7 2RD, England

(*Accepted* 28 *April* 1992)

Abstract—The Stone–Wales (pyracylene) rearrangement is a hypothetical mechanism for interconversion of fullerene isomers, accounting for mobility of the pentagons on the surfaces of these clusters. Two versions of the transformation, involving patches of 4 and 12 rings, respectively, can be defined. Energetic and group-theoretical aspects of the transformation are discussed, symmetry-based selection rules devised, and complete isomerisation maps presented for small fullerenes (C_{20} to C_{40}) and isolated-pentagon isomers (C_{78} to C_{88}). Fullerenes that self-racemise under this transformation include D_2 C_{28}. Limitations on the usefulness of this mechanism as a means of rationalising experimental isomer distributions are briefly discussed.

Key Words—Carbon, fullerene, isomerisation, chirality, symmetry.

1. INTRODUCTION

The fast-growing family of fullerenes now includes a range of chiral and achiral molecules in addition to the C_{60} archetype and its C_{70} satellite. Chromatography of soluble products of graphite evaporation yields C_{60}, C_{70}, C_{76}, C_{78}, C_{82}, C_{84}, C_{90}, C_{96}, . . . [1–6], some in several isomeric forms. This new experimental evidence presents an opportunity to test the qualitative theoretical approaches[7–9[developed in the years since the first observation of C_{60}[10]. Three main targets for a systematic theory of fullerenes are: enumeration of possible isomers, classification of their electronic structures, and estimation of steric factors in overall stability. As argued elsewhere[11], the first two problems are solved by the spiral algorithm[9] and the leapfrog/cylinder magic-number rules [12,13], respectively. Some progress has been made on the steric problem, but more remains to be done.

In the present paper we concentrate on another factor that may help to decide the experimental isomer distributions and stable magic numbers (i.e., the likelihood of interconversion between isomers of a given fullerene). It is now possible to compute a complete solution to the mapping of C_n interconversions for any given mechanism, and here we study the most promising candidate mechanism: the Stone–Wales[14] (or pyracylene[4]) transformation. A recent suggestion links the experimental isomer distribution to thermodynamic equilibrium within closed groups of Stone–Wales interconverting isomers[4]. Isomerisation maps for particular fullerenes (C_{78}, C_{84}, C_{82}) have been published[4,15†,16], but our aim here is to provide a general discussion of the mathematical and chemical characteristics of the transformation. Such a discussion gives a basis for prediction of experimentally isolable isomers within any particular

model of electronic and steric stability. Maps for other higher and lower fullerenes of interest are also presented.

2. THE STONE–WALES TRANSFORMATION

The pyracylene transformation was first proposed in a paper by Stone and Wales on possible nonicosahedral isomers of C_{60}[14], and was taken up by Coulombeau and Rassat[17,18] in a further exploration of low-symmetry C_{60} isomers. Although there is no experimental evidence for stability of any other C_{60} isomer, and characterisation of bulk fullerite has confirmed the primacy of I_h C_{60}, the transformation itself gives a general hypothetical mechanism for interconversion between fullerene isomers and may explain features of the observed isomer distributions for higher carbon cages.

To perform the transformation, we first search the fullerene surface for a patch (I) where two pentagons and two hexagons meet in a pyracylene/pyracene pattern. If such a patch is found, we twist the central bond (the Stone–Wales or SW bond), breaking two neighbouring bonds and making two more, thereby swapping pentagons and hexagons and leading to a "rotated" patch within the same 12-atom perimeter. The product is still a fullerene (in general different from the starting isomer) and any single transformation of this kind is clearly reversible. The "sense" of the twist is immaterial and so the transformation links in a well-defined way two isomers of a given fullerene C_n (or one isomer to its enantiomer, or conceivably one isomer of low symmetry to itself).

There are steric[7] and electronic[18,19,20] grounds for believing that isolated-pentagon isomers are more stable than fullerenes with abutting pentagons. This intuition is now often described as a rule—the isolated-pentagon rule (IPR). Certainly it appears to be obeyed by all fullerenes isolated and characterised so far. It is useful, therefore, to consider a restricted version of the Stone–Wales transformation

---

*Permanent address: Department of Chemistry, St. Patrick's College, Maynooth, Co. Kildare, Ireland.

†In Fig. 2 of this paper the transformation between isomers 9 and 13 is to be deleted.

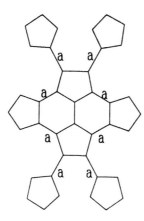

Fig. 5. Changes in the number of Stone–Wales patches can accompany twisting of a single SW bond. All eight bonds marked with an "a" lose their Stone–Wales status upon rotation of the central bond.

enantiomer of B if they are exchanged by a proper operation, but opposite enantiomers of B if they are exchanged by an improper operation. When B is also achiral (case (a)), all patches in the orbit lead to the same product as B is superimposable on its mirror image.

Both examples shown in Fig. 6 are of patches with C_1 site symmetry, which span the regular orbit[24] of the molecular point group. In this orbit every pair of patches is swapped by just one operation and since half of the operations of an achiral group are proper and half improper, the orbit splits into two equal disjoint sets, one for each handedness of the product. If we associate signs + and − with enantiomers of the product, then the pattern of signs transforms as the antisymmetric representation Γ_ϵ of the molecular point group (Γ_ϵ has character +1 under proper and −1 under improper operations). Similar remarks apply to the "half-orbit" of the molecular group

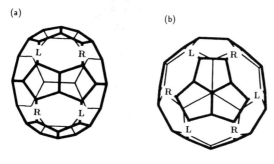

Fig. 6. Transformation of an achiral to a chiral fullerene. Orbits of SW patches for the conversions (a) C_{2v} C_{40} (isomer **8**) ↔ C_1 C_{40} (isomer **6**, L and R forms), (b) C_{3v} C_{38} (isomer **16**) ↔ C_1 C_{38} (isomer **14**, L and R forms). The labels L and R are arbitrarily assigned to the two enantiomers of the product. Within the orbit of SW sites, all patches producing a given enantiomer are related to proper operations ((a) C_2, (b) C_3, C_3^2 about an axis perpendicular to the plane of the paper), and swap with patches of opposite effect under improper operations (σ reflections perpendicular to the paper).

spanned by patches with C_2 site symmetry. The permutation representation of the C_1- and C_2-orbits of an achiral group contains distinct copies of both Γ_ϵ and Γ_0 (the totally symmetric representation).

Case (c) is the conversion of a chiral isomer A to a chiral product B. All patches of an orbit in A in this case are exchanged amongst themselves by proper rotations alone and so all lead to the same enantiomer of B. There are three subcases, depending on whether A is identical with B, enantiomeric to B or simply different from B. A chiral-to-chiral SW transformation may thus produce a null result, self-racemisation (e.g., twisting of one of the two C_2 patches in D_2 C_{28}), or conversion between fixed enantiomers of A and B (e.g., the map 1 D_2 ↔ 2 C_2 ↔ 5 D_2 for C_{84}[15] where two mirror-image triads exist and interconvert one enantiomer each of 1, 2, and 5 without racemisation).

These considerations have direct consequences for the prospects of resolution of optically active fullerenes. If an isomerisation map includes a single achiral structure, then resolution of enantiomers of even the most stable isomer on that map is possible only under conditions where the SW transformation is "frozen out." On the other hand, if the map contains only chiral isomers and has no case of self-racemisation either explicit or implicit in a cycle, then the freezing out of the SW transformation is not a prerequisite for resolution.

5. STONE–WALES TRANSFORMATIONS AND LEAPFROGS

The leapfrog fullerenes form a class of isolated-pentagon polyhedra of particular theoretical interest. They have the formula C_n ($n = 60 + 6k$, $k \neq 1$) and are obtained by omnicapping a fullerene $C_{n/3}$ and taking the dual of the resulting deltahedron. They are the only fullerenes with closed electronic shells and antibonding LUMOs within simple Hückel theory. An equivalent definition can be made in terms of localised bonding[26]: every leapfrog fullerene has a unique Kekulé structure that is found by placing 5 double bonds *exo* to any one pentagon and filling in the single and double bonds over the rest of the structure as forced by valency considerations. Every pentagon in this Kekulé structure has single bonds along its sides and all *exo* bonds double. The SW bonds, if any, on a leapfrog fullerene are therefore formally double. The special Kekulé structure of a leapfrog fullerene has the property of achieving the maximum possible number of benzenoid hexagonal faces ($n/3$)[26].

As Fig. 7 shows, the Stone–Wales transformation on a leapfrog fullerene disrupts this special Kekulé structure. It follows that no two leapfrog fullerene isomers of C_n may be connected by a single SW step, though they may be connected by a sequence of such steps.

Figure 8 shows the genesis and fate of SW patches under the leapfrog operation, showing that every patch in a leapfrog comes from a pair of pentagons

(a)   (b)

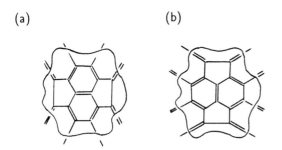

Fig. 7. Stone–Wales transformation and the leapfrog construction. (a) In the special Kekulé structure of the leapfrog fullerene every pentagon has 5 *exo* double bonds. Bond orders of contacts across the boundary of the 4-ring patch are indicated by broken lines. (b) On rotation of the SW bond within the patch a mismatch is produced at the edges, and the transformed fullerene is therefore no longer a leapfrog isomer.

sharing an edge in the parent, and every large patch in the leapfrog from an "anti-patch" where a bond common to two pentagons joins atoms in two hexagons. The figure also shows that one repetition of the leapfrog operation is sufficient to destroy all SW patches. Double-leapfrog fullerenes are thus disconnected from all other isomers on an interconversion map, but as the first double-leapfrog not already disqualified by icosahedral symmetry is C_{216} this fact is unlikely to have chemical relevance in the near future. The smallest leapfrog fullerene with a large SW patch, and therefore an allowed isomerisation to another isolated-pentagon fullerene, is a D_{3h} isomer of C_{78}[27,4].

6. GENERATION OF ISOMERISATION MAPS

It is a finite task to determine all possible SW transformations between isomers C_n. For each n it is relatively straightforward to catalogue all possible fullerene isomers using the spiral algorithm[9]. The symmetry, spectroscopic signature and approximate (topological) atomic coordinates are all available from the adjacency matrix[19] which is itself constructible from a 1D sequence of 12 pentagonal and $(\frac{1}{2}n - 10)$ hexagonal faces in the spiral that encodes each fullerene isomer. For a given isomer, searching of the face adjacency patterns can be used to find SW bonds (in patches of type I or II or both), site symmetries can then be assigned and the effect of the transformation ascertained by diagonalisation of the transformed adjacency matrix. A map of allowed isomerisations can thus be constructed computationally. One informal indication of the completeness of the spiral algorithm is that this Stone–Wales procedure has never yet generated an isomer outside the starting set. Handling of chiral SW transformations of the type discussed earlier requires some care as the coordinates generated from the eigenvectors of the adjacency matrix are of arbitrary handedness[19].

The results of a survey using these programs will now be discussed in two parts. First, we look at the results for small ($n \leq 40$) fullerenes with abutting pentagons. Secondly, the results for the restricted transformation in higher fullerenes are presented.

The isomer counts for fullerene polyhedra C_{20} to C_{40} appear as part of a larger tabulation in[19]. Combining the spiral algorithm with new programs to identify and perform the SW transformation, we have mapped isomerisations for this range of clusters. A classification of the patches is given in Table 2 and maps are shown in Figs. 9–12.

Several general properties of the transformation can be illustrated by reference to the maps. Their most obvious feature is that they are incompletely connected: the isomers of C_n fall into families and are not all interconvertible by repeated transformation. An analogous "factorisation" is found for many rearrangement mechanisms in chemistry. For example, the diamond-square-diamond borane rearrangement is the dual of the Stone–Wales transformation when the four atoms involved are appropriately connected. A concerted multiple DSD rearrangement has been postulated as a mechanism converting (1,2) to (1,7) $C_2B_{10}H_{12}$ via a cuboctahedral transition state[28]. The third isomer, the (1,12) form, cannot be reached by this mechanism from either of the

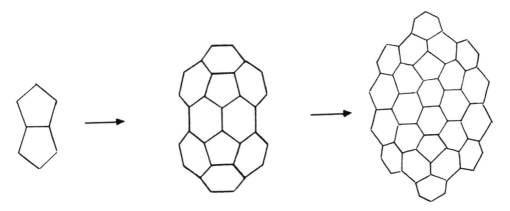

Fig. 8. Generation of a Stone–Wales patch (I) by leapfrogging, and destruction by double leapfrogging. Leapfrogging a pre-existing SW patch destroys it but may create new patches on its periphery.

Table 2. Catalogue of fullerene isomers C_{20} to C_{40}

Isomer	Ring spiral	SW bonds	Orbits
C_{20} **1**(I_h)	555555555555	0	0
C_{24} **1**(D_{6d})	55555565565555	0	0
C_{26} **1**(D_{3h})	555556565655555	0	0
C_{28} **1**(D_2)	5555565665655555	2	$1 \times C_2$
C_{28} **2**(T_d)	5556565655555556	0	0
C_{30} **1**(D_{5h})	55555566666555555	5	$1 \times C_{2v}$
C_{30} **2**(C_{2v})	55555566665655555	3	$1 \times C_{2v}, 1 \times C_s$
C_{30} **3**(C_{2v})	55556656655555556	2	$1 \times C_s$
C_{32} **1**(C_2)	555556566665655555	5	$1 \times C_2, 2 \times C_1$
C_{32} **2**(D_2)	555556656665565555	6	$1 \times C_2, 1 \times C_1$
C_{32} **3**(D_{3d})	555556665665556555	6	$1 \times C_2$
C_{32} **4**(C_2)	555566656555655655	4	$2 \times C_1$
C_{32} **5**(D_{3h})	555566656655655555	0	0
C_{32} **6**(D_3)	555665656556565555	3	$1 \times C_2$
C_{34} **1**(C_2)	5555565666665655555	2	$1 \times C_1$
C_{34} **2**(C_s)	5555566666655555556	6	$3 \times C_1$
C_{34} **3**(C_s)	5555665665565656555	3	$1 \times C_s, 1 \times C_1$
C_{34} **4**(C_2)	5555665666555655565	6	$3 \times C_1$
C_{34} **5**(C_2)	5556565665565655655	5	$1 \times C_2, 2 \times C_1$
C_{34} **6**(C_{3v})	5556565665566556555	3	$1 \times C_s$
C_{36} **1**(C_2)	55555565666666565555	4	$2 \times C_2, 1 \times C_1$
C_{36} **2**(D_2)	55555656666566655555	2	$1 \times C_2$
C_{36} **3**(C_1)	55555666666655655555	6	$6 \times C_1$
C_{36} **4**(C_s)	55555666666556556555	4	$2 \times C_1$
C_{36} **5**(D_2)	55556656655665665555	0	0
C_{36} **6**(D_{2d})	55556656656566565555	8	$1 \times C_s, 1 \times C_2$
C_{36} **7**(C_1)	55556656665565655655	5	$5 \times C_1$
C_{36} **8**(C_s)	55556656665566556555	8	$2 \times C_s, 3 \times C_1$
C_{36} **9**(C_{2v})	55556666665555555566	7	$1 \times C_{2v}, 1 \times C_s, 1 \times C_1$
C_{36} **10**(C_2)	55565656556566565555	2	$1 \times C_1$
C_{36} **11**(C_2)	55565665656566556555	7	$1 \times C_2, 3 \times C_1$
C_{36} **12**(C_2)	55565666655655555656	7	$1 \times C_2, 3 \times C_1$
C_{36} **13**(D_{3h})	55566566655565556565	9	$1 \times C_{2v}, 1 \times C_s$
C_{36} **14**(D_{2d})	56566565556555656565	6	$1 \times C_{2v}, 1 \times C_s$
C_{36} **15**(D_{6h})	56566665556555566565	6	$1 \times C_{2v}$
C_{38} **1**(C_2)	555556566666666565555	2	$1 \times C_1$
C_{38} **2**(D_{3h})	555556665666566655555	0	0
C_{38} **3**(C_1)	555556666665565656555	5	$5 \times C_1$
C_{38} **4**(C_1)	555556666666555655565	4	$4 \times C_1$
C_{38} **5**(C_1)	555556666655665656555	7	$7 \times C_1$
C_{38} **6**(C_2)	555566566665565655655	5	$1 \times C_2, 2 \times C_1$
C_{38} **7**(C_1)	555566566665566556555	3	$3 \times C_1$
C_{38} **8**(C_1)	555566666655556556655	7	$7 \times C_1$
C_{38} **9**(D_3)	555566666655655565565	6	$1 \times C_1$
C_{38} **10**(C_2)	555656566565665656555	9	$1 \times C_2, 4 \times C_1$
C_{38} **11**(C_1)	555656566566656555556	4	$4 \times C_1$
C_{38} **12**(C_{2v})	555656566566565555556	0	0
C_{38} **13**(C_2)	555656666556565566555	9	$1 \times C_2, 4 \times C_1$
C_{38} **14**(C_1)	555656666565655556556	6	$6 \times C_1$
C_{38} **15**(C_{2v})	555656666655655555656	4	$1 \times C_1$
C_{38} **16**(C_{3v})	555666666555555555666	6	$1 \times C_1$
C_{38} **17**(C_2)	565666565565565665655	8	$4 \times C_1$
C_{40} **1**(D_{5d})	5555556666666666555555	0	0
C_{40} **2**(C_2)	5555565666666665655555	2	$1 \times C_1$
C_{40} **3**(D_2)	5555565656666666555555	0	0
C_{40} **4**(C_1)	5555566666665565665555	3	$3 \times C_1$
C_{40} **5**(C_s)	5555566666656566565555	7	$1 \times C_s, 3 \times C_1$
C_{40} **6**(C_1)	5555566666665565655655	3	$3 \times C_1$
C_{40} **7**(C_s)	5555566666665566556555	3	$1 \times C_s, 1 \times C_1$
C_{40} **8**(C_{2v})	5555566666665655655565	4	$1 \times C_{2v}$
C_{40} **9**(C_2)	5555665666556665665555	8	$4 \times C_1$
C_{40} **10**(C_1)	5555665666565665656555	5	$5 \times C_1$
C_{40} **11**(C_2)	5555665666566566556555	1	$1 \times C_2$
C_{40} **12**(C_1)	5555665666656665555556	5	$5 \times C_1$
C_{40} **13**(C_s)	5555665666656656555655	4	$2 \times C_1$
C_{40} **14**(C_s)	5555666666555656566555	9	$1 \times C_s, 4 \times C_1$
C_{40} **15**(C_2)	5555666666556555656565	6	$3 \times C_1$

(Continued)

Table 2. (Continued)

Isomer	Ring spiral	SW bonds	Orbits
C_{40} **16**(C_2)	5555666666556556556556	7	$1 \times C_2, 3 \times C_1$
C_{40} **17**(C_1)	5555666666556556565655	7	$7 \times C_1$
C_{40} **18**(C_2)	5555666666556655665555	4	$2 \times C_1$
C_{40} **19**(C_2)	5555666666565556555656	7	$1 \times C_2, 3 \times C_1$
C_{40} **20**(C_{3v})	5556565656666665555556	3	$1 \times C_s$
C_{40} **21**(C_2)	5556565665665665656555	7	$1 \times C_2, 3 \times C_1$
C_{40} **22**(C_1)	5556565665666656655655	6	$6 \times C_1$
C_{40} **23**(C_2)	5556565665666566556555	3	$1 \times C_2, 1 \times C_1$
C_{40} **24**(C_s)	5556565665666656555556	8	$2 \times C_s, 3 \times C_1$
C_{40} **25**(C_2)	5556566665566555665565	7	$1 \times C_2, 3 \times C_1$
C_{40} **26**(C_1)	5556566665656555656565	8	$8 \times C_1$
C_{40} **27**(C_2)	5556566665656556565655	6	$2 \times C_2, 2 \times C_1$
C_{40} **28**(C_s)	5556566665656655665555	7	$3 \times C_s, 2 \times C_1$
C_{40} **29**(C_2)	5556566665665555655665	11	$1 \times C_2, 5 \times C_1$
C_{40} **30**(C_3)	5556566665665556555656	9	$3 \times C_1$
C_{40} **31**(C_s)	5556566665665655655565	11	$1 \times C_s, 5 \times C_1$
C_{40} **32**(D_2)	5556566666556556556655	2	$1 \times C_2$
C_{40} **33**(D_{2h})	5556566666556655656555	4	$1 \times C_s$
C_{40} **34**(C_1)	5556566666566555565556	4	$4 \times C_1$
C_{40} **35**(C_2)	5565665556556566656555	5	$1 \times C_2, 2 \times C_1$
C_{40} **36**(C_2)	5565665565556565665655	5	$1 \times C_2, 2 \times C_1$
C_{40} **37**(C_{2v})	5565665656556656565556	6	$1 \times C_2, 1 \times C_1$
C_{40} **38**(D_2)	5565665656565656565565	10	$1 \times C_2, 4 \times C_1$
C_{40} **39**(D_{5d})	5565665656565656655655	10	$1 \times C_s$
C_{40} **40**(T_d)	5566656655565555656565	12	$1 \times C_s$

Each spectrally distinct isomer is represented by a ring spiral[9]. For each isomer the total number of Stone–Wales bonds is listed, along with its decomposition into orbits of the point group. Thus, for example, isomer **9** of C_{36} has 7 SW bonds, 4 in a set of C_1 site symmetry, 2 in a set of C_s site symmetry and 1 in a set with the full C_{2v} symmetry. The interconversions are mapped out later in the present paper.

Table 3. Interconversions of the 40 isomers of C_{40}

Reactant isomer	Product isomers
2(C_2)	5(C_1)
4(C_1)	4(C_1),5(C_1),10(C_1)
5(C_s)	2(C_1),4(C_1),7(C_s),9(C_1)
6(C_1)	8(C_1),13(C_1),21(C_1)
7(C_s)	5(C_s),12(C_1)
8(C_{2v})	6(C_1)
9(C_2)	5(C_1),10(C_1),12(C_1),14(C_1)
10(C_1)	4(C_1),9(C_1),10(C_1),21(C_1),22(C_1)
11(C_2)	23(C_2)
12(C_1)	7(C_1),9(C_1),17(C_1),24(C_1),25(C_1)
13(C_s)	6(C_1),22(C_1)
14(C_s)	9(C_1),17(C_1),22(C_1),24(C_s),29(C_1)
15(C_2)	16(C_1),26(C_1),35(C_1)
16(C_2)	15(C_1),17(C_1),19(C_1),29(C_2)
17(C_1)	12(C_1),14(C_1),16(C_1),26(C_1),27(C_1),30(C_1),31(C_1)
18(C_2)	28(C_1),34(C_1)
19(C_2)	16(C_1),19(C_1,C_2),30(C_1)
20(C_{3v})	24(C_s)
21(C_2)	6(C_1),10(C_1),22(C_1),25(C_2)
22(C_1)	10(C_1),13(C_1),14(C_1),21(C_1),26(C_1),27(C_1)
23(C_2)	11(C_2),28(C_1)
24(C_s)	12(C_1),14(C_s),20(C_s),26(C_1),31(C_1)
25(C_2)	12(C_1),21(C_2),26(C_1),34(C_1)
26(C_1)	15(C_1),17(C_1),22(C_1),24(C_1),25(C_1),29(C_1),36(C_1),38(C_1)
27(C_2)	17(C_1),22(C_1),32(C_2),38(C_2)
28(C_s)	18(C_1),23(C_1),33(C_s),37(C_s),39(C_s)
29(C_2)	14(C_1),16(C_2),26(C_1),30(C_1),31(C_1)
30(C_3)	17(C_1),19(C_1),29(C_1)
31(C_s)	17(C_1),24(C_1),29(C_1),38(C_1),40(C_s)
32(D_2)	27(C_2)
33(D_{2h})	28(C_s)

(Continued)

Table 3. (Continued)

Reactant isomer	Product isomers
$34(C_1)$	$18(C_1),25(C_1),36(C_1),37(C_1)$
$35(C_2)$	$15(C_1),35(C_1),36(C_2)$
$36(C_2)$	$26(C_1),35(C_1),35(C_2)$
$37(C_{2v})$	$28(C_s),34(C_1)$
$38(D_2)$	$26(C_1),27(C_2),31(C_1)$
$39(D_{5d})$	$28(C_s)$
$40(T_d)$	$31(C_s)$

Isomers 1 and 3 have no SW bond and therefore no allowed conversion. For each isomer in the numbering scheme of Table 2 and Fig. 12, this table lists products of twisting each distinct type of SW bond.

other two, and the three isomers are thus split into two disjoint families on the transformation map. In this case it is easy to see the reason for the selection rule: the concerted transformation preserves inversion partners and so cannot exchange isomers with two (B,C) antipodal pairs for one with a single (C,C)

pair. In the case of the SW transformation the constant factor linking families is presumably also geometrical, though less easy to see.

Examples of single isomers, self-racemising fullerenes, chiral pathways, and closed loops of transformations are all present in these maps for small fullerenes and show what may be expected in the superficially more complex maps for higher fullerenes.

The only fullerene for which the unrestricted Stone–Wales transformation has received attention in the past has been C_{60}. A full isomerisation map for the 1812 spectrally distinct isomers of this fullerene would be a daunting prospect, and instead the present programs have been used only to check the first few steps away from the canonical icosahedral isomer. Coulombeau and Rassat studied the introduction of Stone–Wales "defects" into the C_{60} structure, confining attention to transformations of those SW bonds

(a)

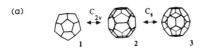

(b)

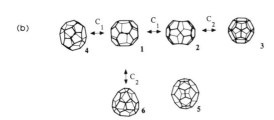

(c)

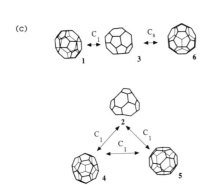

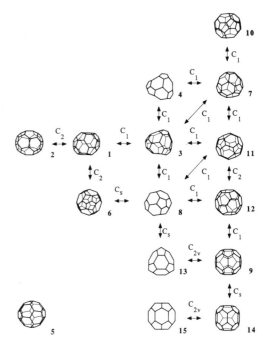

Fig. 9. Stone–Wales interconversion maps for isomers of C_{30} to C_{34}. (a) C_{30} forms a single family, (b) C_{32} forms two families (isomer 5 is disconnected from the rest), (c) C_{34} forms two families of three isomers each. The conventions used here apply to all other maps in the present paper: each isomer is labelled by its place in the list of ring spirals given in Table 2 or 4, enantiomers are not distinguished, the site symmetry of the active bond is indicated on the arrow. Spectrally neutral transformations (i.e., those where the product is enantiomeric with, or identical to, the reactant) are not shown but are listed in the captions. Such transformations occur for C_{32} (isomer 4) and C_{34} (isomers 2, 4, and 5).

Fig. 10. The Stone–Wales interconversion map for C_{36}. Conventions as in Fig. 9. Only isomer 5 is disconnected from the others. Spectrally neutral transformations occur for C_{36} (isomers 3, 7, and 12).

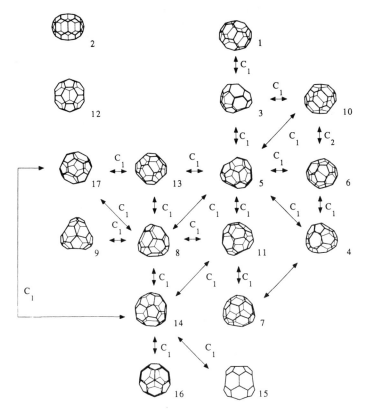

Fig. 11. The Stone–Wales interconversion map for C_{38}. Conventions as in Fig. 9. Isomers 2 and 12 are disconnected from the family that includes the other 15 isomers. Spectrally neutral transformations occur for C_{38} (isomers 3, 4, 7, 10, 13, and 17). 10 and 13 also convert via two C_1 patches.

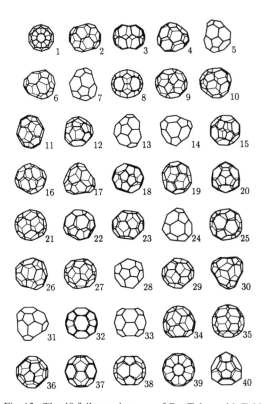

Fig. 12. The 40 fullerene isomers of C_{40}. Taken with Table 3, this allows the construction of the interconversion map. Spectrally neutral transformations occur for C_{40} (isomers 4, 10, 19, and 35).

present in the original structure[17,18]. Lifting this restriction we find the cascade, with only one "1-defect" (C_{2v}) isomer, 6 "2-defect" isomers and 18 "3-defect" forms. It is a general feature of the isomerisation maps that high-symmetry isomers with their few, large orbits tend to lie on the ends of branches whereas low-symmetry isomers are often multiple junction points on the map. It is also notable that, if in the formation process any nonicosahedral isomer of C_{60} is ever produced, the overwhelming dominance of I_h C_{60} indicates that all material must funnel through the C_{2v} bottleneck on the interconversion map if the number of carbon atoms per molecule remains constant.

Of more chemical significance are the isolated-pentagon isomers of the higher fullerenes. It is not yet clear what determines the isomer distribution in every case, but electronic and steric factors are often finely balanced. C_{76} adopts a closed-shell chiral D_2 structure rather than the geodesic but open-shell tetrahedral cage[29,3]; C_{78} occurs as several isomers[27,4], no one of which has a properly closed shell in simple Hückel theory. Isolation of pentagons would appear to be necessary for stability, and the restricted Stone–Wales transformation, linking only isolated-pentagon fullerenes, is plausibly of lower energy than the general transformation (see Fig. 2). The smallest fullerenes to support the larger type II patches are isomers of C_{78}, and a map of C_{78} isomer-

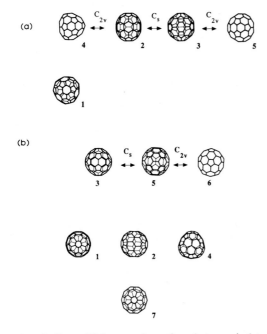

Fig. 13. Stone–Wales transformations between isolated-pentagon isomers of (a) C_{78} (two families), and (b) C_{80} (five families).

isations was presented in [4] as a rationalisation of the experimental observation of two isomers of this cluster. The map splits into two: a linked group of four isomers and an unconnected single isomer. There are difficulties with interpretation—the isomer predicted to be most stable in [4] was not observed at all, though it may be the predominant isomer in independent experiments performed in Tokyo[30], and in any case one might expect a thermodynamic equilibrium between all four isomers on the larger branch of the map.

Maps for isolated-pentagon isomers C_{78} to C_{88} are presented in Figs. 13–15. A classification of their SW patches is given in Table 4. One interesting point that has been emphasised in our recent treatment of C_{82}[16] is that this fullerene appears to be the only one for which all isolated-pentagon isomers form a single SW map. For C_{80} we see the isolation of the icosahedral isomer from all the others. For C_{84} (map plotted in [15]) the two leapfrog isomers belong to distinct families, as do the three leapfrog isomers of C_{90} (map not shown here), whereas for C_{96} 2 of the 6 leapfrog isomers are found in one family.

The partition of isolated-pentagon isomers into distinct families for C_{78} is a consequence of disallow-

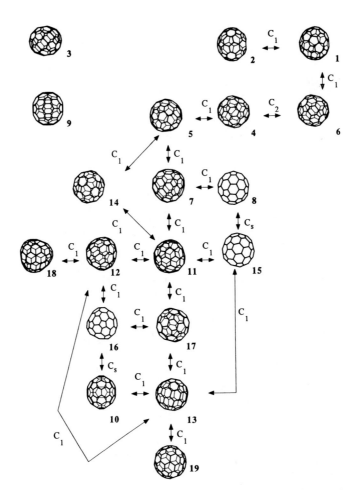

Fig. 14. Stone–Wales transformations between isolated-pentagon isomers of C_{86} (three families).

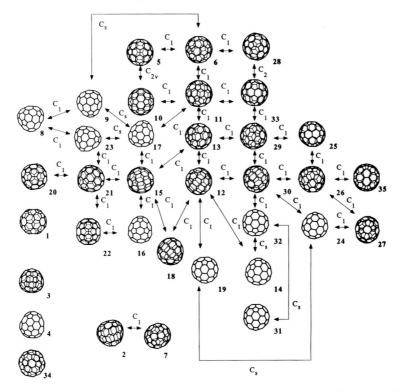

Fig. 15. Stone–Wales transformations between isolated-pentagon isomers of C_{88} (six families).

Table 4. Catalogue of isolated-pentagon fullerene isomers C_{78} to C_{88}

Isomer	Ring spiral	$n(\mathrm{II})$	Orbits
C_{78} **1**(D_3)	56666656565656566666656666566565656565666	0	0
C_{78} **2**(C_{2v})	56666656565656566666666666566566565666565656	3	$1 \times C_{2v}, 1 \times C_s$
C_{78} **3**(C_{2v})	56666656565656656666666666566565656566656666565	3	$1 \times C_{2v}, 1 \times C_s$
C_{78} **4**(D_{3h})	56666656565666566566656656666666566656565	3	$1 \times C_{2v}$
C_{78} **5**(D_{3h})	56666656565665656666665666656565656666565	3	$1 \times C_{2v}$
C_{80} **1**(D_{5d})	56666656565656566666666666666656565656565666665	0	0
C_{80} **2**(D_2)	56666656565656566665666666656666565656565666665	0	0
C_{80} **3**(C_{2v})	56666656565665666666666566665665666565656566	2	$1 \times C_s$
C_{80} **4**(D_3)	56666656565665666666666656666565665656565666	0	0
C_{80} **5**(C_{2v})	56666656565665656666666666566656656566665665	3	$1 \times C_{2v}, 1 \times C_s$
C_{80} **6**(D_{5h})	56666656665656566665666666656566656566665665	5	$1 \times C_{2v}$
C_{80} **7**(I_h)	56666665656565656666666666666565656565666665	0	0
C_{82} **1**(C_2)	56666656565656566666666666566565656566656665	2	$1 \times C_1$
C_{82} **2**(C_s)	56666656565665666666666656566656565666656665	4	$2 \times C_1$
C_{82} **3**(C_2)	56666656565665666666666566565656566565666665	4	$2 \times C_1$
C_{82} **4**(C_s)	56666656565665666666665666565656566656666665	4	$2 \times C_1$
C_{82} **5**(C_2)	56666656566565666666656666656566566565666665	4	$2 \times C_1$
C_{82} **6**(C_s)	56666656566565656666665666666656566565656666665	5	$1 \times C_s, 2 \times C_1$
C_{82} **7**(C_{3v})	56666656566565656666666656666666565656565666	3	$1 \times C_s$
C_{82} **8**(C_{3v})	56666656566565666666656566656565666656666565	3	$1 \times C_s$
C_{82} **9**(C_{2v})	56666656566565656665666666656666565665666665	6	$1 \times C_s, 1 \times C_1$
C_{86} **1**(C_1)	56666656565656566666666665666566565666565666566	2	$2 \times C_1$
C_{86} **2**(C_2)	56666656565656566666665666666666566565656656566	2	$1 \times C_1$
C_{86} **3**(C_2)	56666656565656566666666665666566565666565666565	0	0
C_{86} **4**(C_2)	56666656565656566666665666666566565665666565666566	3	$1 \times C_2, 1 \times C_1$
C_{86} **5**(C_1)	56666656565656566666666666666665665666565665666	3	$3 \times C_1$
C_{86} **6**(C_2)	56666656565656566666666665666566565666565666566	3	$1 \times C_2, 1 \times C_1$
C_{86} **7**(C_1)	56666656565656566666666666566565666566666565666	3	$3 \times C_1$
C_{86} **8**(C_s)	56666656565656566666666665656565666566666565666566	3	$1 \times C_s, 1 \times C_1$
C_{86} **9**(C_{2v})	56666656565656566666665656565665666666566656565	0	0
C_{86} **10**(C_{2v})	56666656565656566666666666565666656656565666665	6	$1 \times C_s, 1 \times C_1$
C_{86} **11**(C_1)	56666656565656566666665656666666566565656565665666	5	$5 \times C_1$
C_{86} **12**(C_1)	56666656565656566666666656666666566565656565656666	5	$5 \times C_1$

(*Continued*)

Table 4. (Continued)

Isomer	Ring spiral	$n(\text{II})$	Orbits
$C_{86}\ 13(C_1)$	56666656566565656666656666666656666565656566566665	6	$6 \times C_1$
$C_{86}\ 14(C_2)$	56666656566565656666656566666656656656565565665666	4	$2 \times C_1$
$C_{86}\ 15(C_s)$	56666656566565656666666666666656565656566666665	5	$1 \times C_s, 2 \times C_1$
$C_{86}\ 16(C_s)$	56666656566565666666656565656666665666656566656656	5	$1 \times C_s, 2 \times C_1$
$C_{86}\ 17(C_2)$	56666656566566666656566565656566656566656566566	6	$3 \times C_1$
$C_{86}\ 18(C_3)$	56666656566565656666656566566566566565665666	3	$1 \times C_1$
$C_{86}\ 19(D_3)$	56666656566565656666656666666656666565656566566665	6	$1 \times C_1$
$C_{88}\ 1(D_2)$	56666656566565666665666666666666666656566565656566	0	0
$C_{88}\ 2(C_1)$	56666656566565666666666665666656566656566665665	1	$1 \times C_1$
$C_{88}\ 3(C_2)$	56666656566565666666666665666656656666656656565	0	0
$C_{88}\ 4(C_s)$	56666656566565666666666666566565656565565665666	0	0
$C_{88}\ 5(C_{2v})$	56666656566565666666666656666566666666565656566665	5	$1 \times C_{2v}, 1 \times C_1$
$C_{88}\ 6(C_1)$	56666656566565666666666656666656566656565656565665	4	$4 \times C_1$
$C_{88}\ 7(C_2)$	56666656566565666666666656666656566566656566665665	2	$1 \times C_1$
$C_{88}\ 8(C_s)$	56666656566565666666666666656565656565656566666	2	$2 \times C_s$
$C_{88}\ 9(C_s)$	56666656566565666666666666656565656566566665	4	$2 \times C_s, 1 \times C_1$
$C_{88}\ 10(C_{2v})$	56666656566565666666666565656566666666656565656	5	$1 \times C_{2v}, 1 \times C_1$
$C_{88}\ 11(C_1)$	56666656566566666666656565656566666666656565656	5	$5 \times C_1$
$C_{88}\ 12(C_1)$	56666656566566666666656566565656666656666565665	5	$5 \times C_1$
$C_{88}\ 13(C_1)$	56666656566566666666656565656566666656666565656	4	$4 \times C_1$
$C_{88}\ 14(C_s)$	56666656566566666666656566565656666666665665665	3	$1 \times C_s, 1 \times C_1$
$C_{88}\ 15(C_1)$	56666656566566666666656565656566656566666566	5	$5 \times C_1$
$C_{88}\ 16(C_s)$	56666656566566666666656565656566566666656566	3	$1 \times C_s, 1 \times C_1$
$C_{88}\ 17(C_s)$	56666656566566666666656565656566566656666566	6	$2 \times C_s, 2 \times C_1$
$C_{88}\ 18(C_1)$	56666656566566666666656565656656666656666565665	2	$2 \times C_1$
$C_{88}\ 19(C_s)$	56666656566566666666656656565656665666665665	3	$1 \times C_s, 1 \times C_1$
$C_{88}\ 20(C_2)$	56666656566566666666656656565656565656566666	2	$1 \times C_1$
$C_{88}\ 21(C_1)$	56666656566566666666666656656566565656565656656	4	$4 \times C_1$
$C_{88}\ 22(C_{2v})$	56666656566566666666666566565656566566565666	6	$1 \times C_s, 1 \times C_1$
$C_{88}\ 23(C_s)$	56666656566566666666666566565656565656566666	4	$2 \times C_s, 1 \times C_1$
$C_{88}\ 24(C_s)$	56666656566565666666656666666566656565656566	5	$1 \times C_s, 2 \times C_1$
$C_{88}\ 25(C_2)$	56666656566565666666666666656656566656566565656	4	$2 \times C_1$
$C_{88}\ 26(C_1)$	56666656566565666666666666656656566565656566	4	$4 \times C_1$
$C_{88}\ 27(C_2)$	56666656566565666666666666656656566656566565656	4	$2 \times C_1$
$C_{88}\ 28(C_2)$	56666656566565666666666666666665665656566665	3	$1 \times C_2, 1 \times C_1$
$C_{88}\ 29(C_1)$	56666656566565666666666666656565656566566665656	4	$4 \times C_1$
$C_{88}\ 30(C_1)$	56666656566565666666666666656565656566566656	5	$5 \times C_1$
$C_{88}\ 31(C_s)$	56666656566565666666666666656565656566566665	1	$1 \times C_s$
$C_{88}\ 32(C_s)$	56666656566565666666666666656565656566566656	4	$2 \times C_s, 1 \times C_1$
$C_{88}\ 33(C_2)$	56666656566565666666656565656666666566565656	5	$1 \times C_2, 2 \times C_1$
$C_{88}\ 34(T)$	56666656566565666666666666656566565656566565656	0	0
$C_{88}\ 35(D_2)$	56666656566565666665666666666666656566565656656	4	$1 \times C_1$

Each spectrally distinct isomer is represented by a ring spiral[9]. A list of spirals for C_{84} is given in [19] and is not repeated here. For each isomer the total number $n(\text{II})$ of type II Stone–Wales bonds is listed, along with their decomposition into orbits of the point group. The interconversions are mapped for C_{78} to C_{88} in the present paper, with the exceptions of C_{82} and C_{84}, for which maps are available elsewhere[15,16].

ing small-patch SW transformations. The D_3 isomer of C_{78} has no type-II patch but 18 type-I patches, and can be converted to the D_{3h} isomer by three consecutive small-patch rotations and thence to any other isolated-pentagon isomer by rotation of newly formed larger patches (Fig. 16).

The role of the restricted SW transformation in rationalising isomer distributions is an intrinsically limited one. Clearly, it cannot be a helpful concept in the range of n below 78, because no such transformations are possible there. It will also become progressively less discriminating in the high n range, since most isolated-pentagon isomers of the very large higher fullerenes will tend to have large separations between their pentagons and hence, no SW patch. The rationalisation proposed in [4] has thus a

naturally finite range of usefulness. It is interesting to speculate on whether all or nearly all isolated-pentagon isomers of the super-high fullerenes will eventually be found, or whether other more complicated rearrangements of pentagons on their surfaces will act to equilibrate the isomers and bring the number of isolable species down to a manageable level. Given the enormous number of isolated-pentagon isomers for $n = 120$ and above[19], the absence of super-Stone–Wales or other rearrangements would make the chemist's life a difficult one.

Prediction of the isomer distribution even within the Stone–Wales regime depends upon assumptions about the relative stability and activation energies across the isomerisation map. At one extreme, under the assumption of freely converting isolated-penta-

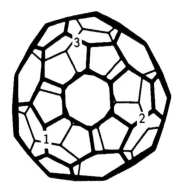

Fig. 16. Conversion between families of isolated-pentagon isomers. The D_3 isomer of C_{78} is in a family by itself and has no type II patch. Successive twisting of the marked small patches leads to a D_{3h} isomer and hence to the second family of isolated-pentagon isomers.

fullerene isomers and to list all their interconversions. Computer algorithms capable of carrying out both of these jobs have been implemented and described here. An atlas of all spectroscopic signatures and of all isomer families, complete to any desired carbon number n is now within reach. The next step in this task of classification is the generalisation from the two-dimensional maps for constant n to the three-dimensional ladders connecting fullerenes C_n and $C_{n\pm2}$, $C_{n\pm4}$, ... accessible by a combination of isomerisation and C_2 ingestion/extrusion. Programs for the systematic treatment of this step are under development.

Acknowledgements—R.P.R. thanks the Irish-American Partnership and AGB Scientific Ltd. for financial support. Help from Robin Batten (Exeter) with computer graphics is gratefully acknowledged.

gon isomers of equal stability, it can be shown that the equilibrium mole fraction of each isomer in any given map is inversely proportional to the order of its molecular point group. (Proof: Consider a SW transformation A ↔ B in which the conserved site group of the transforming patch has order O_S, the molecular point group of A has order O_A, and the molecular point group of B has order O_B. The number of SW patches in A which give B on transformation is $n_{A\to B} = O_A/O_S$, and the number of patches in B which give A is $n_{A\leftarrow B} = O_B/O_S$. The equilibrium concentrations of A and B are related by microscopic reversibility, so that $K_c = [B]_e/[A]_e = k_1/k_{-1}$, where k_1 and k_{-1} are the total forward and reverse reaction rate constants. Since ΔH for the transformation is assumed to be zero, these total rate constants are simply $k_1 = n_{A\to B}k$ and $k_{-1} = n_{A\leftarrow B}k$, where k is the generic rate constant of a single A ↔ B transformation. Hence, $[B]_e/[A]_e = n_{A\to B}/n_{A\leftarrow B} = O_A/O_B$, as claimed. The resulting distribution can be interpreted as entropic, the molar statistical entropy of each isomer X in the map being given by $S_X = R \ln W_X$ with W_X being proportional to $1/O_X$.) Under the opposite assumption of significant differences in stability between rapidly interconverting isomers, a distribution governed by enthalpy (rather than entropy) differences is to be expected. In particular, if one isomer is very much more stable than any other, this thermodynamic distribution may further simplify to one observable isomer per map. Experimental results for C_{78} do not decide conclusively between the two extreme pictures ([4,30], see discussion in [16]).

7. CONCLUSIONS

The focus of this paper has been mainly mathematical and taxonomic. We have been discussing the systematics of a particular mechanism of isomerisation, because we believe that this is an important preliminary to detailed modelling. In order to test ideas about fullerene stability and isomerisation it is necessary, at least in principle, to be able to generate all

REFERENCES

1. W. Krätschmer, L. D. Lamb, K. Fostiropoulos, and D. R. Huffman, *Nature* **347**, 354 (1990).
2. R. Taylor, J. P. Hare, A. K. Abdul-Sala, and H. W. Kroto, *J. Chem. Soc. Chem. Commun.* 1423 (1990).
3. R. Ettl, I. Chao, F. Diederich, and R. L. Whetten, *Nature* **353**, 149 (1991).
4. F. Diederich, R. L. Whetten, C. Thilgen, R. Ettl, I. Chao, and M. Alvarez, *Science* **254**, 1768 (1991).
5. F. Diederich, R. Ettl, Y. Rubin, R. L. Whetten, R. Beck, M. Alvarez, S. Anz, D. Sensharma, F. Wudl, K. C. Khemani, and A. Koch, *Science* **252**, 548 (1991).
6. K. Kikuchi, N. Nakahara, T. Wakabayasi, M. Honda, H. Matsumiya, T. Moriwaki, S. Suzuki, H. Shiromaru, K. Saito, K. Yamauchi, I. Ikemoto, and Y. Achiba, *Chem. Phys. Lett.* **188**, 177 (1992).
7. H. W. Kroto, *Nature* **329**, 529 (1987).
8. P. W. Fowler, *Chem. Phys. Lett.* **131**, 444 (1986).
9. D. E. Manolopoulos, J. C. May, and S. E. Down, *Chem. Phys. Lett.* **181**, 105 (1991).
10. H. W. Kroto, J. R. Heath, J. C. O'Brien, R. F. Curl, and R. E. Smalley, *Nature* **318**, 162 (1985).
11. P. W. Fowler and D. E. Manolopoulos, *Nature* **355**, 162 (1985).
12. P. W. Fowler and J. I. Steer, *J. Chem. Soc. Chem. Commun.* 1403 (1987).
13. P. W. Fowler, *J. Chem. Soc. Faraday* **86**, 2073 (1990).
14. A. J. Stone and D. J. Wales, *Chem. Phys. Lett.* **128**, 501 (1986).
15. P. W. Fowler, D. E. Manolopoulos, and R. P. Ryan, *J. Chem. Soc. Chem. Commun.* 408 (1992).
16. D. E. Manolopoulos, P. W. Fowler, and R. P. Ryan, *J. Chem. Soc. Faraday,* **88**, 1225 (1992).
17. C. Coulombeau and A. Rassat, *J. Chim. Phys.* **88**, 173 (1991).
18. C. Coulombeau and A. Rassat, *J. Chim. Phys.* **88**, 665 (1991).
19. D. E. Manolopoulos and P. W. Fowler, *J. Chem. Phys.,* **96**, 7603 (1992).
20. T. G. Schmalz, W. A. Seitz, D. J. Klein, and G. E. Hite, *Chem. Phys. Lett.* **130**, 203 (1986).
21. S. C. O'Brien, J. R. Heath, R. F. Curl, and R. E. Smalley, *J. Chem. Phys.* **88**, 220 (1988).
22. M. Endo and H. W. Kroto, to be published.
23. A list of 36 possible groups is given in P. W. Fowler, J. E. Cremona, and J. I. Steer, *Theor. Chim. Acta* **73**, 1 (1988). More detailed considerations (P. W. Fowler, D. E. Manolopoulos, C. M. Quinn, D. B. Redmond, and R. P. Ryan, to be published) show that the symmetry

criterion can be tightened because groups S_{2n}, C_n, C_{nh}, C_{nv} (n = 5,6) cannot be realised as fullerenes.

24. P. W. Fowler and C. M. Quinn, *Theor. Chim. Acta* **70**, 333 (1986).

25. P. W. Fowler, *J. Chem. Soc. Faraday* **87**, 1945 (1991).

26. P. W. Fowler, *J. Chem. Soc. Perkin* **2**, 145 (1992).

27. P. W. Fowler, D. E. Manolopoulos, and R. C. Batten, *J. Chem. Soc. Faraday* **87**, 3103 (1991).

28. See for example, F. A. Cotton and G. Wilkinson, *Advanced Inorganic Chemistry,* 3rd Ed. Wiley, New York (1972).

29. D. E. Manolopoulos, *J. Chem. Soc. Faraday* **87**, 2861 (1991).

30. K. Kikuchi, N. Nakahara, T. Wakabayasi, S. Suzuki, H. Shiromaru, Y. Miyake, K. Saito, I. Ikemoto, M. Kianosho, and Y. Achiba, *Nature* **357**, 142 (1992).

Table 3. Properties for different negatively curved graphite structures

Structure	ρ	ΔE	B	a	a Å	N
Schwarzite $D7$	1.15	0.18	9.4	17.39	24.7	216
Schwarzite $P7$	1.02	0.20	7.5	11.4	16.2	216
Schwarzite $P8$	1.16	0.19	10.3	10.47	14.9	192
Schwarzite $D7'$	1.28	0.22	11.5	15.42	21.9	168
f.c.c. C_{60}	1.71	0.42	1.4	10.0	14.12	60
Diamond	3.52	0.02	44.3	2.31	3.5595	2

ρ is the density in g/cm^3.
ΔE is the total energy relative to graphite in eV/atom.
B is the bulk modulus in units of 10^{11} dyne/cm^2.
a is the cubic unit cell size in units of the bond length.
a Å is the cubic unit cell size in Ångstroms.
N is the number of C atoms per primitive unit cell (the f.c.c. cell contains four primitive cells).

incorporating heptagons. $D7'$ is the D- parallel surface found by Vanderbilt and Tersoff. $P8$, $D8$, and $G8$ are the symmetrical P-, D-, and G-surfaces incorporating octagons found by Mackay and Terrones.

Tables 4–6, give the coordinates of the corresponding structures.

Vanderbilt and Tersoff[6] have proposed a graphite structure based on a surface parallel to the D-surface, so that the two subspaces are not equivalent (Fig. 12). A tetrahedral joint is built out of 84 atoms in hexagons and heptagons so that each point is a member of two hexagons and one heptagon ($6^2.7$) and is topologically (but not metrically) equivalent to every other. There are thus 2×84 atoms per primitive unit cell with space group $Fd3$ and $8 \times 84 = 672$ per cubic unit cell with $a = 21.8$ Å (assuming graphite-type bonds). The fractional coordinates of the nonequivalent atoms are given in Table 7.

The density is expected to be 1.29 g/cm^3. The authors calculate the energy of formation to be 0.11 eV/atom as compared with 0.67 eV/atom for C_{60}, but their calculation is not exactly the same as that of Lenosky.

Table 4. Fractional coordinates of the $P8$ surface with origin at the centre of the cubic cell ($Im\overline{3}m$)

Atom	x	y	z
1	0.286	0.173	0.286
2	0.314	0.0908	0.314
3	0.387	0.0494	0.2729

Table 5. Fractional coordinates of the G8 surface with origin at one of the corners of the cubic cell ($Ia\overline{3}d$)

Atom	x	y	z
1	0.1903	0.2117	−0.2265
2	0.1348	0.1703	−0.1988
3	−0.0844	0.2014	−0.1511
4	−0.0279	0.1520	−0.1217

Table 6. Fractional coordinates of the D8 surface with origin at one of the corners of the cubic cell ($Fd\overline{3}m$)

Atom	x	y	z
1	−0.0416	0.0	0.54163
2	−0.0831	0.0	0.58315
3	−0.1668	0.0	0.6668
4	−0.2083	0.0	0.7083
5	−0.0967	0.05172	0.61032
6	−0.1532	0.05172	0.63966

Table 7. Fractional coordinates of Vanderbilt's $D7'$ structure

Atom	x	y	z
1	0.1134	0.0050	0.2415
2	−0.0109	0.0303	0.2200
3	0.0272	0.0756	0.2435
4	−0.0984	0.0826	0.1704
5	−0.0662	0.1382	0.1542
6	−0.0743	0.0402	0.2146
7	−0.1915	0.0177	0.1649

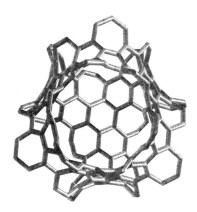

Fig. 12. Structure proposed by Vanderbilt and Tersoff ($D7'$).

side the focus of this article.[13,17] Here we simply recite the main results of this work:

Single value for T_c. For each alkali metal, only a single value for T_c is observed on annealed samples, regardless of the sample composition. This implies that there exists a single, stoichiometric phase that is superconducting, thereby ruling out the possibility of a continuous variation in T_c with "doping."

The composition of this phase is A_3C_{60}. For both pure metals, and for alkali alloys, the maximum shielding observed occurs near $x = 3$.

Phase segregation. The observation of sizeable shielding diamagnetism fractions at compositions away from the stoichiometric one points to a very strong tendency toward phase segregation. Recall that for any shielding to be observed, the superconducting domains must have a thickness comparable to the penetration depth, necessitating the formation of large single-phase regions. This is also supported by x-ray diffraction evidences.[13,22]

2.3 *Structure of the A_3C_{60} compounds*

The existence of the first single-phase samples of K_3C_{60} materials in May 1991 made it possible to determine the crystal structure of this compound,[22] which has since served as the archetype for all other A_3C_{60} compounds.[10–16] The x-ray diffraction measurements were carried out on a set of single-phase pressed samples at the NSLS beamline (Brookhaven), from which the structure illustrated in Fig. 5 was deduced. The diffraction patterns and optimized structure, obtained from Rietveld refinement, indicate the following:

1. The pattern can be indexed to an *fcc* structure, in which the lattice parameter ($a = 14.24$ Å) is expanded slightly from that of *fcc* C_{60} (14.11 Å).

2. The alkali ions occupy the octahedral and tetrahedral vacancies of the host C_{60} lattice, as in the well known cryolite structure of $(A^+)_3(M^{3-})$, where M^{3-} is a complex ion with octahedral symmetry. This causes the diffraction intensities to be drastically altered from that of C_{60} fullerite: The scattering centers in fullerite C_{60} are located not on the *fcc* lattice points, but rather displaced radially to a distance coincidentally near $a/4$, leading to dramatic near-cancellations for some lines; the presence of the alkali ions as strong scatterers drastically modifies this condition. From the refined fit, the thermal amplitude of the octahedral alkali ion is found to be substantially larger than the tetrahedral alkali ions, in accord with geometrical considerations.

3. The orientations of the C_{60} with respect to the alkali ions and toward each other can be determined, even at ambient temperature, and is as illustrated in Fig. 5. The C_{60} molecules are oriented so that the Cartesian axes of the cubic lattice bisect pyracylene bonds, so that the octahedral A^+ ions are six-coordinated to these electron-rich regions of the molecule. The tetrahedral A^+ ions are four-coordinated to the centers of six-membered rings. The corresponding space group is Fm3m, which is higher than allowed

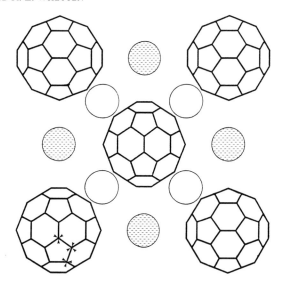

Fig. 5. Structure of the A_3C_{60} superconductors. The open and hatched spheres represent the alkali atom at tetrahedral and octahedral sites, respectively. Note the two possible orientations of the C_{60} molecules and the three inequivalent carbon sites (marked by arrows on the molecule in the lower left corner) as described in the text. After P. Stephens *et al.*[22]

by the icosahedral symmetry of the molecule, but can be explained by a 50% occupancy of the two possible orientations (related by 90° rotation).

4. The size of the C_{60} molecule is essentially unchanged from that of the undoped material (7.08 Å diameter between opposing nuclei), and there are three distinct carbon-atom sites.

The 2:1 occupancy of the tetrahedral and octahedral sites is directly observed in the ^{39}K-NMR experiment (Fig. 6),[46] which shows distinct chemical shifts for these sites, and the two peaks integrate to give the appropriate ratio.

The temperature variation in a is 0.7% for K_3C_{60} from 11 K to 300 K, and is 0.9% for pristine C_{60}[16]. Using these thermal expansion coefficients and the known compressibility of C_{60}, one can extract from standard thermodynamic relations an estimate for the compressibility for K_3C_{60} of 1.1 to 1.4 × 10^{-2} GPa^{-1} (see discussion below).

The diffraction results for the Rb_3C_{60} compound were analyzed similarly[11,12,15] and its structure was determined to be in every way analogous to the K_3C_{60} structure, with the main difference being that the lattice is expanded, $a = 14.43$ Å, with the result that the interball distance is now larger. The crystal structure of mixed-alkali A_3C_{60} superconductors such as Rb_2Cs, K_2Rb, and KRb_2 has also been determined[7,13] to be analogous, with varying lattice parameters. Finally, the structure of Na_3C_{60}, for which no superconducting transition has been found, is also analogous. All the parameters are given in Table 1.

The determination of these structures has been used as input to a large number of band-structure calculations. Most of these seem to indicate that the

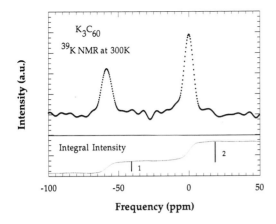

Fig. 6. Ambient temperature ³⁹K-NMR spectrum of the K_3C_{60} superconductor referenced to potassium-fluoride in water solution, demonstrating the distinct chemical shifts and occupation of the tetrahedral and octahedral lattice sites in the superconducting compound (Fig. 5). After ref. [46].

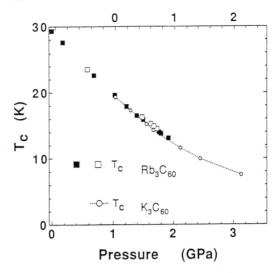

Fig. 7. Dependence of T_c on applied pressure for Rb_3C_{60} and K_3C_{60}. Note that the upper scale displaced by 10,6 GPa serves for the K_3C_{60} data (open circles). After Sparn et al.[19]

charge transfer is very nearly complete, $[A^+]_3 \cdot [C_{60}^{3-}]$, that the electrons are used to half-fill the t_{1u} band, so that the conductivity occurs among the fullerene anions with the alkali ions as spectators. Experimentally this is indicated by the small diamagnetic shift (i.e., the absence of positive (Knight) shift) of the ³⁹K resonance. From the lattice constant and this single-band picture with three electrons, the total carrier density is three times the molecule number density (i.e., $n \approx 4 \times 10^{21}$ cm⁻³. The structure shows that the direct path is from ball to ball (along 110 axes), and that the alkali-alkali distance is very large. The retention of cubic symmetry about each C_{60} acts to maintain the degeneracy of the three t_{1u} orbitals, and no indications of other symmetry breaking (Jahn-Teller effect) have been found, even at temperatures near or below T_c. However, the thermal expansion coefficient is indicative of a material much softer than normal ionic crystals.

Our focus here has been on the structure of the A_3C_{60} compounds and its relation to the *fcc* C_{60}. It is noteworthy that so far only the A_3C_{60} *fcc* structure seems to result in superconductivity; other compositions (i.e., A_4C_{60}, A_6C_{60}) form in different symmetry structures and the ground state is (suspected) insulating. The reader is referred to other articles in this volume.

3. HOW TO CHANGE T_c? DEPENDENCE ON LATTICE CONSTANT AND ALKALI-ION IDENTITY

3.1 *Pressure-dependence of T_c, lattice-constant dependence via compressibility*

The effect of applied pressure on the superconductivity in A_3C_{60} compounds (A = K, Rb) was investigated by Sparn et al.[18,19] on single-phase pressed pellet samples using susceptibility measurements in the range up to 2 GPa. Figure 7 shows the combined trends in T_c obtained over this range for both compounds, plotted in such a way that the two curves are essentially superimposed. The important results and conclusions are the following.

The effect of pressure is very large and negative, with T_c varying from 8–19 K in K_3C_{60} and from 13 to 30 K in Rb_3C_{60}, for a combined range in transition temperature of 8 to 30 K. The effect is monotonic (neglecting the inflection near 1.5 GPa in Rb_3C_{60}) so, in the absence of evidence for any structural phase changes, it is interpreted to reflect only the monotonic decrease in the lattice constant with increasing pressure.

Although the curves $T_c(P)$ are not very linear, for the purposes of comparison one can extract the pres-

Table 1. Structural properties of A_3C_{60} superconductors

Property	C_{60}	K_3C_{60}	Rb_3C_{60}
Space group	Pa$\bar{3}$ *(fcc)*	Fm$\bar{3}$m *(fcc)*	Fm$\bar{3}$m *(fcc)*
Lattice parameter (nm)	1.418	1.424	1.443
C_{60}-C_{60} distance (nm)	1.003	1.006	1.020
Volume per C_{60} (nm³)	0.713	0.722	0.750
Carrier density (10^{21} cm⁻³) (3 valence e^- per C_{60})	—	4.155	4.200
Compressibility (GPa⁻¹) ($\delta \ln a / \delta P$)	0.018	0.012	0.015
Thermal expansion: $(a_{300} - a_{11})/a_{300}$	0.009	0.007	—
A–C_{60} closest distance (nm)	—	0.327	0.333
A thermal factor, $B^{1/2}$ (nm)(tetrahedral, octahedral)	—	0.065,0.16	0.020,0.15

sure coefficient $(\delta T_c/\delta P)_{P\ =\ 0}$ from these plots, to find for A = Rb and K the values -9.7 and -7.8 K/GPa, respectively. These are among the largest, in absolute value, known. For comparison with high-T_c materials, it is more useful to express the effect in terms of the reduced coefficient $(1/T_c)(\delta T_c/\delta P)_{P\ =\ 0}$, in which case both compounds have the same value, ~ -0.35 GPa^{-1}.

It was noticed[19] that the curves for the two different materials can be overlapped in the common region of T_c simply by displacing the pressure scale of one with respect to the other by about 1 GPa (i.e., $T_c(0)_K\ =\ T_c(1\text{GPa})_{Rb}$ and $(\delta T_c/\delta P)_{P\ =\ 0,K}\ =\ (\delta T_c/\delta P)_{P\ =\ 1.0,Rb}$. The simplest hypothesis to account for this coincidence is that T_c is a function only of the lattice parameter, in which case the lattice parameter of the K and Rb materials would need to be equal when the pressure difference is 1 GPa. This is equivalent to saying that T_c is a function only of the interball distance and not the identity of the alkali, so that the role of the alkali ions would be reduced to that of electron donor and intermolecular "spacer."

The missing information needed to evaluate this hypothesis is the lattice parameter of the two materials as a function of pressure. In the absence of this information, Sparn et al.[19] calculated that the (linear) compressibility of the Rb$_3$C$_{60}$ would need to be $-\delta(\ln a)/\delta P\ =\ 0.012$ GPa^{-1}, as compared to the C$_{60}$ fullerite value of 0.018 GPa^{-1}. This was subsequently validated by Zhou et al.,[48] who measured the pressure-dependence of the lattice constant, $a(P)$, for both Rb and K compounds, and reported $-\delta(\ln a)\delta P\ =\ 0.015$ and 0.012, respectively.

The combination of the pressure dependence experiments on the T_c and lattice parameter thus allows the curve $T_c(a)$, that every theorist should love, to be determined over the range $a\ =\ 13.9\ -\ 14.5$ Å, corresponding to the center-to-center distances $2^{-1/2}a\ =\ d\ =\ 9.858$ to 10.284. Although this variation may not seem too large, one has to remember that variation of the distance between the carbon atoms of neighboring C$_{60}$ molecules is relevant. To obtain that, one has to substract the fixed diameter of the molecules (as the molecule is practically incompressible) 7.1 Å, resulting in a variation from 2.76 to 3.18 Å of the free space between the molecules. As the covalent C—C bond length of 1.5 Å sets a practical lower limit, the excess distance is varied from 1.2 to 1.6 Å—a considerable range of relative change.

3.2 Ternary compounds, and non-superconducting A_3C_{60} compounds

The relation between T_c and lattice parameter was independently deduced by the investigation of the ternary compounds, generic formula A$_2$A$'$C$_{60}$, where A and A$'$ are different elements from K, Rb, and Cs. These have also been described[12] somewhat contradictorily as continuous (alloy) compounds, for the case of Rb$_{3-x}$K$_x$C$_{60}$. An approximately linear dependence of T_c on the lattice parameter was found for these compounds, but the range of T_c spanned is 18–

32 K, in contrast to the 5–30 K range covered by pressure measurements. Although we are not aware of structure refinements, these are materials with the same *fcc* structure, indicating again the importance of the interball spacing and insignificance of the other alkali-ion characteristics.

It is important, therefore, that an exception has been found in the form of Na$_3$C$_{60}$,[47] which assumes the *fcc* structure but shows no indications of a superconducting phase. More recently, a series of Na$_{3-x}$A$_x$C$_{60}$ has been described that shows deviations from the simple relationship.[15,16]

Beyond the alkali metals, there have also been reports of superconductivity in large fractions of samples using Ga or Ca,[49,50] both claiming an *fcc* structure. The former is said to have the formula Ga$_3$C$_{60}$, suggesting that perhaps only one electron is transfered from each Ga atom, while the latter is Ca$_5$C$_{60}$, with three ions in the octahedral site. The characterization of the superconducting properties of these materials remains incomplete at the time of writing.

4. SUPERCONDUCTING STATE PARAMETERS

4.1 Critical-field measurements; evaluation of λ_L and ξ_0

The parameters characterizing the superconducting state are the (London) penetration length λ_L, the coherence length ξ_o, and the gap, Δ. Values of the lower and upper critical fields, H$_{c1}(T)$ and H$_{c2}(T)$, extrapolated to zero temperature, can be used to obtain values for $\lambda(T\ =\ 0)$ and ξ_o. In combination with other known quantities such as $T_c(H\ =\ 0)$ and the carrier number density, n, these values can be used within the assumptions of the classical formulas to calculate estimates for other properties. Measurements of the critical fields in K$_3$C$_{60}$ and Rb$_3$C$_{60}$ single-phase pressed powders were reported by Holczer et al.[20] and by Sparn et al.[19], respectively, and for K$_3$C$_{60}$ films by Palstra et al.[51] Because the upper critical fields in these materials are very high, measurements at high field strengths are required to determine H$_{c2}(T\ =\ 0)$, and these were recently carried out, up to 30 Tesla, on pressed-powder samples of both binary superconductors by Jiang and co-workers.[21]

The methods of measurement (*dc* magnetic susceptibility) and of extracting the critical fields follow standard procedures. The H$_{c1}$ values, extrapolated to $T\ =\ 0$, are 13 mT for K$_3$C$_{60}$ and 26 mT for Rb$_3$C$_{60}$. Because of the granular morphology of the material and the substantial vortex pinning observed, the experimentally determined H$_{c1}$ should be regarded as an upper bound. (H$_{c1}\ \sim\ 30$ mT is about the field strength where the number of vortices per grain is near unity.) Therefore only a lower bound for λ_L is obtained using eqn (1) below.

The upper critical fields are plotted as a function of temperature in Fig. 8, taken from refs. [20] and [21]; an early estimate from the slopes in the 0–5 T

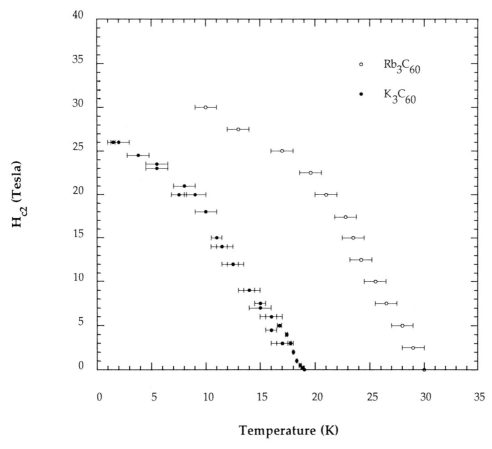

Fig. 8. The H_{c2} vs T phase diagram of the superconductors K_3C_{60} and Rb_3C_{60}, combined from refs. [19–21]. The coherence lengths deduced from $H_{c2}(T=0)$ are in Table 2.

range and using the formula from Werthamer-Helfand-Hohenberg,[52] gave extreme high values, H_{c2} = 49 and 78 T for the K and Rb compounds, respectively. These are both in excess of the Pauli limit (1.84 T_c in Tesla units, for weak-coupling BCS), where the Zeeman energy $\mu_B \cdot H$ of the electrons approaches the superconducting gap Δ. The recent high-field measurements, shown in Fig. 8 in the form of a phase diagram (H-T plane), complete the earlier results, and show a more substantial downward curvature than the Werthamer-Helfand-Hohenberg theory, which omits the effect of paramagnetism. The empirical H_{c2} values at $T = 0$ K are therefore much lower than the earlier extrapolations, near 27 and 34 T for K_3C_{60} and Rb_3C_{60}.

From $H_{c2}(0)$ and $H_{c1}(0)$, the zero-temperature superconducting coherence length ξ_0 and London penetration depth λ_L, were evaluated using the relations[14]

$$H_{c2}(0) = \frac{\Phi_0}{2\pi\xi_o^2}, \; H_{c1}(0) = \left(\frac{\Phi_0}{4\pi\lambda_L^2}\right) \text{Log}\left(\frac{\lambda_L}{\xi_o}\right), \quad (1)$$

where Φ_0 is the flux quantum. Strictly speaking, these formulas apply when ξ_0 is much smaller than the elec-

tronic mean-free-path (so-called clean limit). These values are collected in Table 2.

The most important points are, first, the estimated values for ξ_o, 2.0 nm and 2.6 nm for Rb_3C_{60} and K_3C_{60} are very small, just a few times the nearest-neighbor C_{60} distance $d = 1.0$ nm. Second, the penetration depths, $\lambda_L = 168$ nm and 240 nm, are large, a good fraction of a grain size in typical powders. Third, the ratio of these quantities, the dimensionless Landau-Ginzburg parameter $\kappa = \lambda_L/\xi_o$, exceeds 10^2 in both materials, showing that the A_3C_{60} family is a second class of extreme type-II superconducting compounds.

Table 2. Superconductivity parameters in A_3C_{60} compounds

Parameter	A = Rb	A = K
T_c (K)	29.6	19.3
$(\delta T_c/\delta P)_{P=0}$ (K-GPa^{-1})	−0.97	−0.78
H_{c1} (mT)	26	13
H_{c2} (T)	78, 34	49, 26
$J_c(10^6$ A-cm^{-2})	1.5	0.12
ξ_0 (nm)	2.0, 3.0	2.6, 3.5
λ_L (nm)	168, 440	240, 480
$\kappa = \lambda_L/\xi_0$	84	92

In principle, the values for λ_L and ξ_o parameters can be used directly to calculate many intrinsic properties of the materials in their normal state, including the effective mass (density of states) and so on. A critical analysis of the validity of this approach is given in Section 6.

4.2 Evaluation of $\lambda_L(T)$ from μSR measurements

The μSR method has been extensively applied to the study of the penetration depth in type-II superconductors. In the case of the A_3C_{60} compounds, the results of Uemura et al.[23,24] obtained at the TRIUMF facility on large ($\sim$0.2 g) pressed powders of single-phase superconductors, were evaluated to give the form of $\lambda_L(T)$ over the range $T = 3$ K to T_c. Results on K_3C_{60} and Rb_3C_{60} are shown in Fig. 9, where the muon spin relaxation rate is plotted vs temperature for both materials and compared to predictions of theoretical curves.[24] The important results are the following:

1. The shapes of $\lambda_L(T)$ curves can be explained without invoking any exotic pairing mechanisms, but rather are fully consistent with s-wave, singlet pairing. This can be seen immediately from the flat region over the temperature range $T = 0$ to about $T_c/2$.

2. The zero-temperature extrapolated values, $\lambda_L(0)$ or simply λ_L in Table 1, are about twice as large as estimated from the H_{c1} measurements, at 440 nm (Rb)[24] and 480 nm (K)[23] (see Table 2 for comparison). The reason for this discrepancy is not clear; in the critical field measurements, systematic errors can arise from the procedure used to extract H_{c1}; in the μSR measurements the main uncertainty arises from the need to simulate the vortex lattice structure. In either case, the conclusion $\lambda_L^{(K)} > \lambda_L^{(Rb)}$ is unaffected by such systematic errors. (An estimate from ^{13}C NMR line broadening[27] gives the same trend, $\lambda_L^{(K)} \approx 600$ nm and $^{(Rb)} \approx 460$ nm.)

Uemura et al.[23] have also discussed how these values place the fulleride superconductors in relation to other high-T_c and molecular superconductors.

4.3 NMR results: Coherence effects; estimate for the superconducting gap Δ

In a conventional metal, the principal source of nuclear spin relaxation is the Pauli paramagnetism of the conduction electrons. In the superconducting state, the unpaired electron density, reflecting thermally activated quasiparticle excitations, continues to dominate the relaxation rate below the normal-to-superconducting transition, allowing for the determination of its temperature dependence, $1/T_1 = \exp(-\Delta/T)$, where Δ is the superconducting gap. Figure 12 shows an Arrhenius plot for ^{13}C relaxation in K_3C_{60}.[28] The activation energy, Δ, derived from the linear low-temperature region, is 35 ± 3 K. Tycko et al.[27] gave the values $\Delta = 21$ and 47 K for K_3C_{60} and Rb_3C_{60}, respectively. The relevant quantity is the dimensionless parameter $2\Delta/k_B T_c$. The weak-coupling

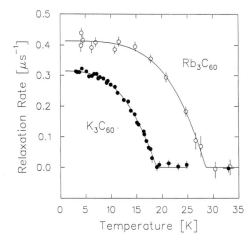

Fig. 9. Temperature-dependence of the muon spin relaxation rate for K_3C_{60} and Rb_3C_{60}; the evaluated λ_L values are in Table 2. After refs. [23,24]; figure by courtesy of Y. Uemura.

BCS mechanism gives 3.52 for this quantity, which is compatible with the results of ref. [27] for K_3C_{60}.

Just below T_c in Fig. 10, there is a small enhancement of the relaxation rate over the Arrhenius law. In traditional superconductors the NMR relaxation rate shows a pronounced maximum below T_c known as a coherence (or Hebel-Slichter) peak, which has been taken to be a phenomenon characteristic of isotropic-gap singlet-pairing superconductors.[33] The high-T_c compounds fail to show the expected coherence peak in the NMR relaxation rate, but show instead a fast (exponential) decrease in $1/T_1$ starting right at T_c. As seen from Fig. 10, measurements in high field do not show such an abrupt decrease, making it seem that the coherence peak amplitude is strongly attenuated, as might arise from a short quasiparticle lifetime and/or clean-limit behavior.

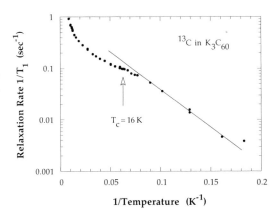

Fig. 10. Arrhenius plot for ^{13}C relaxation rate in K_3C_{60}, demonstrating the temperature dependence, $1/T_1 \approx \exp(-\Delta/T)$, where Δ is the superconducting gap. The activation energy, Δ, derived from the linear low-temperature region, is 35 ± 3 K. Tycko et al. gave the values, $\Delta = 21$ and 47 K for K_3C_{60} and Rb_3C_{60}, respectively. After ref. [28].

The superconducting gap Δ can also be extracted directly from measurements of the *ac* conductivity, from the equivalent optical (reflectance) spectroscopy,[53] or from *dc*-tunneling current-voltage curves. From the latter method, Zhang *et al.*[54] obtain a value, $2\Delta = 5.5\ k_BT_c$ higher than the values inferred by the NMR results, $\Delta = 5.5\ k_BT_c$, which would argue for a strong coupling limit within a BCS model.

5. NORMAL-STATE PROPERTIES OF A$_3$C$_{60}$ SUPERCONDUCTORS

5.1 *Magnetic susceptibility and ESR spectroscopy*

In the absence of transport property measurements, magnetic measurements have been most valuable in revealing the properties of the normal state of these superconductors. Static magnetic susceptibility measurements[25]aiming to evaluate the spin susceptibility contain contributions from the core diamagnetism and orbital para- and diamagnetism as well, whose separation is in principle a difficult task, particularly when the electron density is so low, no more than 4×10^{21} cm^{-3}, two orders of magnitude below that of most metals. Fortunately, electron spin resonance (ESR) in these materials[25,55] shows an easily observable signal even at room temperature, allowing the spin susceptibility to be evaluated separately. Figure 11 shows an ambient temperature spectrum of K$_3$C$_{60}$, consisting of a single Lorentzian line with a *g*-value of 2.0002 (as compared to the free-electron value of 2.0023), which is a characteristic *g*-value observed so far in all C$_{60}$ radical anions,[56] leaving little doubt that the spin is located on the C$_{60}$.

Evidence that this signal comes from the metallic phase, and not from some sensitively detected impurity phase, comes from the relatively large intensities observed and also from the invariance of the ambient temperature line width with field (ESR frequencies of 10 GHz and 35 GHz have been

used[55]), as for cases when the main source of line width is the intrinsic lifetime of conduction electrons.

The integrated intensity for the K$_3$C$_{60}$ at ambient temperature corresponds to 0.2–0.5 *free* spins per C$_{60}$ molecule, a remarkably high value, although the total susceptibility is still small due to the low C$_{60}$ number density, $n_o \approx 1.4 \times 10^{21}$ cm^{-3}. Figure 12 shows the temperature dependence of the integrated ESR line intensity, which when converted to a molar susceptibility gives the value $10 \pm 2 \times 10^{-4}$ emu/mol. This value seems to be temperature-independent, suggesting Pauli paramagnetism as in a normal metal; the density-of-states that would be evaluated from it would be 31 eV^{-1} per C$_{60}$ site indicating an extremely narrow band. However, the following should be noted:

First, the susceptibility value, interpreted as a Pauli susceptibility, would imply a Fermi temperature around 300 K, which in turn implies that there should be a temperature dependence (i.e., a decreasing susceptibility toward high temperature) in the temperature range shown on the figure. Because this is not observed, this interpretation is doubtful (see also the next section). Furthermore, the ambient-temperature magnitude of the susceptibility from the ESR experiments turns out to be a factor of two to four times smaller for the Rb$_3$C$_{60}$ than for the K$_3$C$_{60}$ compound, opposite the tendency expected from the naïve density-of-states argument.

More importantly—or confusingly—below ~ 60 K the peak-to-peak line width of Rb$_3$C$_{60}$ starts to increase, as shown on Fig. 13. This is accompanied by significant line shape changes, also observable on the K$_3$C$_{60}$ at low temperatures, so the signal can no longer be regarded as a single Lorentzian line.[55] A detailed study of the line shapes at 35 and at 10 GHz, combined with pulsed experiments (i.e., direct measurements of T_1 and T_2 on both echo and free-induction decay signals) show the increasing weight of a slightly higher *g*-value, broad line of about 30 Gauss. Neither a clear separation of this additional line from the high-temperature signal as a second phase, nor an understanding of the new line shape was possible up to

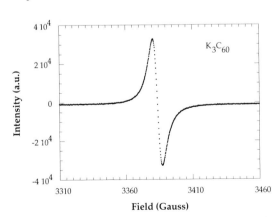

Fig. 11. Derivative ESR line shape of K$_3$C$_{60}$ superconductor obtained at ambient temperature. After ref. [55].

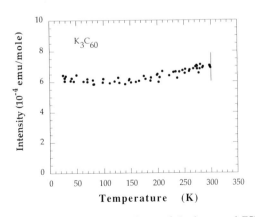

Fig. 12. Temperature dependence of the integrated ESR signal of K$_3$C$_{60}$. After ref. [25].

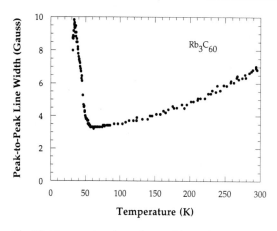

Fig. 13. Temperature dependence of the peak-to-peak ESR line width of Rb_3C_{60}. After ref. [55].

now on the available powder samples. The possibility of an exotic spin state at low temperature cannot be ruled out.

An electron spin echo signal was observed in both K_3C_{60} and Rb_3C_{60} at low temperature showing a Curie low temperature dependence.[55,57] The intensity of this roughly 1 Gauss width electron spin echo signal corresponds to a small concentration of the order of 1 spin per $10^5 C_{60}$. This is the only clearly separable "second" phase so far identified, most probable isolated C_{60}^{1-} ions (the g-factor is the smallest[56]). This part of the signal cannot be associated to any conduction electron spin, but to some kind of localized (at least to a C_{60} molecule) spin state characterized by a phase-memory time of $T_M \sim 2\,\mu s$ and T_1 of the order of 200 μs. This is further proved by the echo amplitude modulation (ESEEM) observed on this signal arising from hyperfine coupling to natural abundance ^{13}C nuclei.[57] Though the nature of these spin states are clearly different from the majority signal (therefore not representative of the conducting/superconducting electrons) and may correspond to a different oxidation state of C_{60}, it is interesting to note the following:

- As the amplitude of echo modulation is proportional to the hyperfine coupling anisotropy (the scalar coupling affects only its frequency), the observation of the modulation effect shows that the dipolar part of the electron-nuclear coupling is comparable to, if not bigger than, that of the scalar part.
- The observed modulation depths are smaller in the K compound than in the Rb one, indicating a smaller dipolar hyperfine coupling in the former.

All these observations, listed in the last three paragraphs, raise doubts on a simple Pauli susceptibility interpretation. Some of these effects can be connected to material quality problems hindering a more definite approach at present. A careful study of their

composition dependence will hopefully lead to a more detailed understanding. Even if the (unlikely) scenario of the copper-oxide superconductors would repeat, that is, all signals seen by ESR turn out not related to the clean superconducting phase, and would be all of material defect origin, the remarkable tendency of local moment formation on the C_{60} anions observed is a fundamental character to remember. Note that forming a monoclinic anisotropic structure, the all organic compound TDAE-C_{60} shows a magnetic ordering transition at 16 K.[58]

The different character signals, whose intensities are summed in the temperature dependence shown in Fig. 12, are nevertheless the principal source of the ^{13}C NMR relaxation. This topic is addressed in the next section.

5.2 Normal-state nuclear spin relaxation

^{13}C nuclear spin-lattice relaxation rates ($1/T_1$) measurements for K_3C_{60} and Rb_3C_{60} were published by Tycho et al.[27] and were interpreted as in conventional metals (i.e., Korringa law), giving a density-of-states of 20–25 per molecule per eV in K_3C_{60} and "about 40 percent higher in Rb_3C_{60}," in agreement with the expectations of the simple density of state argument. Their analysis is based on the assumption that relaxation is dominated by contact (scalar) hyperfine coupling to the conduction electrons. This is supported by calculations showing that the nonplanarity of the C_{60} molecule leads to a substantial carbon 2s orbital component in the conduction electron wave functions. The measured Knight-shift values were notedly inconsistent with this interpretation. It was also noted that the estimated value of the scalar hyperfine coupling is rather uncertain, and the dipolar and orbital contribution may not be negligible. The absolute values of the derived density of states were taken with caution, but the relative value found for the two materials was claimed to provide quantitative support for the density of state change being the principal source of the different $T_c - s$. Tycho et al. conclude that their results favor low-frequency phonons being responsible for the superconductivity, unless electron-electron interactions are important, in which case $T_1 T$ would not be proportional to $\rho(\epsilon_F)^{-2}$.

The ^{13}C line is narrow and homogeneous only at high temperature (>250 K for the K_3C_{60} and >300 for the Rb_3C_{60}), where the ratcheting motion of the molecules is still sufficiently fast. Once this effect disappears, the line becomes inhomogeneous as a chemical shift (incompletely averaged by the slowing down motion) anisotropy of 200 ppm emerges. One can find an exponential relaxation only in the high temperature range.[28]

The T_2 relaxation varies strongly below 250 K, and the longitudinal (T_1) relaxation curves encountered are clearly non-exponential.[28] Tycho et al.[27] used a stretched exponential fit to recover an average relaxation time. As the shape of the relaxation curve varies little in the temperature range 10–

200K, an investigation of the temperature dependence of an average T_1 (no matter how it is defined) is hoped to provide relevant information.

The A_3C_{60} structure (see Fig. 5.) implies three distinct carbon atom sites, allowing for different chemical shifts and electron densities on the different carbon sites (i.e., for different relaxation rates as well). Band structure calculations[35,36] indeed suggest substantial difference in the conduction electron density on the different carbon sites, namely a maximum on the 12 carbon atoms whose pyracylane bonds are perpendicular to the Cartesian axes of the cubic cell (facing the octahedral Alkali atoms) and minimum on the hexagons facing the tetrahedral Alkali atoms. Recent ^{13}C Magic Angle Rotation spectra by Yannoni et al.[59] on Rb_3C_{60} indeed show three distinct lines with intensity ratios close to that corresponding to the 12-24-24 different carbon sites, confirming this assessment. One can therefore argue that the spin-lattice relaxation is not exponential due to the different electron density on the molecular level, or the value of the hyperfine coupling is different on the different carbon sites (for instance, due to a slight deformation of the C_{60} shape in the structure), resulting in a temperature-independent relaxation shape,[27,28] but the overall temperature dependence of the "average" T_1 is still representative to the band formed by the molecular orbitals.

Figure 14 is a plot of $(T_1T)^{-1}$ vs T measured at Orsay,[28] and it is in overall good agreement with those from AT&T.[27] For a quasi-free electron band (a degenerate Fermi gas), this quantity is expected to be temperature-independent. Even if one assumes that the temperature dependence above 200 K is strongly influenced by other processes (as a result of molecular motion), limiting the range of "normal state" to $T_c < T < 200$ K, this "constant" increases by close to 30% in this temperature range. This temperature dependence seems too large to explain as a simple thermal dilatation effect on the band, and any other conventional temperature effect, like $T \sim T_F$ or $\rho(\epsilon)$ sharply picking around ϵ_F would result in variations of opposite sign. It seems difficult to explain the observations without evoking correlation effects.

Owing to the complications in the measured electron spin susceptibility[55] (as discussed above) and the uncertainties in the value and the nature of the electron-nuclear coupling, quantitative conclusions based on T_1T have to be considered with cautions at present. We expect that the situation will gradually clarify as a result of the ongoing effort of different groups.

6. DISCUSSION: AN EARLY COMPARISON WITH THEORY

Sections 4 and 5 contain a set of quantitative values for the fundamental properties of the superconducting state and some trends in observed properties of the corresponding normal states. Some of these data are assembled in Table 2. One can easily introduce these quantities into the appropriate location of a number of formulas, thereby interrelating the normal and superconducting properties, and also leading to extractable values for theoretically predicted or observable quantities. Among these, the favorites are the closely related set that includes the width of the conduction band (or ϵ_f), density-of-states at the Fermi energy, $\rho(\epsilon_F)$, or the effective (band) mass m*, and plasma frequency ω_p.

However, all these closed formulas assume an isotropic (spherical) Fermi surface and gap, and a single conduction band. In addition, the formulas within the BCS framework require a weak pairing attraction compared to the bandwidth. Referring to the introduction, it seems possible that some of these assumptions will not turn out to be satisfied, or may even break down altogether (see next subsection for one scenario). (The formulas also rely on other quantities, such as the carrier density, n, assumed to be either 1.4 $\times$ 10^{21} cm^{-3}—one electron per C_{60}—or 4.2 $\times$ 10^{21} cm^{-3}—three electrons).

As a first example, we reproduce here the estimates of the conduction-band parameters from the measured superconductivity parameters, using the relations:

$$\lambda_L = c \cdot \omega_p^{-1}, \text{ where } \omega_p = (4\pi n/m^*)^{1/2}$$

and

$$\xi_o = hv_F/2\pi^2\Delta.$$

Inserting the carrier density into the first of these allows the calculation from λ_L of the effective mass of the carrier (m*) and from this to obtain the Fermi energy. Using the λ_L values for K_3C_{60} from μSR measurements,[23,24] with one electron per C_{60} unit, gives an effective mass of m* = 11 m_e, corresponding to a T_F = 470 K (i.e., an extreme narrow band, ca.

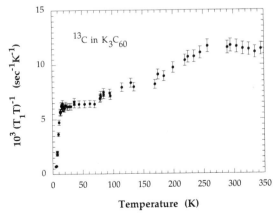

Fig. 14. The ^{13}C relaxation rates for K_3C_{60} plotted as $(T_1T)^{-1}$ vs T. Note that the relaxation shape is exponential only about 250 K; below that an 'average' relaxation rate is evaluated (see text). After ref. [28].

0.1 eV bandwidth). Using the same assumptions for Rb_3C_{60} gives a slightly smaller value for m^* and a *larger* value by 20% for T_F. (Assuming three electrons per C_{60} unit reduces these mass values by $\sqrt{3}$ and increases the T_F values by the same amount.) *This trend is opposite to what would be expected on the basis of the popular density-of-states arguments, which would assign a narrower band (smaller T_F) to the expanded compound Rb_3C_{60}.*

A similar result applies on the basis of the ξ_o values,[19–21] from which the Fermi velocity can be estimated. To get an idea of the magnitudes, first assume the weak coupling BCS relation $\Delta = 1.76 \, k_B \, T_c$ holds, in which case the $v_F = 3.6 \times 10^6$ cm/s is found for K_3C_{60} using the values extrapolated from the linear region of H_c (a 40% higher value would be found from the high-field measurements). The same analysis for Rb_3C_{60} gives a slightly *larger* value for v_F, near 4.1×10^6 cm/s. This is again contradictory to the anticipated trend, which is made even worse when the high-field results are used.

One can argue that the evaluations presented above are based on clean limit formulas, therefore certainly numerically missleading.[51] At present, although some single crystal transport measurements[30] start to appear in the literature, we think that it will take some time until a well established experimental value of the mean free path will emerge (implying that the applicability of the concept will also be accepted). Only the ^{13}C NMR relaxation data seemed at first site interpretable,[27] providing support to the observed T_c-lattice parameter relation as a consequence of changes in $\rho(\epsilon_F)$. We have discussed in detail the different eventual reasons in chapter 5.2 that may lead to a re-interpretation of these results.[28]

Problems have been encountered also in determining the magnetic susceptibility of normal state,[25] especially the interpretation and assignment of the observed ESR signals[55] as discussed in chapter 5.1. Facing the novelty of the material chemistry, spurious and intrinsic effects are still to be separated; at present (even if one considers it unlikely) one can not experimentally rule out the existence of some exotic spin states on the molecular level. Continuing efforts to improve the sample quality and a comparative study of different compounds will hopefully resolve this problem.

The most fundamental question is:

- What is the mechanism of superconductivity in these materials?
- In other words, are we dealing with something fundamentally new, or can the conventional BCS theory of superconductivity—with more or less small modifications/adaptations—adequately describe the properties of these materials? has still no experimental answer.

Several theories[60,61] have been proposed on the assumption of electron-phonon coupling being responsible for the superconductivity in these materials, differing in the choice of relevant phonon modes (low energy acoustic and libronic, high energy vibronic, etc.) and the coupling strength required to reflect the observed T_c values. Inelastic neutron scattering studies[62] (see also K. Prassides *et al.* in this volume) indeed show evidences for strong electron-intramolecular coupling; whether this is a simple consequence of charging of the molecule or the fundamental cause of pair formation, remains to be proved.

A fundamentally different root has been proposed by Chakravarty and Kivelson[63] by pointing out that electron-electron correlation alone can result in electron pairing on the C_{60} molecule, leading to superconductivity of a different nature than BCS. This fascinating possibility at present cannot be either proved or disproved on the bases of the presently available experimental data—even the observed isotope effect[64,65] on T_c can be accounted for in this way as well.[66] This model seems to provide a more natural starting point for the understanding of the observed tendency of local moment formation[55] (independently of whether these momenta are intrinsic of the superconducting compound or a character of some close-lying 'second phase' to be identified) and the magnetic transition[58] observed in the TDAE C_{60}.

The small value of superconducting coherence length ξ_o, of the order of the molecule size, never found before in isotrope superconductors, the relatively large normal state resistivity[30] (so far) obtained (~ 5 mΩ), higher than the maximum metallic resistivity estimated by Mott, lead to a near concensus in the literature, that the attractive interaction responsible for the pairing of the electrons is most probably connected to a molecular property: either an intramolecular phonon mode coupled to the electrons[60,61] or a more intimate nature of the electron-electron correlation on the molecule.[63] Either way, the intermolecular hopping (bandwidth) plays only a secondary role in establishing superconductivity, and the observed T_c lattice parameter relation can be equally explained, at least qualitatively. With the range of T_c − s expanding by the discoveries of new materials, some specific models may work better than others. In view of this unusual situation, more should be learned about the excitations of the individual molecule in its different anionic states, $[C_{60}]^{n-}$, $n = 1$–4, and how these excitations influence the observed macroscopical properties. Experiments probing the materials on a more microscopic level (like NMR and ESR) and transport measurements on single crystals will hopefully bring deeper insight.

7. CONCLUSIONS

The little more than two years passed since the new form of carbon, especially C_{60}, has been available have provided an unexpected gift for the solid state physicist—a new family of superconducting materials. A growing number of simple compounds based on C_{60} with mono- and divalent metals challenge our understanding of superconductivity. The hope of separating carbon cage molecules containing different atoms (endohedral fullerenes) promises more exciting materials to come.

We have attempted to provide an early summary of the properties of A_3C_{60} superconducting compounds. The most fundamental question, whether there is a need or not to introduce special elementary excitations, quasi-particles, at the level of the C_{60} molecules and compose special statistics of them to explain the macroscopical superconductivity of these materials is still an open question. As we have discussed in the introduction and illustrated throughout the paper, the conventionally separate energy scales are certainly sufficiently overlapping that it is realistic to think about a different set of quantum numbers than the traditional ones (inter- and intra-molecular phonons, electron bands described in k space, etc.) to be more appropriate to describe the properties of these fascinating materials.

Acknowledgements—We would like to thank R. Peccei, G. Grüner, M. Héritier, and H. Alloul for their continuing support and interest. This review paper was born out of a long-standing cooperation with our colleagues from UCLA-LANL, namely: S.-M. Huang, R. B. Kaner, F. Diederich, G. Grüner, J. D. Thompson, G. Sparn, W. G. Clark, F. Ettl, J. Wiley, O. Klein, S. Donovan, and H.-W. Jiang. We are also thankful for cooperation with several groups, lead by: F. Wudl, Y. Uemura, P. Stephens, L. Mihály, H. Alloul, P. Petit, J.-J. André, and permission to quote from our common work prior to its publication. We gained from discussions with many colleagues, especially P. Monod, S. Kivelson, S. Chakravarty, J. Fischer, R. L. Green, R. C. Haddon, A. Hebard, R. Tycko, G. Kriza, and C. S. Yannoni. K. H. is especially thankful to P. Petit for his help in preparing the present manuscript.

REFERENCES

1. W. Krätschmer, L. D. Lamb, K. Fostiropoulos, and D. R. Huffman, *Nature* **347**, 354–358 (1990).
2. H. W. Kroto, J. R. Heath, S. C. O'Brien, R. F. Curl, R. E. Smalley, *Nature* **318**, 162 (1985).
3. *Mat. Res. Soc. Symp. Proc.* **206** (1991).
4. *Acc. Chem. Res.* **25** (1991).
5. F. Diederich and R. L. Whetten, *Acc. Chem. Res.* **25**, 121 (1991).
6. A. F. Hebard, M. J. Rosseinsky, R. C. Haddon, D. W. Murphy, S. H. Glarum, T. T. M. Palstra, A. P. Ramirez, and A. R. Kortan, *Nature* **350**, 600 (1991). R. C. Haddon, A. F. Hebard, M. J. Rosseinsky, D. W. Murphy, S. J. Duclos, K. B. Lyons, B. Miller, J. M. Rosamilia, R. M. Fleming, A. R. Kortan, S. H. Glarum, A. V. Makhija, A. J. Muller, R. H. Eick, S. M. Zahurak, R. Tycko, G. Dabbagh, and F. A. Thiel, *Nature* **350**, 320 (1991).
7. D. W. Murphy, M. J. Rosseinsky, R. M. Fleming, R. Tycko, A. P. Ramirez, R. C. Haddon, T. Siegrist, G. Dabbagh, J. C. Tully, and R. E. Walstedt, *J. Chem. Phys. Solids* (in press).
8. M. J. Rosseinsky *et al.*, *Phys. Rev. Lett.* **66**, 2830 (1991).
9. K. Holczer, O. Klein, S.-M. Huang, R. B. Kaner, K.-J. Fu, R. L. Whetten, and F. Diederich, *Science* **252**, 1154 (1991).
10. K. Tanigaki *et al. Nature* **352**, 222 (1991).
11. C-C. Chen, S. P. Kelty, and C. M. Lieber, *Science* **253**, 886 (1991).
12. R. M. Fleming *et al., Nature* **352**, 787 (1991).
13. K. Holczer, G. R. Chalmers, J. B. Wiley, S.-M. Huang, R. B. Kaner, C. E. Strouse, F. Diederich, R. L. Whetten, *Synthetic Metals* (in press).
14. P. W. Stephens, L. Mihály, J. B. Wiley, S.-M. Huang, R. B. Kaner, F. Diederich, R. L. Whetten, and K. Holczer, *Phys. Rev. B* **45**, 543 (1992).
15. M. J. Rosseinsky, D. W. Murphy, R. M. Fleming, R. Tycko, A. P. Ramirez, T. Siegrist, G. Dabbagh, and S. E. Barratt, *Nature* **356**, 417 (1992).
16. K. Tanigaki, I. Hirosawa, T. W. Ebbesen, J. Mizuki, Y. Shimakawa, Y. Kubo, J. S. Tsai and S. Kuroshima, *Nature* **356**, 419 (1992).
17. Q. Zhu, O. Zhou, N. Coustel, G. Vaughan, B. M. Gavin, J. P. McCauley, W. J. Romanow, J. E. Fischer, and A. B. Smith, *Science* **254**, 545 (1991).
18. G. Sparn, J. D. Thompson, S.-M. Huang, R. B. Kaner, F. Diederich, R. L. Whetten, G. Gruner, K. Holczer, *Science* **252**, 1829 (1991).
19. G. Sparn, J. D. Thompson, R. L. Whetten, S.-M. Huang, R. B. Kaner, F. Diederich, G. Grüner, and K. Holczer, *Phys. Rev. Lett.* **68**, 1228 (1992).
20. K. Holczer, O. Klein, G. Gruner, J. D. Thompson, F. Diederich, R. L. Whetten, *Phys. Rev. Lett.* **67**, 271 (1991).
21. C. E. Johnson, H. W. Jiang, K. Holczer, R. B. Kaner, R. L. Whetten, and F. Diederich, *Phys. Rev. B.* (in press).
22. P. W. Stephens, L. Mihály, P. L. Lee, R. L. Whetten, S.-W. Huang, R. Kaner, F. Diederich, and K. Holczer, *Nature* **351**, 632 (1991).
23. Y. J. Uemura, A. Keren, L. P. Le, G. M. Luke, B. J. Sternlieb, W. D. Wu, J. H. Brewer, R. L. Whetten, S. M. Huang, S. Lin, R. B. Kaner, F. Diederich, S. Donovan, G. Grüner, and K. Holczer, *Nature* **352**, 605 (1991).
24. Y. Uemura *et al. Proceedings of ICSM '92*, Göteborg (submitted for publication).
25. W. H. Wong, M. Hanson, W. G. Clark, G. Grüner, J. D. Thompson, R. L. Whetten, S.-M. Huang, R. B. Kaner, F. Diederich, P. Petit, J.-J. Andre, and K. Holczer *Europhys. Lett.* **18**, 79 (1992).
26. R. Tycko, *et al. Science* **253**, 884 (1991).
27. R. Tycko, G. Dabbagh, M. J. Rosseinsky, D. W. Murphy, A. P. Ramirez, and R. M. Fleming, *Phys. Rev. Lett.* **68**, 1912 (1992).
28. K. Holczer, O. Klein, F. Hippert, H. Alloul, S.-M. Huang, R. B. Kaner, R. L. Whetten (submitted for publication).
29. G. P. Kochanski, A. F. Hebard, R. C. Haddon, and A. T. Fiory, *Science* **255**, 184 (1992); T. T. M. Palmstra, R. C. Haddon, A. F. Hebard, and J. Zaanan, *Phys. Rev. Lett.* **68**, 1054 (1992).
30. X.-D. Xiang, *et al., Science* **256**, 1190, (1992).
31. J. G. Bednorz and K. A. Müller, *Z. Phys.* **B64**, 189, (1986).
32. *Organic superconductors* (Edited by V. Kresin and W. A Little). Planum Press, New York (1991).
33. P. G. deGennes *Superconductivity of metals and alloys.* Benjamin (1966). M. Tinkham, *Introduction to superconductivity.* McGraw-Hill, (1975).
34. P. W. Anderson, *Science* **235**, 1196 (1987); *Mat. Res. Bull.* **8**, 153 (1973).
35. S. Saito and A. Oshiyama, *Phys. Rev. Lett.* **66**, 2637 (1991).
36. S. Satpathy, V. P. Antropov, O. K. Andersen, O. Jepsen, O. Gunnarsson, and A. I. Lichtenstein, *Phys. Rev. B*, (1992).
37. R. C. Haddon, L. E. Brus, and K. Raghavachari, *Chem. Phys. Lett.* **125**, 459 (1986). **131**, 165 (1986).
38. C. Reber, L. Yee, J. McKiernan, J. I. Zink, R. S. Williams, W. M. Tong, D. A. Ohlberg, R. L. Whetten, and F. Diederich, *J. Phys. Chem.* **95**, 2127 (1991). A. Y. Saito, H. Shinohara, M. Kato, H. Nagashima, M.

Ohkohchi, and Y. Ando, *Chem. Phys. Lett.* **189**, 236 (1992).

39. A. Tokmakoff, D. R. Haynes, and S. M. George, *Chem. Phys. Lett.* **186**, 450 (1991).
40. S. Duclos et al., *Nature* **351**, 380 (1991).
41. C. T. Chen et al., *Nature* **352**, 603, (1991).
42. R. Fleming, *et al., Mat. Res. Soc. Symp. Proc.* **206**, (1991).
43. W. I. F. David, R. M. Ibberson, J. C. Matthewman, K. Prassides, T. J. Dennis, J. P. Hare, H. W. Kroto, R. Taylor, and D. R. M. Walton, *Nature* **353**, 147 (1991).
44. D. A. Neumann, J. R. D. Copley, W. A. Kamitakahara, J. J. Rush, R. L. Cappelletti, N. Coustel, J. P. McCauley, J. E. Fischer, A. B. Smith, K. M. Creegan, and D. M. Cox, *J. Chem. Phys.* (submitted). W. P. Beyerman, M. F. Hundley, J. D. Thompson, F. N. Diederich, and G. Grüner, *Phys. Rev. Lett.* **68**, 2046 (1992).
45. S. Pekker, G. Faigel, K. Foder-Csorba, L. Granasy, E. Jakab, and M. Thegze, *Solide State Commun.* (in press).
46. K. Holczer et al., (submitted for publication).
47. K. Holczer et al., (unpublished manuscript).
48. O. Zhou, G. B. M. Vaughan, Q. Zhu, J. E. Fischer, P. A. Heiney, N. Coustel, J. P. McCauley, and A. B. Smith, *Science* **255**, 833 (1992).
49. S. S. Xie, Z. B. Zhang, W. H. Yang, and B. Ren, *Solid State Commun.* (in press).
50. A. R. Kortan, N. Kopylov, S. Glarum, E. M. Gyorgy, A. P. Ramirez, R. M. Fleming, F. A. Thiel, and R. C. Haddon, *Nature* **355**, 529 (1992).
51. T. T. M. Palstra, R. C. Haddon, A. F. Hebard, and J. Zaanen, *Phys. Rev. Lett.* **69**, 1054 (1992).
52. N. R. Werthamer, E. Helfand, and P. C. Hohenberg, *Phys. Rev.* **147**, 295 (1966).
53. L. D. Rotter, Z. Schlesinger, J. P. McCauley, N. Coustel, J. E. Fischer, and A. B. Smith, *Nature* **355**, 532 (1992).
54. Z. Zhang, C.-C. Chen, and C. M. Lieber, *Science* **254**, 1619 (1991).
55. K. Holczer, P. Petit, and P. Höfer (submitted for publication).
56. P.-M. Allemand, G. Srdanov, A. Koch, K. Khemani, F. Wudl, Y. Rubin, F. Diederich, M. M. Alvarez, S. J. Anz, and R. L. Whetten, *J. Am. Chem. Soc.* **113**, 2780 (1991).
57. P. Höfer, P. Petit, and K. Holczer (submitted for publication).
58. P.-M. Allemand et al. *Science* **253**, 301 (1991). P. Stephens et al. *Nature* **355**, 331 (1992).
59. C. S. Yannoni et al. *Synthetic Metals* (in press).
60. M. Schluter et al., *Phys. Rev. Lett.* **68**, 526, (1992).
61. C. M. Varma, J. Zaanen, and K. Raghavachari, *Science* **254**, 989 (1991).
62. K. Prassides, J. Tomkinson, C. Christides, M. J. Rosseinsky, D. W. Murphy, and R. C. Haddon, *Nature* **354**, 462 (1991).
63. S. Chakravarty and S. Kivelson, *Europhys. Lett.* **16**, 751 (1991). S. Chakravarty, M. Gelfand, and S. Kivelson, *Science* **254**, 970 (1991).
64. Ebbesen et al. *Nature* **355**, 620 (1992).
65. A. P. Ramirez, A. R. Kortan, M. J. Rosseinsky, S. J. Duclos, A. M. Mujsce, R. C. Haddon, D. W. Murphy, A. V. Makhija, S. M. Zahurak, and K. B. Lyons, *Phys. Rev. Lett.* **68**, 1058 (1992).
66. S. Chakravarty, S. A. Kivelson, M. I. Salkola, and S. Tewari, *Science* **256**, 1306 (1992).

FULLERENES AND FULLERIDES IN THE SOLID STATE: NEUTRON SCATTERING STUDIES

K. Prassides, H. W. Kroto, R. Taylor and D. R. M. Walton
School of Chemistry and Molecular Sciences, University of Sussex, Brighton BN1 9QJ, U.K.

W. I. F. David, J. Tomkinson and R. C. Haddon
Rutherford Appleton Laboratory, Didcot, Oxon OX11 0QX, U.K.

M. J. Rosseinsky and D. W. Murphy
AT&T Bell Laboratories, Murray Hill, NJ 07974, U.S.A.

(*Accepted* 28 *April* 1992)

Abstract—Fullerene-60 shows a high molecular symmetry, consistent with the icosahedral point group I_h. At low temperature in the solid state, high-resolution powder neutron diffraction reveals that crystalline C_{60} (local symmetry S_6) adopts a simple cubic crystal structure whose stability is driven by optimisation of the intermolecular electrostatic interactions. Above 90 K, molecular motion is no longer frozen and the molecules shuffle between nearly degenerate orientations, differing in energy by 11.4(3) meV. At 260 K, a first-order phase transition leads to a face-centred cubic structure, characterised by rapid isotropic reorientational motion of the molecules. The phonon spectra of pristine fullerene, superconducting K_3C_{60} and saturation-doped Rb_6C_{60} measured by inelastic neutron scattering in the energy range 20–2000 cm^{-1}, reveal substantial broadening of fivefold degenerate H_g intramolecular vibrational modes both in the low-energy radial and the high-energy tangential part of the spectrum. This provides strong evidence for a traditional phonon-mediated mechanism of superconductivity in the fullerides but with an electron–phonon coupling strength distributed over a wide range of energies (33–195 meV) as a result of the finite curvature of the fullerene spherical cage.

Key Words—Fullerenes, neutron scattering, orientational order, superconductivity, electron–phonon coupling.

1. INTRODUCTION

The isolation of crystalline fullerene samples from arc-processed graphite[1] and their successful chromatographic separation[2] led almost overnight to an exciting new era of carbon chemistry, physics, and materials science. Fullerenes ceased to be the scientific curiosities, originally proposed to account for the intense features observed in the mass spectra of the products of the laser vapourisation of graphite[3]. The early research efforts naturally focussed in characterising the materials and confirming the originally proposed structures: truncated icosahedral (I_h) for fullerene-60 and ellipsoidal (D_{5h}) for fullerene-70. Soon the availability of significant quantities of fullerenes opened the way towards the synthesis of numerous derivatives with equally exciting properties. For instance, intercalation of solid C_{60} with electron donors, like the alkali and alkaline earth metals, can lead to metallic compositions, which become superconducting[4–6] at critical temperatures as high as 33 K, surpassed only by the high T_c superconducting cuprates.

In this paper, we review our recent neutron scattering studies on solid fullerene-60 and the alkali-metal doped fullerides K_3C_{60} and Rb_6C_{60}, performed at the Rutherford Appleton Laboratory, UK. A review of neutron scattering studies on solid C_{60} and its compounds emphasizing work performed at the National Institute of Standards and Technology is also soon to appear[7]. Neutron diffraction measure-ments allowed us to extend earlier X-ray work[8] on the structural properties of solid C_{60}, leading to a more complete understanding of the low-temperature crystal structure, its intermolecular bonding implications and the structural phase transitions accompanying orientational ordering of the C_{60} molecules. Inelastic neutron scattering measurements, on the other hand, have provided a detailed picture of the vibrational spectra of C_{60}, superconducting K_3C_{60} and insulating Rb_6C_{60}, revealing a plethora of additional features to those observed by optical spectroscopy. Such knowledge of the experimental vibrational properties of $C_{60}{}^{n-}$ ($n = 0,3,6$) is very important for both the theoretical description of the electronic structure of fullerenes and their implications towards the possible mechanism for superconductivity in the fullerides, providing evidence for or against the participation of phonons in the pairing interaction.

2. EXPERIMENTAL

2.1 *Neutron scattering*

When a neutron impinges on a nucleus, several interactions lead to a variety of different scattering processes[9–11]. The neutron may be transmitted, scattered, or absorbed. The scattering process may be either elastic or inelastic [i.e., it may involve transfer of energy to (neutron energy gain) or from the scattered neutron (neutron energy loss)]. The above processes can be either coherent, arising by scattering in a regular way from the lattice, or incoherent, arising from nonperiodic fluctuations in scattering power. The static structure of a solid gives rise to elastic scattering; the wavelength of a thermal neutron, $\lambda = h/m_n v$ is of the same order of magnitude as the interatomic distances in solids and neutron diffraction may be used for the study of crystal structures. The dynamical behaviour of solids gives rise to inelastic neutron scattering; because of their finite mass, the nuclei gain or lose energy when neutrons collide with them; such energy is transferred from or to crystal vibrations. The use of neutrons in the study of molecular and crystal vibrations stems from the fact that their energies are of the same order of magnitude as those of the phonon modes in solids. In the scattering process, both the exchange of energy, $\hbar\omega$ and momentum, $\hbar Q$ can be observed. From the conservation laws for energy and momentum, we have:

$$\hbar\omega = E - E', \qquad \mathbf{Q} = \mathbf{k} - \mathbf{k}', \qquad (1)$$

where $\mathbf{k}$ and $\mathbf{k}'$ are the wavevectors of incident and scattered neutrons, respectively, E and E' are the initial and final neutron energies, respectively, and $\mathbf{Q}$ is the scattering vector.

Information about the phonon modes in solids may be derived from either optical or neutron spectroscopic techniques. IR and Raman spectroscopies are characterised by excellent energy resolution; however, they are confined to modes close to the Brillouin zone centre ($\mathbf{k}' \approx \mathbf{k}$, $\mathbf{Q} \approx 0$). In neutron spectroscopy, even though the energy resolution is considerably poorer, information across the Brillouin zone is obtained routinely and the observed intensities may be related to atomic displacements in a straightforward manner. Furthermore, no selection rules are present so that optically inactive modes are also observed. For anisotropic oscillators in powder samples, the scattering law $S(Q,\omega)$ may be expressed as[12]:

$$S(Q,\omega)_\nu \propto (\tfrac{1}{3})(Q^2 Tr\mathbf{B}_\nu) \exp(-Q^2\alpha_\nu), \qquad (2)$$
$$\alpha_\nu \approx (\tfrac{1}{5})\{Tr\mathbf{A} + 2(\mathbf{B}_\nu : \mathbf{A}/Tr\mathbf{B}_\nu)\},$$

where the total mean square displacement is given by $\mathbf{A} = \Sigma_\nu \mathbf{B}_\nu$, $\mathbf{B}_\nu$ is the mean square displacement tensor of the scattering atom in internal mode ν, and $Q = |\mathbf{Q}|$ is the momentum transferred during scattering. For a simple harmonic isotropic oscillator, the scattering law may be simplified as: $S(Q,\omega) = (Q^2B) \exp(-Q^2B)$, with the mean square displacement of the oscillator given by: $B = (\hbar/2\mu\omega)$ and μ the reduced mass and ω the fundamental frequency.

2.2 Instrumental details

Diffraction profiles of pristine C_{60} were recorded on the high-resolution powder diffractometer (HRPD) at the ISIS pulsed neutron source, Rutherford Appleton Laboratory, UK over a time-of-flight range 30,000–230,000 μs (d-spacing range: 0.6–3.2 Å, instrumental resolution, $\Delta d/d \approx 8 \cdot 10^{-4}$). A series of shorter runs was also recorded in 10-K steps

between 5 and 320 K (tof range: 40,000–115,000 μs). Data analyses were performed using the standard TF12LS powder diffraction package. Inelastic neutron scattering spectra of C_{60}, K_3C_{60}, and Rb_6C_{60} at low temperatures were recorded in the energy range 2.5–200 meV (1 meV = 8.066 cm^{-1}) on the time-focused crystal analyser spectrometer (TFXA) at ISIS (instrumental resolution, $\Delta\omega/\omega \approx 2\%$). In order to achieve such a resolution, the spectrometer uses a very low final neutron energy ($E' \approx 4$ meV); consequently, eqn (1) results in:

$$\hbar\omega \approx E, \qquad \mathbf{Q} \approx \mathbf{k} \text{ for large incident energies.}$$

Since $E \propto k^2$, this leads to $Q^2 \propto \omega$. The consequence of these instrumental characteristics is that even for principally coherent scatterers like carbon, the incoherent approximation may be used for energy transfers $\hbar\omega > 25$ meV; however, at high energy transfers (>150 meV), the resulting Debye–Waller factors are so large that they tend to wash out some of the vibrational features.

3. STRUCTURE OF PRISTINE C_{60}

Early X-ray diffraction work[1,8,13] revealed that at room temperature, the C_{60} crystalline powder consisted of spheroidal molecules of diameter 7.1 Å, forming a random mixture of hexagonal close-packed (hcp) and face-centred cubic (fcc) arrays. However, careful elimination, by sublimation, of solvent molecules trapped in interstitial cavities leads only to a fcc crystal structure in which each C_{60} molecule is orientationally disordered. The powder neutron[14] and X-ray[8] diffraction data are consistent with the space group $Fm\bar{3}m$, with all four quasi-spherical C_{60} units being symmetry equivalent; the cubic lattice constant at 290 K is 14.1569(5) Å. A variety of experimental techniques, including NMR[15–17], quasielastic neutron scattering[18], and μSR[19] measurements, revealed that in this plastic "rotator" phase, the C_{60} molecules rotate freely, effectively randomly with no time-averaged preferred orientation in space. Rotational correlation times τ_{ROT} of 9–12 ps at room temperature are only three to four times longer than expected for unhindered gas phase rotation and faster than any other known solid state rotor[16–18]; furthermore, C_{60} reorients even faster in the plastic phase than in tetrachloroethane solution ($\tau \approx 15$ ps). Thus the high temperature description of pristine C_{60} that has emerged is of a prototypical plastic crystal (cf. adamantane, norbornane, etc) with a well-defined translational order and a smooth rotational potential with many shallow minima, leading to continuous small-angle molecular motions.

As the temperature is lowered, the rotational properties of the C_{60} molecules become more complicated. In analogy with adamantane which shows a phase transition at 208.6 K from a high-temperature fcc $Fm\bar{3}m$ to an orientationally tetragonal $P\bar{4}2_1 c$

structure[20], differential scanning calorimetry (DSC) measurements of C_{60}[21] first established the existence of a phase transition near 250 K that was subsequently confirmed by means of X-ray diffraction[8]. New reflections appear indexing on a simple cubic unit cell. The results were interpreted as manifesting the orientational ordering of the four C_{60} molecules in the unit cell which ceased to be symmetry equivalent, as orientational long-range order developed. Sachidanandam and Harris[22] proposed the cubic $Pa\overline{3}$ space group for the low-temperature crystal structure of C_{60}; starting from an ideal $Fm\overline{3}$ space group with the four $\langle 111 \rangle$ axes coinciding with threefold and the $\langle 100 \rangle$ axes with twofold symmetry axes of the molecule, the equilibrium configuration was found for a $\phi \approx 94°$ rigid anticlockwise rotation about the [111] direction. This finding was in strong disagreement with theoretical calculations[23], employing van der Waals interactions and graphitic-type potentials describing the intracage bonding; a lowering in symmetry to orthorhombic was predicted upon orientational ordering at low temperatures.

High-resolution neutron diffraction measurements on a highly crystalline, solvent-free C_{60} sample were employed for the structure determination at 5 K[24]. Rietveld profile refinement essentially confirmed the $Pa\overline{3}$ space group (a = 10.0408(1) Å) and resulted in an improved value for the rotation angle about [111] of $\phi \approx 98°$. Closer inspection of the relative orientations of neighbouring C_{60} molecules (of S_6 local symmetry) in the lattice (Fig. 1), readily reveals the reason for which theoretical attempts to model the intermolecular interactions had failed. C_{60} is indeed a quasispherical molecule and consequently, the intermolecular energy is dominated by the van der Waals contributions. However, a fundamental difference from graphite that forms the *raison d' être* of C_{60} itself[3] is the presence of isolated pentagonal units on the surface of the sphere. The result

is that there exist two types of C–C bonds, long (fusions of a five- and a six-membered ring) and short (fusions of two six-membered rings), of differing π order, leading to an anisotropic electronic distribution. Thus bonding models developed for the graphitic sheets ought to be modified accordingly in order to be applicable to the fullerenes. Figure 2a shows the intermolecular C_{60}–C_{60} nearest contact as viewed along the centre-to-centre direction; a high π-order six–six ring fusion is positioned almost parallel ($\approx 179.6°$) to the pentagonal face which is the most "electron-deficient" region of the surface of the fullerene cage, leading to a subtle optimisation of the molecular electrostatic potential. The seemingly arbitrary rotation angle of 98° in $Pa\overline{3}$ thus ensures all twelve nearest-neighbour interactions are optimised with six of the 30 short bonds and six of the 12 pentagonal faces of each molecular unit taking part in the intermolecular interactions. The C–C bonds range from 1.379(10) to 1.485(10) Å with average values for the six–six ring fusions of 1.404(10) Å and for the five–six ring fusions of 1.448(10) Å, in excellent agreement with the values derived from NMR data[24] of 1.400(15) and 1.450(15) Å, respectively. More refined theoretical calculations, based on our proposed model for the intermolecular interactions have now appeared[26–28]. Explicit consideration of the electrostatic interaction is taken into account by assigning different partial bond charges to the high ($q \approx -0.54e$) and low ($q \approx 0.27e$) π-order C–C bonds. The low temperature global energy minimum is now found to be the experimentally observed $Pa\overline{3}$ crystal structure with a rotation angle of $\phi \approx 98.7°$[26]. Alternatively, Sprik et al.[28] arrive in a similar global minimum by assigning a negative charge of $-0.35e$ to the high π-order bonds and positive charges (0.175e) to the carbon atoms. More recent experimental diffraction work has also provided further support for the $Pa\overline{3}$ structure[29,30].

NMR studies[16,17] of the rotational dynamics of C_{60} revealed that at the phase transition there is a dramatic change in the nature of the dynamics: the molecular motion in the low-temperature phase was described as jumps between symmetry-equivalent orientations. Figure 3 shows the temperature evolution of the lattice constant of the cubic unit cell, as derived from high-resolution neutron diffraction measurements[14]. The signature of the first-order phase transition, accompanying the change in the rotational dynamics is apparent at 260 K, where a sudden change of 0.344(8)% in the lattice dimensions occurs. Coexistence of the two phases is also observed at T_c with the present sample showing a fcc:sc ratio of $\approx 45{:}55$ (Fig. 4). The lattice constant smoothly decreases below 260 K, albeit with a distinct curvature until at 90 K, a well-defined cusp in its variation appears. The smooth contraction then continues down to 5 K. The cusp at 90 K is the signature of a higher-order phase transition, arising from the freezing of the jump motion of the C_{60} molecules. Since both experimental diffraction[24] and theoretical[26–28]

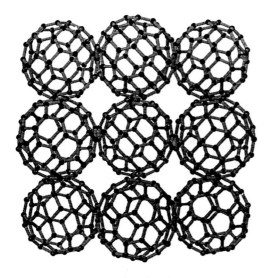

Fig. 1. Basal plane projection of the low temperature structure of C_{60} (space group $Pa3$).

markable agreement with the neutron diffraction results. However, the present two-state model is certainly only an approximation of the description of the rotational dynamics of C_{60} below 260 K; the existence of other local minima (e.g., configurations resulting from 36° jumps[26] about the twofold [110] axes) with comparable energies necessitates a more complicated picture. Similarly the performance by the molecules of small angular oscillations[33] about their equilibrium orientations should be taken into account to explain the presence of diffuse scattering (peaking at $Q \approx 3.5$ and 5.8 Å^{-1}) even at 5 K. Using a Monte-Carlo simulation, Copley *et al.*[7] find that a rms librational amplitude of 1.9° and a fraction $p \approx 0.833$ of the majority phase can account well for both the observed Bragg and diffuse scattering at 5 K.

4. INELASTIC NEUTRON SCATTERING AND SUPERCONDUCTIVITY

The appearance of superconductivity in alkali-doped A_3C_{60}[4–6] compounds naturally led to considerable efforts in attempting to understand the origin of the high T_c's in these materials (cf. the much smaller T_c's in alkali-doped graphite intercalates) and the mechanism responsible for the pair formation. The fullerides are potentially simpler materials than the high-T_c cuprates, and thus more easily amenable to theoretical modelling. Theoretical treatments of superconductivity in A_3C_{60} have considered purely electronic[34–36] as well as phonon-induced pairing, involving either low-energy ($\approx$10-20 meV) A^+-C_{60}^{3-} optic modes[37] or the high-energy intramolecular modes[38–40]. The electronic mechanisms for pair-binding derive from the resonating valence bond (RVB) picture of Pauling and Anderson and focus on the strong intrafullerene electron–electron repulsions. On the other hand, phonon-induced mechanisms[37–42] follow the classic route by ascribing the effective attractive interaction between electrons to their strong scattering near the Fermi surface by the vibrations of the ions. Experimental evidence derived from the relation between lattice dimensions and superconducting transition temperatures in A_3C_{60} both at ambient[43] and at high pressures[44] is consistent with T_c being modulated by the density-of-states at the Fermi level $N(\epsilon_F)$. Furthermore, a strong alkali-specific effect is not observed, supporting the importance of the intramolecular fullerene rather than the external intercalate-fullerene phonons. From a BCS-type relation,

$$T_c \propto \hbar\omega_{ph} \exp\left[-\frac{1}{VN(\epsilon_F)}\right]$$
$$= \hbar\omega_{ph} \exp(-1/\lambda) \quad (4)$$

the observed high T_c's may be understood in terms of a high average phonon frequency $\hbar\omega_{ph}$, resulting from the light carbon mass and the large force constants associated with the intramolecular modes, a

high DOS at ϵ_F, resulting from the weak intermolecular interactions and strongly scattering intramolecular modes. Strong support for a phonon-mediated pair-binding interaction has now come from isotope-effect measurements which find isotope exponents α equal to 0.37 ± 0.05 for Rb_3C_{60}[45] and 0.30 ± 0.06 for K_3C_{60}[46] in the expression that governs the T_c variation with ionic mass M, $T_c \propto M^{-\alpha}$. These exponents are reduced from the ideal BCS value of $\alpha \approx 0.5$, presumably because of significant Coulomb interaction effects.

For a phonon-mediated superconductor, the superconducting properties may be derived from a knowledge of the Eliashberg spectral function $\alpha^2F(\omega)$, where $F(\omega)$ is the phonon density-of-states and α^2 is an effective electron–phonon coupling strength. This spectral function can be expressed in terms of the phonon linewidths $\gamma_{\nu Q}$, in principle experimentally available through neutron scattering measurements at low temperatures[47]:

$$\alpha^2F(\omega) \propto N(\epsilon_F)\omega \, \Sigma_{\nu Q} \, \gamma_{\nu Q} \, \delta(\omega - \omega_{\nu Q}). \quad (5)$$

The molecular nature of the superconducting fullerides results in fairly dispersionless phonon branches, and consequently, the electron–phonon Hamiltonian[38,39] has a particularly simple form. Then the total electron–phonon coupling constant λ may be simply expressed as a sum of partial contributions λ_ν, associated with the each mode ν mediating the pairing interaction[39], $\lambda = \Sigma_\nu\lambda_\nu$. The situation can be further simplified if we focus our attention on the t_{1u} electronic states which are occupied on electron-doping of the C_{60} fullerene. Using simple symmetry arguments[38,39], it can be shown that in icosahedral symmetry the only relevant intramolecular modes that can couple to the t_{1u} conduction electrons ($t_{1u} \otimes t_{1u} = A_g \oplus T_{1g} \oplus H_g$) are of H_g symmetry, while the totally symmetric A_g modes become active only for finite wavevectors ($Q \neq 0$). Strong electron–phonon coupling should produce substantial broadening and softening of the affected intramolecular H_g modes in the superconducting fullerides, compared to pristine C_{60}. Quantitatively such effects on the position and width may be estimated using the expressions:

$$\Delta\omega_\nu \approx -(\lambda_\nu/5)\omega_\nu, \qquad \gamma_\nu \approx (\pi/5)N(\epsilon_F)\lambda_\nu\omega_\nu^2, \quad (6)$$

where $\Delta\omega_\nu$ is the change in frequency upon reduction to C_{60}^{3-} and γ_ν is the increase in full-width at half-maximum of the fivefold degenerate νth phonon.

The INS spectra of superconducting K_3C_{60}[48] and Rb_3C_{60}[49], insulating C_{60}[50–52] and over-doped Rb_6C_{60}[53]have all been measured at low temperatures (Fig. 6). The vibrational spectra of C_{60} and its reduced forms C_{60}^{3-} and C_{60}^{6-}, are extremely rich in structure, as expected for such a large 60-atom moiety. There are 174 intramolecular modes giving rise to 46 fundamentals which can be classified, using the I_h point group, as follows:

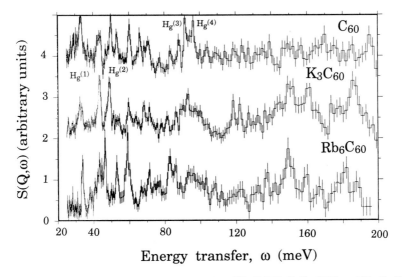

Fig. 6. INS vibrational spectra of the intramolecular modes of C_{60} (20 K), K_3C_{60} (5 K), and Rb_6C_{60} (20 K) in the energy region 25–200 meV.

$$\Gamma_{vib} = 2A_g + A_u + 3T_{1g} + 4T_{1u} + 4T_{2g} + 5T_{2u} + 6G_g + 6G_u + 8H_g + 7H_u. \quad (7)$$

From these, four are infrared active (T_{1u}) and ten are Raman active (A_g, H_g), leaving 32 optically inactive modes. INS measurements are not restricted by the usual optical selection rules and consequently, the full vibrational spectrum of $C_{60}{}^{n-}$ ($n = 0, 3, 6$), containing information on all 46 normal modes of vibration may be recorded. In addition, there is information on the intermolecular (external) low-energy modes. The intramolecular modes are not expected to show significant dispersion (except for the low-energy modes, which may show dispersion up to 2.5 meV[54]), permitting direct comparisons between the INS data and the available zone-centre theoretical and experimental data to be made. Furthermore, considerations of reduced optical penetration depth in K_3C_{60}, important in Raman and IR spectroscopy, are not relevant here.

The vibrational spectra may be divided into two principal regions: 0–30 meV (lattice modes) and 30–210 (intramolecular modes). There is an energy gap of the order of ≈ 10–15 meV separating inter- from intramolecular modes. The intramolecular part of the spectrum extends smoothly from 30 to 210 meV. C_{60} lacks a central atom and consequently, radial forces are weak and the cage vibrations may be distinguished into radial (≈ 30–110 meV) and tangential (≈ 110–200 meV) ones. As a result of the carbon cage reduction $C_{60} \rightarrow C_{60}{}^{3-} \rightarrow C_{60}{}^{6-}$, the high-energy cutoff in the vibrational DOS softens substantially from 205.0 meV through 198.0 meV (3.4%) to 190.2 meV (7.2%), whereas the low-energy onset is barely affected, showing a hardening (0.8 meV) for $C_{60}{}^{6-}$; this reflects the pronounced effect of reduction on the vibrational modes involving substantial stretching components, particularly of the high π-order six-six ring fusions. Table 1 summarises positions and (tentative) assignments of some vibrational modes. We find excellent agreement with the optical results. Inactive modes are assigned to specific vibrations, using the results of theoretical calculations. In particular, the quantum chemical calculations of Negri et al.[55] are in respectable agreement with the experimental results.

Phonon modes are well separated in the low-energy (30–90 meV) range and changes in position and width can be followed with confidence as a function of the reduction level of the fullerene cage. The INS data provide evidence of strong electron–phonon

Table 1. Vibrational spectra of C_{60} (20 K), K_3C_{60} (5 K), and Rb_6C_{60} (20 K) with tentative assignments. All energies are given in meV (1 meV = 8.066 cm^{-1}). Asterisks denote optically active (IR, Raman) vibrational modes. Assignments of the optically inactive modes were made using the results of the quantum chemical calculations of Negri et al.[55]

Assignment	C_{60} Ref. [50]	K_3C_{60} Ref. [48]	Rb_6C_{60} Ref. [53]
Lattice modes	—	4.3, 14.0	6.4, 8.4, 13.1
$H_g^{(1)*}$	33.1	32.7	33.9
T_{2u}, G_u	42.7, 44.3	43.7	43.7
H_u	50.0	49.2	46.6
$H_g^{(2)*}$	53.2	52.5	53.1
G_g, A_g^*	60.3	59.9	59.1
T_{1u}^*, $(G_g)(?)$	66.1	63.6, 68.5	66.7, 68.2
$(T_{1g}, H_u)(?)$, T_{1u}^*	(68.6), 70.7	71.6	71.2
$T_{2u}(?)$	83.4	81.4	82.7
$H_g^{(3)*}$	88.2	86.5	85.9
$H_u(?)$	91.9	—	—
$H_g^{(4)*}$	96.2	94.0	—
Energy cutoff	205.0	198.0	190.2

$H_g^{(1)}$ $H_g^{(2)}$

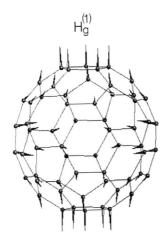   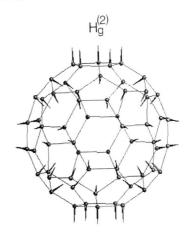

Fig. 7. Schematic representation of the two lowest-energy H_g modes of the C_{60} cage, using the eigenvectors of ref. [54].

coupling to the H_g modes. Indeed the most remarkable feature when comparing the K_3C_{60} and C_{60} INS spectra is the virtual disappearance, due to a 4.3 meV increase in FWHM, of the $H_g^{(2)}$ mode (Fig. 7) at 53.2 meV, consistent with strong electron–phonon coupling (Table 2). Using eqn (6), we find an electron–phonon coupling constant for this mode of $\lambda^{(2)} \approx 0.17$, approximately 30% of the total electron–phonon coupling constant λ. Note that by coincidence the $H_g^{(2)}$ mode occurs in a region of the spectrum where there are no neighbouring or overlapping peaks, making our conclusions on the effects of doping on its width and position totally *unambiguous*. The dominant role of this mode has also been confirmed by Raman studies on ultrathin Rb_xC_{60} ($x = 0$–3) films[56]. Strong coupling is also evident for the well-separated $H_g^{(1)}$ mode which broadens by 1.0 meV, resulting in an estimated $\lambda^{(1)}$ contribution to the total λ of 0.10; Raman studies also show that this mode shows a Fano lineshape[57]. $H_g^{(4)}$ is more weakly coupled and we estimate an increase in γ of at least 3.1 meV, corresponding to a lower limit for $\lambda^{(4)}$ of 0.04. On the other hand, $H_g^{(3)}$ shows a virtually unchanged FWHM, even though it is somewhat softer. Information on the λ_ν's may be also extracted through the softening of phonon frequencies due to electron–phonon coupling, eqn (6). However,

changes in phonon linewidths are more reliable measures of electron–phonon interactions as energy changes also arise as a result of the cage reduction and depend on details of the electronic structure of the cluster.

Interpretation of the INS spectra in the tangential-mode regime (120–200 meV) is not unambiguous because of both a worsening in resolution and a substantial intensity attenuation due to the large Debye–Waller factors. Furthermore, broadening scales with ω^2, making identification of the affected phonons very difficult. Nonetheless, we can see that $H_g^{(8)}$ should couple strongly as there is in K_3C_{60} a hole in the spectrum at the position where it was observed for C_{60}. A definitive statement about the effect of intercalation on the $H_g^{(5-7)}$ modes at 136, 155, and 177 meV in C_{60} is difficult because they are weak, and there are significant changes in the spectra of K_3C_{60} in these regions. However, our experimental results uniquely confirm the important conclusion that electron–phonon coupling strength should be distributed between both buckling and tangential modes. Using the LDA frozen-phonon calculated λ_ν values of Schluter et al.[38] for the tangential and combining it with our estimates for the radial modes, we find

$$\lambda = 0.64.$$

This may be combined with estimates of the density-of-states $N(0) \approx 14$ states/C_{60}/eV/spin and the Coulomb pseudopotential $\mu^* \approx 0.15$[45] to give, using McMillan's formula,

$$T_c = \frac{\omega_{\log}}{1.2} \exp\left[-\frac{1.04(1 + \lambda)}{\lambda - \mu^*(1 + 0.62\lambda)} \right], \quad (8)$$

an estimate of $T_c \approx 17$ K (for $\omega_{\log} = 1072$ K), in good agreement with the experimentally observed values. A crucial point revealed by our results is the contrast with alkali-intercalated graphite, where the corresponding buckling modes do not couple to the π elec-

Table 2. Observed positions, ω_ν of the four radial H_g modes in C_{60} and their broadenings, γ_ν on reduction to C_{60}^{3-} together with calculated electron–phonon coupling strengths, V_ν in meV for a density-of-states $N(0) = 14$ states/eV/spin/C_{60}. We also include for comparison the values of γ_ν, estimated by the LDA frozen phonon calculations of Schluter et al.[38]

Vibrational mode	$H_g^{(1)}$	$H_g^{(2)}$	$H_g^{(3)}$	$H_g^{(4)}$
ω_ν/meV	33.1	53.2	88.1	96.2
γ_ν/cm^{-1}	8	35	1	25
$\gamma_{\nu,\text{calc}}$/cm^{-1}	7	15	23	56
V_ν/meV	7.4	12.4	0.1	2.7

trons because of symmetry restrictions. It appears that coupling to the radial modes, because of the finite curvature of the fullerene cage, accounts for roughly half the total electron–phonon coupling strength in K_3C_{60} and leads to a substantially increased T_c, compared to the lamellar graphite intercalates.

Further insight into the origin of phonon broadening in K_3C_{60} is provided by the INS spectra of insulating Rb_6C_{60}. The $H_g^{(2)}$ mode with the strongest contribution to the total electron–phonon coupling strength reappears very sharp at 53.1 meV with only a 0.1-meV change in FWHM, compared to C_{60}. The $H_g^{(1)}$ mode, also strongly coupled, similarly sharpens substantially compared to K_3C_{60} and shows an even smaller FWHM than in C_{60}. As a result, we can exclude both K^+ disorder effects and effects due to reduction, as being responsible for the remarkable broadening of certain intramolecular modes in K_3C_{60}. Strong electron–phonon coupling is an excellent candidate for these experimental observations.

5. POSTSCRIPT

The discovery of the family of the fullerene molecules has led to the development of a new world of round carbon chemistry, physics, and materials science, currently growing at an explosive rate. Papers dealing with novel aspects of fullerenes and their derivatives appear continuously. Neutron scattering experiments have provided, and no doubt will continue to provide, a unique insight into the properties of these remarkable materials. Globular C_{60} is a prototypical isotropic rotor, spinning at a very fast rate even in the solid state. The nature of the molecular motion changes drastically at 260 K to one of jump reorientations among nearly degenerate configurations, differing in energy by 11.4 meV. Below 90 K, the solid freezes into an orientational glass with substantial disorder down to 5 K. Crucial to the understanding of the details of fullerene packing in the solid state is that in contrast to the graphitic sheets, C_{60} shows an anisotropic electronic distribution, as a result of the presence of the twelve pentagonal faces. The unusually high T_c's observed in the superconducting fullerides can be explained without invoking any unusual or exotic interactions; the INS results are consistent with weak coupling superconductivity, in agreement with isotope-effect measurements and intramolecular electron–phonon coupling theories. Both radial and tangential vibrational modes contribute significantly to the electron–phonon coupling strength, as a result of the curvature of the spherical fullerene skeleton. This may be contrasted with the graphite intercalates which show much lower T_c's and no coupling is allowed by symmetry for the low-energy radial modes.

Acknowledgements—We thank our co-workers (C. Christides, T. J. S. Dennis, J. P. Hare, R. M. Ibberson) for their contributions to this work, and M. Schluter, J. R. D. Copley, S. F. J. Cox, J. P. Lu, G. Onida, and F. Zerbetto for valuable discussions. K.P. acknowledges support from the SERC and the Royal Society.

REFERENCES

1. W. Krätschmer, L. D. Lamb, K. Fostiropoulos, and D. R. Huffman, *Nature* **347**, 354 (1990).
2. R. Taylor, J. P. Hare, A. Abdul-Sada, and H. W. Kroto, *J. Chem. Soc., Chem. Commun.* 1423 (1990).
3. H. W. Kroto, J. R. Heath, S. C. O'Brien, R. F. Curl, and R. E. Smalley, *Nature* **318**, 162 (1985).
4. A. F. Hebard, M. J. Rosseinsky, R. C. Haddon, D. W. Murphy, S. H. Glarum, T. T. M. Palstra, A. P. Ramirez, and A. R. Kortan, *Nature* **350**, 660 (1991).
5. M. J. Rosseinsky, A. P. Ramirez, S. H. Glarum, D. W. Murphy, R. C. Haddon, A. F. Hebard, T. T. M. Palstra, A. R. Kortan, S. M. Zahurak, and A. V. Makhija, *Phys. Rev. Lett.* **66**, 2830 (1991).
6. K. Holczer, O. Klein, S. M. Huang, R. B. Kaner, K. J. Fu, R. L. Whetten, and F. Diederich, *Science* **252**, 1154 (1991).
7. J. R. D. Copley, D. A. Neumann, R. L. Cappelletti, and W. A. Kamitakahara, *J. Phys. Chem. Solids*, in press.
8. P. A. Heiney, J. E. Fischer, A. R. McGhie, W. J. Romanow, A. M. Denenstein, J. P. McCauley Jr., A. P. Smith III, and D. E. Cox, *Phys. Rev. Lett.* **66**, 2911 (1991).
9. G. E. Bacon, *Neutron Diffraction.* Oxford University Press, Oxford (1975).
10. S. W. Lovesey, *Theory of Neutron Scattering from Condensed Matter.* Clarendon Press, Oxford (1984).
11. G. L. Squires, *Introduction to the Theory of Thermal Neutron Scattering.* Cambridge University Press, Cambridge (1978).
12. J. Tomkinson, M. Warner, and A. D. Taylor, *Mol. Phys.* **51**, 381 (1984).
13. H. W. Kroto, A. W. Allaf, and S. P. Balm, *Chem. Rev.* **91**, 1213 (1991).
14. W. I. F. David, R. M. Ibberson, T. J. S. Dennis, J. P. Hare, and K. Prassides, *Europhys. Lett.* **18**, 225 (1992).
15. C. S. Yannoni, R. D. Johnson, G. Meijer, D. S. Bethune, and J. R. Salem, *J. Phys. Chem.* **95**, 9 (1991).
16. R. Tycko, G. Dabbagh, R. M. Fleming, R. C. Haddon, A. V. Makhija, and S. M. Zahurak, *Phys. Rev. Lett.* **67**, 1886 (1991).
17. R. D. Johnson, C. S. Yannoni, H. C. Dorn, J. R. Salem, and D. S. Bethune, *Science* **255**, 1235 ((1992).
18. D. A. Neumann, J. R. D. Copley, R. L. Cappelletti, W. A. Kamitakahara, R. M. Lindstrom, K. M. Creegan, D. M. Cox, W. J. Romanow, N. Coustel, J. P. McCauley, N. C. Maliszewskyj, J. E. Fischer, and A. B. Smith, III, *Phys. Rev. Lett.* **67**, 3808 (1991).
19. R. F. Kiefl *et al., Phys. Rev. Lett.* **68**, (1992).
20. C. E. Nordmann and D. L. Schmitkons, *Acta Cryst.* **18**, 764 (1965); J. P. Amourex, M. Bée, and J. C. Damien, *Acta Cryst. B* **36**, 2633 (1980).
21. A. Dworkin, H. Szwarc, S. Leach, J. P. Hare, T. J. S. Dennis, H. W. Kroto, R. Taylor, and D. R. M. Walton, *C.R. Acad. Sci. Paris II* **312**, 979 (1991).
22. R. Sachidanandam and A. B. Harris, *Phys. Rev. Lett.* **67**, 1467 (1991).
23. Y. Guo, N. Karasawa, and W. A. Goddard, III, *Nature* **351**, 464 (1991).
24. W. I. F. David, R. M. Ibberson, J. C. Matthewman, K. Prassides, T. J. S. Dennis, J. P. Hare, H. W. Kroto, R. Taylor, and D. R. M. Walton, *Nature* **353**, 147 (1991).
25. C. S. Yannoni, P. P. Bernier, D. S. Bethune, G. Meijer, and J. R. Salem, *J. Am. Chem. Soc.* **113**, 3190 (1991).
26. J. P. Lu, X. P. Li, and R. M. Martin, *Phys. Rev. Lett.* **68**, 1551 (1992).
27. X. P. Li, J. P. Lu, and R. M. Martin, to be published.
28. M. Sprik, A. Cheng, and M. L. Klein, *J. Phys. Chem.* **96**, 2027 (1992).

29. S. Liu, Y. Lu, M. M. Kappes, and J. A. Ibers, *Science* **254**, 408 (1991).

30. J. R. D. Copley *et al., Physica B,* **180**, 706 (1992).

31. X. D. Shi, A. R. Kortan, J. M. Williams, A. M. Kini, B. M. Savall, and P. M. Chaikin, *Phys. Rev. Lett.* **68**, 827 (1992).

32. R. C. Yu, N. Tea, M. B. Salamon, D. Lorents, and R. Malhotra, *Phys. Rev. Lett.* **68**, 2050 (1992).

33. D. A. Neumann *et al., J. Chem. Phys.,* **96**, 8631 (1992).

34. S. Chakravarty and S. Kivelson, *Europhys. Lett.* **16**, 751 (1991); S. Chakravarty, M. P. Gelfand, and S. Kivelson, *Science* **254**, 970 (1991).

35. P. W. Anderson, in press.

36. G. Baskaran and E. Tossati, *Curr. Sci.* **61**, 33 (1991).

37. F. C. Zhang, M. Ogata, and T. M. Rice, *Phys. Rev. Lett.* **67**, 3452 (1991).

38. M. A. Schluter, M. Lannoo, M. Needels, G. A. Baraff, and D. Tomanek, *Phys. Rev. Lett.* **68**, 526 (1992).

39. C. M. Varma, J. Zaanen, and K. Raghavachari, *Science* **254**, 989 (1991).

40. R. A. Jishi and M. S. Dresselhaus, *Phys. Rev. B* **45**, 2597 (1992).

41. L. Pietronero, *Europhys. Lett.* **17**, 365 (1992).

42. I. I. Mazin *et al., Phys. Rev. B* **45**, 5114 (1992).

43. R. M. Fleming, A. P. Ramirez, M. J. Rosseinsky, D. W. Murphy, R. C. Haddon, S. M. Zahurak, and A. V. Makhija, *Nature* **352**, 787 (1991).

44. O. Zhou *et al., Science* **255**, 833 (1992).

45. A. P. Ramirez *et al., Phys. Rev. Lett.* **68**, 1058 (1992).

46. C. C. Chen and C. M. Lieber, *J. Am. Chem. Soc.,* **114**, 3141 (1992).

47. P. B. Allen, *Phys. Rev. B* **6**, 2577 (1972).

48. K. Prassides, J. Tomkinson, C. Christides, M. J. Rosseinsky, D. W. Murphy, and R. C. Haddon, *Nature* **354**, 462 (1991).

49. J. W. White, G. Lindsell, L. Pang, A. Palmisano, D. S. Sivia, and J. Tomkinson, *Chem. Phys. Lett.,* **191**, 92 (1992).

50. K. Prassides, T. J. S. Dennis, J. P. Hare, J. Tomkinson, H. W. Kroto, R. Taylor, and D. R. M. Walton, *Chem. Phys. Lett.* **187**, 455 (1991).

51. R. L. Cappelletti, J. R. D. Copley, W. A. Kamitakahara, F. Li, J. S. Lannin, and D. Ramage, *Phys. Rev. Lett.* **66**, 3261 (1991).

52. C. Coulombeau, H. Jobic, P. Bernier, C. Fabre, D. Schütz, and A. Rassat, *J. Phys. Chem.* **96**, 22 (1992).

53. K. Prassides, C. Christides, M. J. Rosseinsky, J. Tomkinson, D. W. Murphy, and R. C. Haddon, *Europhys. Lett,* **19**, 629 (1992).

54. G. Onida and G. Benedek, *Europhys. Lett.,* **18**, 403 (1992).

55. F. Negri, G. Orlandi, and F. Zerbetto, *Chem. Phys. Lett.* **144**, 31 (1988); *Chem. Phys. Lett.* **190**, 174 (1992).

56. M. G. Mitch, S. J. Chase, and J. S. Lannin, *Phys. Rev. Lett.* **68**, 883 (1992).

57. P. Zhou, K. A. Wang, A. M. Rao, P. C. Eklund, G. Dresselhaus, and M. S. Dresselhaus, *Phys. Rev. B.* **46**, 2595 (1992).

Part Two

Physics and Chemistry of Fullerene-Based Solids

INTRODUCTION

The vitality of solid state chemistry and physics has once again been demonstrated by the extraordinary response to the landmark discovery of Krätschmer, Huffman and coworkers of a method for synthesizing macroscopic quantities of fullerenes. Coming less than 4 years after the Bednorz–Muller discovery of superconducting cuprates, this breakthrough had three important consequences. First, it provided for the first time enough material for experiments confirming the truncated icosahedral geometry of the C_{60} molecule proposed 5 years earlier by Kroto, Smalley and colleagues. Second, it stimulated synthetic organic chemists to envision C_{60} as a starting point for a new family of derivative molecules, much as the discovery of benzene gave birth to the petrochemical, plastics and polymer industries. Finally, almost lost in the initial excitement, was that this discovery represented the creation of a new family of solid phases, unusual and interesting in their own right but also extremely versatile as host lattices for synthesizing new intercalation compounds and other solid state derivatives. This special issue of the *Journal of Physics and Chemistry of Solids* presents a state-of-the-art review of progress in the synthesis and characterization of these new materials.

Three articles deal with synthesis and characterization of pristine and intercalated phases of solid C_{60}. Murphy *et al.* describe the solid state chemistry of alkali metal-intercalated C_{60} with emphasis on superconducting phases. Zhou and Cox present a systematic review of crystal structures obtained by Rietveld refinement of X-ray power diffraction profiles. Gensterblum and co-workers show that under appropriate conditions one can obtain high-quality epitaxial thin films of C_{60} which should eventually broaden the scope of possible experiments on solid state derivatives.

Four papers describe how the electronic and vibrational properties of solid C_{60} are derived from the corresponding molecular properties, and how these vary with doping. Eklund *et al.* discuss vibrational properties from the standpoint of Raman and infrared spectroscopy and symmetry arguments. Lucas and Weaver review our knowledge of single particle and collective electronic spectra based on energy loss and photoemission spectroscopies, respectively. Oshiyama summarizes the results of band structure calculations and discusses how the electron spectra evolve with intercalation.

Four papers focus on manifestations of the novel physics and chemistry of fullerene-derived solids. Heiney, and Copley *et al.*, apply X-ray and neutron scattering techniques, respectively, to study the consequences of the coupling between intramolecular dynamics and average intermolecular long-range order in solid C_{60}. Schlüter and co-workers address the mechanism of superconductivity in the alkali-intercalated $M_3 C_{60}$ phases, and give a rationale for the much higher T_cs as compared to the chemically-similar graphite intercalation compounds. Finally, Wudl and Thompson describe a compound which exhibits soft ferromagnetic behavior, obtained by reacting solid C_{60} with an organic donor molecule.

The reviewing process of these 11 manuscripts was completed in the spring of 1992. We heartily thank the authors for taking the time to contribute to this project, which hopefully will serve as a timely review of this rapidly expanding new field.

John E. Fischer
David E. Cox

SYNTHESIS AND CHARACTERIZATION OF ALKALI METAL FULLERIDES: A_xC_{60}

D. W. Murphy, M. J. Rosseinsky, R. M. Fleming, R. Tycko, A. P. Ramirez, R. C. Haddon, T. Siegrist, G. Dabbagh, J. C. Tully and R. E. Walstedt

AT&T Bell Laboratories, Murray Hill, NJ 07974-2070, U.S.A.

(Received 4 February 1992)

Abstract—Alkali metal fullerides (A_xC_{60}) are a subject of considerable current interest because of the occurrence of superconductivity for A_3C_{60} at temperatures surpassed only by the high T_c copper oxides. The preparation and characterization of A_xC_{60} (A = alkali metal, $x = 2,3,4,6$) by powder X-ray diffraction, NMR (^{13}C, ^{23}Na and ^{87}Rb), and d.c. magnetization are reported. The structures are described as intercalation compounds of the FCC structure of pristine C_{60} or of hypothetical BCC or BCT structures. The structures and phase diagrams can be rationalized on the basis of ion size and electrostatic considerations. Only the A_3C_{60} compounds are metallic (and superconducting). The superconducting T_c increases nearly linearly with unit cell size. EHT (Extended Hückel Theory) calculations and ^{13}C NMR relaxation measurements indicate higher densities of states for the higher T_c compositions.

Keywords: Superconductivity, Fullerenes, carbon, structures.

INTRODUCTION

Since the discovery of [1] of C_{60} and higher fullerenes there has been considerable interest in the chemistry and physical properties of these molecular forms of carbon. The isolation of macroscopic quantities of C_{60} by the spark erosion technique [2] added the possibility of investigating solid state properties. The molecular fullerene solids are held together by relatively weak van der Waal's forces similar to those between layers in graphite. A rich variety of physical properties have already been identified for solid C_{60} and its compounds [3]. The discovery of superconductivity in K_xC_{60} [4] and identification of K_3C_{60} as the superconducting phase [5, 6] has led to a broad class of superconductors [4–11] of composition A_3C_{60} (A = alkali metals) with superconducting critical temperatures, T_c, surpassed only by the copper oxides. The occurrence of superconductivity has focused considerable interest on A_xC_{60} and a rich variety of A_xC_{60} have already been reported with compositions at $x = 2$ [11, 12], 3 [6, 8, 11], 4 [13] and 6 [11, 14] depending on A. Alkali metal intercalation in the fullerenes already appears to be more diverse than the widely investigated graphite analogs [15]. In this paper we expand on our previous reports on bulk, polycrystalline A_xC_{60} and integrate these with the results of others to present an overall picture of the A_xC_{60} compounds.

EXPERIMENTAL

The C_{60} used in this work was purified by standard extraction and chromatography of soot prepared in-house by the spark erosion technique [2] or purchased from Texas Fullerenes. Extract was also purchased from MER. We have observed no differences in the materials prepared from any of these sources.

A variety of techniques have been employed to synthesize A_xC_{60} with the choice of technique dependent on both A and x. A_6C_{60} (A = K,Rb,Cs), is best prepared by direct reaction with excess alkali metal vapor at 250–350°C for 10–15 days in sealed pyrex tubes [14], and for A = Na at 350–400°C for 6 days in stainless steel tubes [11]. The A_3C_{60} and A_4C_{60} (A = K,Rb,Cs) phases can be prepared by several methods, but the most reliable way to control stoichiometry is reaction of appropriate amounts of A_6C_{60} and C_{60} at 200–250°C for 10–15 days [16]. For A = Na, Na_2C_{60} is best prepared by reaction of NaH or Na_5Hg_2 with C_{60} at 350°C for nine days with two intermediate regrindings [11]. Na_3C_{60} is best synthesized by reaction of Na_5Hg_2 with C_{60} at 330°C for 10–15 days followed by distillation of Hg away from the product [11]. A_3C_{60} with mixed alkalis are prepared either by reaction of appropriate quantities of the different A_6C_{60}s with C_{60}, or by reaction of A_6C_{60}, C_{60} and NaH for Na containing compositions. In each of these methods the reaction times are dependent on amount and crystallite size of the C_{60} and the size and ambient pressure in the reaction tube. Reactions in this study used from 10 to 500 mg C_{60}. Appropriate physical properties (e.g. X-ray diffraction, NMR, magnetization) must be monitored to insure phase purity. Even brief exposure to air results in serious degradation of the superconducting properties of A_3C_{60}. They should be maintained in an inert atmosphere at all times.

Powder X-ray diffraction patterns were taken using CuK_α radiation from a high intensity rotating anode source with a singly bent pyrolytic graphite (PG) focusing monochromator and a flat PG analyser. The resolution was enhanced over that provided by the 0.5° acceptance of PG by placing slits before the monochromator and the analyser. This gives an instrument with a moderate angular resolution ($\sim 0.25°$ 2θ) but a high dynamic range ($> 10^3$ for powder diffraction). Samples for X-ray diffraction were sealed in Debye–Scherrer capillaries to protect them from air. The capillaries were then mounted in a vacuum chamber with a beryllium window to eliminate scattering from air at low angles. LAZY/PULVERIX [17] and the NRCVAX [18] package were used in refinement and modeling of intensity data after modification of the source code to allow a form factor for C_{60}, $\propto \sin QR/QR$, which models C_{60} as a hollow spherical shell of radius R containing 360 electrons. R was fixed at 3.55 Å as determined from our previous single crystal analysis [19]. The value of R was manually varied for various A_xC_{60} and found to be constant for each. The NMR spectra were obtained at 9.39T (100.5 MHz for ^{13}C). Samples for NMR were sealed in pyrex under a He atmosphere, except for ^{23}Na which were sealed in quartz to minimize background signals. Superconducting shielding measurements were taken on a Quantum Design SQUID at fields of 5–10 Oe. DSC data were taken on a DuPont 1090 Thermal Analyser. Electronic structure calculations were carried out with the EHT (Extended Hückel Theory) band structure programs.

STRUCTURES

The room temperature solid state structure of C_{60} may be described as FCC packing of identical, spherical C_{60} molecules. The observed X-ray powder data are modeled well by using uniform shells of electron density (radius 3.55 Å) for the C_{60} molecules at the $4a$ positions of a FCC cubic cell of symmetry $Fm\bar{3}m$ [19, 20]. A van der Waal's radius of 5.01 Å is calculated from the crystallographic unit cell. Motional narrowing of the ^{13}C NMR spectra shows that the molecules rotate rapidly compared to the NMR timescale ($\approx 100\,\mu s$) [21, 22]. At 260 K there is a first order phase transition to a simple cubic cell [20] of symmetry $Pa\bar{3}$, corresponding to rotational ordering of the molecules. The $Pa\bar{3}$ structure has been refined by X-ray [20] and neutron diffraction [23] on powders and X-ray diffraction on twinned single crystals [24]. Continued motional narrowing of the ^{13}C NMR spectra below 260 K leads to the conclusion that C_{60} molecules continue to rapidly 'ratchet' between symmetry equivalent orientations above about 140 K [25].

The structures of all the alkali metal fullerides characterized to date may be regarded as intercalation compounds of the pristine FCC structure or of hypothetical C_{60} structures with BCC (body centered cubic) or BCT (body centered tetragonal) ball packing. The ideal sizes and coordination numbers of interstitial sites for FCC and BCC packings of rigid spheres of radii 5.01 Å are summarized in Table 1 along with the ionic radii of the alkali metal ions. The octahedral and tetrahedral interstices of FCC are fixed on special positions. The octahedral site is larger than any alkali metal ion, and the tetrahedral site is closest in size to Na^+. (Ions can generally occupy sites smaller than their hard sphere sizes would indicate.) These are the only interstitial sites generally considered for FCC packing, but the large size of the C_{60} molecules makes consideration of other sites reasonable. The most likely of these is (x,x,x) which for $x = 1/3$ gives a trigonal site of appropriate size for small cations (e.g. Li^+). For BCC packing, the

Table 1. Interstitial sites in C_{60}

Structure type	Unit cell (Å)	Interstitial sites				Alkali metals	
		(x,y,z)	No. per C_{60}	CN	Radius (Å)	A^+	Radius (Å)
FCC (s.g. $Fm\bar{3}$)	14.17	$(\frac{1}{2}, \frac{1}{2}, \frac{1}{2})$	1	6	2.06	Cs	1.70
		$(\frac{1}{4}, \frac{1}{4}, \frac{1}{4})$	2	4	1.12	Rb	1.49
		$(x, x, x)\, x = 1/3$	8	3	0.78	K	1.38
						Na	1.02
BCC (s.g. $Im\bar{3}$)	11.57†	$(0, \frac{1}{2}, z)\, z = \frac{1}{4}$	6	4	1.46	Li	0.69
		$(x, y, z)\, x = y = 1/8, z = \frac{1}{2}$	6	3	1.13		

†Calculated lattice parameter for spheres of radius 5.01 Å.

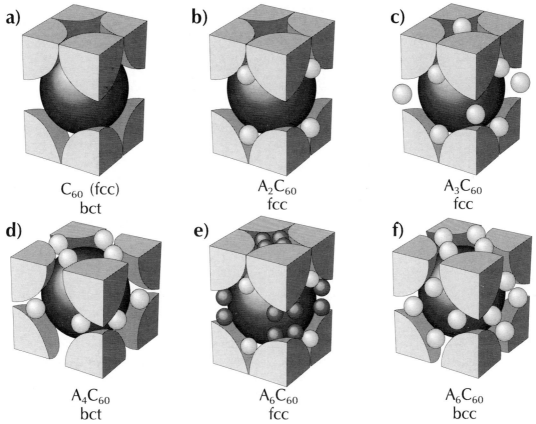

Fig. 1. Schematic structures of C_{60} and A_xC_{60} with C_{60}s as large spheres and A as the smaller spheres. (a) FCC C_{60} drawn in an equivalent BCT representation. (b) The structure of Na_2C_{60} with Na ions in tetrahedral interstices. (c) A_3C_{60} with A ions in both tetrahedral and octahedral interstices. (d) The A_4C_{60} structure exhibited by K, Rb and Cs. (e) The FCC A_6C_{60} structure (A = Na, Ca) with the darker Nas 50% occupied. (f) The BCC A_6C_{60} structure of K, Rb and Cs.

interstitial sites located at $(0,1/2,z)$ are tabulated for $z = 0.25$, which is equidistant from four C_{60}s. The sites for BCT packing are more complex, depending both on x and c/a, but are approximately the size of those for BCC.

Schematic representations of the known A_xC_{60} structures are illustrated in Fig. 1. Tables 2 and 3 list unit cell data for the various observed phases. The A_3C_{60} structure was first refined by Stephens *et al.* [6] for K_3C_{60}. For the C_{60} to be completely ordered and

Table 2. Unit cell and T_cs for FCC A_xC_{60}

	Lattice parameter(s) (Å)	T_c(K) (%)†	^{13}C shift (ppm vs TMS)
Na_2C_{60}‡	14.189 (1)	—	173 (2)
Na_3C_{60}	14.183 (3)	—	175 (2)
Na_6C_{60}	14.380 (8)	—	167 (2)
Na_2KC_{60}	14.120 (4)	—	
Na_2RbC_{60}	14.091 (6)	—	180 (2)
$Na_2Rb_{0.5}Cs_{0.5}C_{60}$	14.148 (3)	8.0 (3)	
Na_2CsC_{60} No. 1§	14.132 (2)	10.5 (8)	182 (2)
Na_2CsC_{60} No. 2§	14.176 (9)	14.0 (9)	
K_3C_{60}	14.253 (3)	19.3 (30)	183 (2)
K_2RbC_{60}	14.299 (2)	21.8 (32)	
Rb_2KC_{60} No. 1§	14.336 (1)	24.4 (34)	
Rb_2KC_{60} No. 2§	14.364 (5)	26.4 (32)	
Rb_3C_{60}	14.436 (2)	29.4 (35)	171 (5)
Rb_2CsC_{60}	14.493 (2)	31.3 (48)	

†T_cs and shielding fractions (%) were measured by d.c. magnetization.
‡Sample is simple cubic.
§Samples labeled No. 1 and No. 2 have the same nominal composition.

Table 3. Unit cell and alkali sites for A_4C_{60} and A_6C_{60}

	Phase	$a(\text{Å})$	$c(\text{Å})$	c/a	$r_A\text{Å}\dagger$ (avg)	$r_\square(\text{Å})\ddagger$ (avg)	[13]C shift (ppm vs TMS)
K_4C_{60}	BCT	11.886 (7)	10.774 (6)	0.906	1.48 / 1.32 } (1.40)	1.39 / 1.65 } (1.53)	
Rb_4C_{60}	BCT	11.962 (2)	11.022 (2)	0.921	1.52 / 1.44 } (1.48)	1.44 / 1.72 } (1.58)	181 (2)
Cs_4C_{60}	BCT	12.057 (18)	11.443 (18)	0.949	1.58 / 1.63 } (1.60)	1.52 / 1.82 } (1.67)	
K_6C_{60}	BCC	11.390§		1.00	1.52 / 1.21 } (1.36)		
Rb_6C_{60}	BCC	11.548§		1.00	1.61 / 1.30 } (1.45)		154 (5)
Cs_6C_{60}	BCC	11.790§		1.00	1.75 / 1.43 } (1.59)		

†Distance from center of A ion site to center of C_{60} minus the van der Waal's radius of C_{60} (5.01 Å).
‡The □ denotes a vacancy.
§Data from Ref. 14.

remain FCC, one expects the space group $Fm\bar{3}$. In this packing all molecules are identical when viewed down ⟨100⟩ directions. Stephens *et al.* [6] found that the observed space groups is $Fm\bar{3}m$ indicating that the molecules are disordered over two orientations differing by $\pi/2$ rotations about ⟨100⟩. All A_3C_{60}s that have been structurally characterized exhibit FCC structures at room temperature.

Powder X-ray diffraction patterns for Na_xC_{60} are shown in Fig. 2. Refinement of the patterns [11] indicate that Na_2C_{60} has Na in tetrahedral sites of a primitive cubic lattice. The Na saturated phase has a composition near Na_6C_{60}, which refines well for two

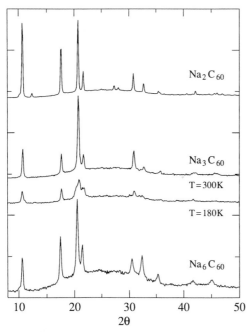

Fig. 2. Powder diffraction patterns of Na_xC_{60}. Intensities are normalized between samples, but the 300K and 180K patterns of Na_3C_{60} are on the same scale.

Na on the tetrahedral sites and 50% occupancy of a cubic cluster of sites at 32f positions (0.43, 0.43, 0.43) centered on the octahedral site. An arrangement of Na atoms consistent with half occupancy of these sites is the presence of Na_4 tetrahedra formed by Na atoms at opposite corners of the cube. These tetrahedra would then be disordered between parent octahedral sites. This structure has also been proposed for Ca_xC_{60} [26]. The diffraction pattern of Na_3C_{60} at room temperature fits the normal FCC A_3C_{60} structure. Although Na_3C_{60} refines well for Na on both octahedral and tetrahedral sites, the possibility of Na displacement to a more general position (e.g. 32f) is likely. Below 260 K there is a structural transition in Na_3C_{60}, resulting in an X-ray pattern (Fig. 2) that appears to be a two-phase mixture arising from disproportionation into compositions with lattice parameters close to those of Na_2C_{60} and Na_6C_{60}. Remarkably this transition is reversible with temperature. Since it is hard to imagine facile Na motion over large distances to produce a macroscopic two phase material, we envisage a shorter range diffusion of Na from the tetrahedral site to form clusters about the octahedral site. Preferred compositions (still to be determined) such as Na_4C_{60} with all Na clustered about the octahedral site could account for formation of two phases.

The BCC A_6C_{60} (A = K, Rb, Cs) structure, first reported by Zhou *et al.* [14], is based on BCC packing of C_{60} molecules with A ions occupying four coordinate interstices between rotationally ordered C_{60} molecules. For A = Cs, refinement located A at (0,1/2,0.28), which may be described as a distorted tetrahedron with A closer to two C_{60}s. Assuming the same fractional coordinate for other A, the radii of the sites are given in Table 3. The average of the four distances to C_{60}s is close to the A^+ ionic radii. The

A_4C_{60} structure, described by Fleming *et al.* [13], has a BCT structure and was refined in the $I/4mmm$ space group with C_{60} as a shell of electron density. The true symmetry must be lower if the C_{60}s are completely ordered due to the absence of a four-fold axis in the C_{60} molecule. The sizes of the A^+ sites in A_4C_{60} are given in Table 3. The sizes of the A^+ ion sites in both A_4C_{60} and A_6C_{60} closely match the ionic radii of the A^+ ions. The A_4C_{60} structure may be viewed as a defect A_6C_{60} structure. The six identical sites in the BCC structure disproportionate into four slightly smaller sites and two larger ones for BCT structures with $c/a < 1.0$. The A ions occupy the smaller sites.

The observed structures can all be rationalized on the basis of alkali ion size and electrostatic considerations. A plot of the unit cell volumes of the A_xC_{60} normalized per C_{60} molecule versus the total cation volume calculated from the ionic radii, shown in Fig. 3, makes a useful comparison both within and between structure types based on alkali ion sizes. For the FCC series, compositions exist with both larger and smaller cell volumes than for the pristine host. This is also true for the BCC series, based on the estimated volume of a hypothetical 'pristine BCC host' with spheres of the same radius as C_{60}. For A_6C_{60} the cell is contracted for A = K ($r_{K+} = 1.38$ Å) which is smaller than the hard sphere site size ($r = 1.49$ Å), whereas Rb ($r_{Rb+} = 1.49$ Å) is nearly a perfect match and Cs ($r_{Cs+} = 1.70$ Å) causes a lattice expansion. The A_3C_{60} data are more complicated because there are two different types of sites. The smaller tetrahedral site is 'lattice expanding' for ions larger than Na, and the octahedral site is 'lattice contracting' for all A ions, giving rise to a minimum in the lattice parameter as a function of cation size. Refinements of X-ray powder intensity data for mixed alkali A_3C_{60} show distinct site ordering only

between ions with the largest size differences, i.e. Na_2CsC_{60} and Na_2RbC_{60} [11]. It should be noted that the large size of the octahedral site compared to that of the A^+ ions is reflected in large thermal factors in the structure refinements of A_3C_{60}. The larger unit cells of Na_3C_{60} and Na_2KC_{60} compared to Na_2RbC_{60} are attributed to Na^+ occupation of sites displaced from the centers of the octahedral sites. The larger volume of A_4C_{60} compared to the A_6C_{60} of the same A reflects the disproportionation of site sizes in the former with the vacant sites being larger as shown in Table 3. The difference is most pronounced for the smallest ion, K^+.

Electrostatic calculations shed light on the existence of the observed phases and their detailed structures [13, 27]. The calculations assume point charges for A^+ and C_{60}^{x-}, experimental lattice parameters for known phases, and an electrostatic energy of putting x electrons on C_{60} of $E_x = -E_A + C_x/R$, where the electron affinity, E_A, is taken as 2.6 eV. For C_{60} with radius $R = 3.5$ Å, C_x/R is the minimum Coulomb repulsion of x point charges on the sphere. For Rb_xC_{60} the known phases ($x = 3$, 4 and 6) are the ones found to be most stable with these simple assumptions and have electrostatic energies within 0.1 eV of each other. Two other structures, Rb_2C_{60} (Rb in tetrahedral sites) and Rb_3C_{60} with the A15 structure (Fig. 4) are less stable by only about 0.2 eV. Other calculations suggest that the A15 structure is energetically comparable [28]. The A_2C_{60} phase is experimentally observed for A = Na, but the A15 structure has yet to be observed. The most likely place to find an A15 structure would be for A = Cs,

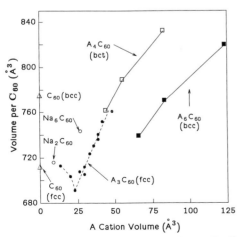

Fig. 3. The crystallographic volume of A_xC_{60} per C_{60} (from Tables 1 and 2) versus the sum of the A^+ ion volume (per C_{60}). The lines are a guide to the eye.

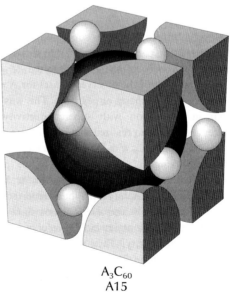

A_3C_{60}
A15

Fig. 4. A hypothetical A_3C_{60} with the A15 structure. The structure can be seen to be an ordered defect structure of A_6C_{60}.

noteworthy that graphite reacts with KH to form the
ternaries $KH_{0.8}C_8$ and $KH_{0.8}C_4$ [39]. No tendency to
form ternary hydrides was observed for C_{60} under our
conditions, but Wudl *et al.* [37] have shown that alkyl
borohydrides add H to C_{60}. These routes are particularly attractive for Na since its low vapor pressure
and reactivity with pyrex present problems for the
direct reaction of Na and C_{60}. Kelty *et al.* [10] used
Hg amalgams to prepare A_xC_{60} with K, Rb, Cs and
mixed alkalis. We have also used the amalgam
Na_5Hg_2 to prepare $Na_xC_{60}s$. Analogous reactions
with graphite again lead to ternary phases [39] which
we have not observed for C_{60}.

NMR

^{13}C NMR has proven a useful spectroscopy for
fullerenes, affording information on the structure,
dynamics and electron structure [21, 22, 25, 27, 40].
For pristine C_{60}, the room temperature ^{13}C spectrum
consists of only one motionally narrowed line at 143
ppm corresponding to freely rotating molecules
[21, 22, 25]. Transition to the ordered $Pa\bar{3}$ structure
at 260 K is accompanied by a sudden increase in the
spin–lattice relaxation rate, $(1/T_1)$ [25], but continued
motional narrowing indicates that molecular rotation
remains fast on the NMR timescale down to ~ 140 K.
Analysis of the temperature dependence of the relaxation rate indicates energy barriers to molecular
reorientation of 42 meV above 260 K and 250 meV
below 260 K [25].

Room temperature ^{13}C NMR spectra of a variety
of A_xC_{60} compounds are shown in Fig. 6. The positions of the centers of gravity (isotropic shifts) of the
NMR lines are listed in Tables 2 and 3. The ^{13}C
resonances of the A_xC_{60} are shifted downfield from
that of pure C_{60}. The overall shifts are the sum of
chemical shifts and (for the conducting phases)
Knight shifts. The magnitudes and signs of the
Knight shifts are currently under study. It is apparent
from the ^{13}C shifts that there is no simple monotonic
correlation of the shifts with the formal charge on
C_{60}. At room temperature, the ^{13}C spectra of Na_2C_{60},
Na_3C_{60}, K_3C_{60}, Na_2CsC_{60}, Rb_4C_{60} and Na_6C_{60} show
significant motional narrowing, indicating that C_{60}^{n-}
anions reorient on a timescale less than $\sim 100 \mu s$. At
low temperatures, where molecular rotation is very
slow, crystallographically inequivalent carbons generally have different ^{13}C chemical and Knight shift
tensors. The NMR spectra are therefore superposi-
tions of powder patterns from the inequivalent sites.
Differences in the shift tensors for different sites lead
to a residual line broadening in A_xC_{60} (e.g. compared
with pure C_{60}) even when molecular reorientation is
rapid. The various chemical shifts and lineshapes of

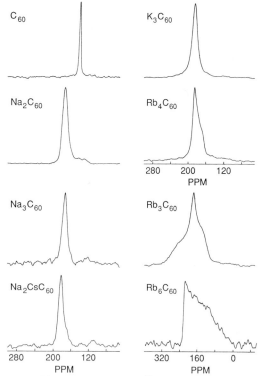

Fig. 6. Representative ^{13}C NMR spectra
for A_xC_{60} (100.5 MHz). Shifts are relative to TMS.

the A_xC_{60} make NMR a useful tool for evaluating
phase purity.

In addition to ^{13}C, we have used ^{23}Na and ^{87}Rb
spectra to gain information about the alkali sites. The
resonances for both ^{23}Na and ^{87}Rb are very close to
the values for the ions in aqueous solution in agreement with the description of these compounds as
$A_n^+ C_{60}^{n-}$. For Rb_3C_{60} there are two resonances at
room temperature with an approximately 2:1 ratio at
-5 and $-130 \pm$ ppm, which we assign to the tetrahedral and octahedral sites, respectively. Powder X-ray
refinements indicate that K and Rb are disordered in
$K_{1.5}Rb_{1.5}C_{60}$, but ^{87}Rb NMR finds approximately
equal amounts of Rb on both sides, indicating that
there is substantially more than a statistical occupation of the octahedral site by Rb as expected based
on the larger size of Rb^+ than K^+.

ELECTRONIC PROPERTIES

The overriding interest in A_xC_{60} is the occurrence
of superconductivity for compounds with $x = 3$.
Superconducting transitions in A_3C_{60} are readily
measured by d.c. magnetization. The susceptibility
measured on warming, after cooling in zero field,
gives a diamagnetic shielding. Representative shielding curves for superconducting A_3C_{60} are shown in
Fig. 7 and the T_cs are tabulated in Table 2. The
fraction of observed diamagnetism to that expected

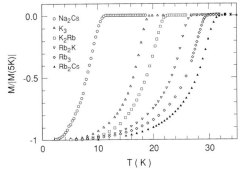

Fig. 7. Shielding measurements for A_3C_{60} normalized to the shielding at 5K.

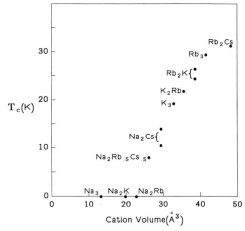

Fig. 9. Superconducting T_cs plotted against calculated cation volume for A_3C_{60}.

for the same volume of perfect diamagnetism is the shielding fraction, or volume fraction. The shielding fraction is a quantitative determination of the phase purity of a superconductor only if the sample morphology is known. For example, anomalously large shielding fractions can be obtained if a closed shell of superconductor surrounds a non-superconductor. For C_{60} there is another effect, related to the finite size of the crystallites. Low values for the shielding fraction are obtained unless the grain size of the superconductor is >10 times the London penetration depth. For these materials the particle size is of the order of $1\,\mu$ and the penetration depth of K_3C_{60} is ≈ 4800 Å [41]. A two fluid model suggests the maximum shielding for a loose powder to be near 35% for these parameters [42]. The superconducting T_cs of A_3C_{60}s are plotted against lattice parameter in Fig. 8 along with T_cs [43–45] and lattice constants [46] for K_3C_{60} and Rb_3C_{60} determined under pressure. For $a > 14.2$ Å the 1 atm and pressure data are in excellent agreement, with substantial differences only for the smallest cations where structural anomalies have

been noted above. A plot of the T_cs against the intercalated cation volume (Fig. 9) extrapolates to $T_c = 0$ near 20–22 Å [3] corresponding to the compositions Na_2KC_{60}–Na_2RbC_{60}. T_c extrapolates to zero for about the same cation size for which the A_3C_{60} unit cell reaches a minimum (Fig. 3). Furthermore, compositions on the low cation volume branch of Fig. 3 are also apparently non-metallic (based on ^{13}C T_1T measurements). Disordered displacements of Na^+ ions on octahedral sites could contribute to electron localization for the low cation volume compositions. In addition, the Na_6C_{60} structure suggests covalent interactions and the possibility of only partial charge transfer. If partial charge transfer does play a role it appears to be more significant for Na on the octahedral site, since the T_c of Na_2CsC_{60} is as expected based on extrapolated values. It is interesting to note that the cell volumes of the A_4C_{60} has a very similar dependence on intercalant volume to the A_3C_{60}s, leading to the possibility of higher T_cs for this series if the electron count could be manipulated to half-fill the t_{1u} band (three electrons per C_{60}).

In order to understand the origin of the superconductivity in A_3C_{60} and the magnitude of the T_cs, a more complete picture of the electronic properties is needed both above and below T_c. In addition, properties of the other A_xC_{60} are of intrinsic interest and may add insight into superconductivity. The unavailability of single crystals, or even dense pellets, combined with the air sensitivity of the compounds have limited the number and variety of experiments relevant to the elucidation of electronic properties. Conductivity measurements have been reported only for thin films [47–51]. These measurements show small increases in the resistivity with decreasing temperature for K_3C_{60} and Rb_3C_{60} with depressed T_cs attributed to granularity [50]. It has further been

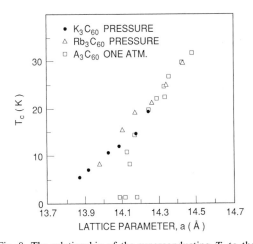

Fig. 8. The relationship of the superconducting T_c to the unit cell size of A_3C_{60}. □ is data from Table 2. ● and △ are data from K_3C_{60} and Rb_3C_{60}, respectively, under pressure as given in Ref. 46.

suggested that the conductivity is near the Mott minimum metallic limit [49]. These studies also show that the A_6C_{60}s are insulating. A number of other techniques such as ^{13}C NMR [27, 40], bulk magnetic susceptibility [51], EPR spin susceptibility [51, 52] and photoemission [12, 53–56] have been used to address the question of metallic conductivity and density of states. There is a general consensus from all of these techniques that the A_3C_{60}s (with the exception of those with Na on octahedral sites) are metallic with the other structures being non-metallic. Calculation of densities of states for A_3C_{60} from each of these techniques, however, involve approximations and vary considerably from each other.

The EHT band structures [57] calculated for FCC C_{60}, $Pa\bar{3}$ C_{60} and A_xC_{60} ($x = 3, 4, 6$) are shown in Fig. 10. As discussed previously [47], the EHT calculations provide a qualitatively correct description of the conduction bands of C_{60}. The band structure calculations indicate that the DOS distributions of the t_{1u-} and t_{1g-} derived bands are relatively constant across the different phases. The band widths and energy spectra remain qualitatively unchanged, and thus the molecular features of C_{60} dominate the electronic structure of the doped phases. It is clear from Fig. 10 that the A_6C_{60} phases will be insulators (filled conduction band), and that phases with $x = 3$, 4 and (presumably) 2 are all predicted to have partially filled bands. Of significance to the density of states, the band width for A_3C_{60} decreases with increasing unit cell parameter resulting in an increase in the density of states of $\approx 15\%$ for Rb_3C_{60} over K_3C_{60} [8].

Information on metallic conductivity and density of states may be obtained from NMR using the Korringa relationship. The Korringa relaxation mechanism predicts that T_1T should be constant for a metal [58]. Both K_3C_{60} and Rb_3C_{60} exhibit ^{13}C T_1T values that are only weakly dependent on temperature [27, 40]. Just above T_c, $T_1T = 100$K-s for Rb_3C_{60} and 165K-s for K_3C_{60}. Assuming relaxation by a contact hyperfine interaction between nuclei and non-interacting electrons, with a magnitude of 0.6 Gauss, the T_1T values imply densities of states at the Fermi energy of roughly 22 eV^{-1} and 17 eV^{-1} for Rb_3C_{60} and K_3C_{60} (per C_{60} per spin state). For Na_3C_{60} T_1T increases by a factor of six with decreasing temperature from 373 K to 93 K and that of Na_2RbC_{60} increases by more than a factor of two over this temperature range, suggesting loss of metallic conductivity for the A_3C_{60}s with the smallest cations. For the A_2C_{60}, A_4C_{60} and A_6C_{60} phases examined to date, T_1T is strongly dependent on temperature, indicating that they are not metallic. For $x = 6$ the insulating behavior is expected from filling of the t_{1u} band. The

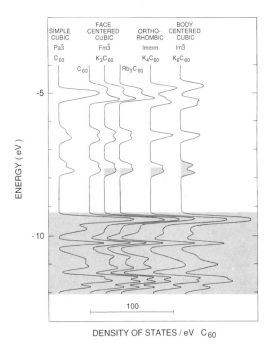

Fig. 10. EHT band structures of A_xC_{60} from Ref. 57.

origin of insulating behavior for $x = 2$ or 4 is not known, but could arise from a localized Jahn–Teller distortion on C_{60}^{x-}, a collective charge density wave, a Mott–Hubbard insulator or carrier concentrations below the Mott limit.

A useful basis for discussion of the magnitude and range of T_c in the A_3C_{60} and of possible pair mediating excitations is the McMillan equation [59]

$$T_c = h\frac{\omega}{1.2k_B}\exp\left[\frac{-1.04(1+\lambda)}{\lambda - \mu^*(1+0.62\lambda)}\right],$$

where $\lambda = N(E_F)V$, V is the strength of the coupling of the conduction electrons to the mediating excitation of frequency ω and μ^* is the renormalized Coulomb repulsion between conduction electrons. The relationship is valid for weak or intermediate coupling ($\lambda \leq 1.5$). Superconductivity in the A_3C_{60}s has been attributed to phonon mediated pairing either via purely intramolecular C_{60} modes [60–62] (≈ 1000–2000 K) or low energy A-C_{60} optic modes [63] (≈ 200 K). Purely electronic coupling has also been proposed [64–66]. The modes calculated to have strong coupling to the conduction electrons are Raman active. In thin film samples, these modes [67] (predominantly H_g symmetry) become unobservable on formation of A_3C_{60}, then reappear on formation of A_6C_{60}. Inelastic neutron scattering on bulk samples is consistent with this observation [68]. The fact that both K_3C_{60} and Rb_3C_{60} have virtually the same T_cs when they have the same lattice constant (under pressure) is strong evidence against involvement of modes containing the A ions. ^{13}C isotope effect experiments, on the other hand, show the involvement of C based modes. Three experiments have been reported with αs $(T_c \propto M^{-\alpha})$ of 1.4 ± 0.5 [69] and 0.37 ± 0.05 [70] for Rb_3C_{60} and 0.30 ± 0.06 [71] for K_3C_{60}. In our experiment [70] a decrease of T_c by 0.65 K for Rb_3C_{60} ($\alpha = 0.37 \pm 0.05$) was found for a sample with 75% enrichment with ^{13}C. Measurements of the superconducting gap are similarly divergent: values of 5.3 have been suggested from STM point contact tunneling [72, 73] although optical data [74] suggest a gap of 3–$5kT_c$ and NMR data suggest 3–$4kT_c$ [40]. Definitive and reproducible measurements of physical properties critical to a complete understanding of superconductivity in A_3C_{60} are still needed.

SUMMARY

In this paper we have described the synthesis, structures and physical properties of A_xC_{60}. Most of the compounds can be synthesized by direct reaction of C_{60} with alkali metals at 200–300°C, although a wider variety of chemistry is available with the exception of the FCC A_6C_{60} structure. The compounds are best described as salts of C_{60}^{n-} and have structures consistent with those predicted on the basis of ion size and electrostatic considerations. The A_3C_{60}s form a class of isostructural superconductors with T_c scaling with unit cell size. The important parameters needed to understand superconductivity in the A_3C_{60}s are emerging, but there is considerable diversity in data from different sources and experiments that is typical of materials at this stage of development.

Acknowledgements—We would like to thank A. Makhija and S. M. Zahurak for purification of C_{60}, and S. H. Glarum, M. Grabow, A. F. Hebard, A. R. Kortan, T. T. M. Palstra, K. Rabe and M. A. Schlüter for stimulating discussions. We also wish to thank those who have shared their preprints with us.

REFERENCES

1. Kroto H. W., Heath J. R., O'Brien S. C., Curl R. F. and Smalley R. E., *Nature* 318, 162 (1985).
2. Krätschmer W., Lamb L. D., Fostiropoulos K. and Huffman D. R., *Nature* 347, 354 (1990).
3. See, for example, (a) *Fullerenes*, ACS Symp. Ser. 481 (Edited by Hammond G. S. and Kuck V. J.). American Chemical Society, Washington D.C. (1992); (b) *Clusters and Cluster-assembled Materials* (Edited by Averback R. S., Bernholc J. and Nelson D. L.). Mater. Res. Soc. Symp. Proc., Vol. 206, Materials Research Society, Boston (1991).
4. Hebard A. F., Rosseinsky M. J., Haddon R. C., Murphy D. W., Glarum S. H., Palstra T. T. M., Ramirez A. P. and Kortan A. R., *Nature* 350, 600 (1991).
5. Holczer K., Klein O., Huang S.-M., Kaner R. B., Fu K.-J., Whetten R. L. and Diederich F., *Science* 252, 1154 (1991).
6. Stephens P. W., Mihaly L., Lee P. L., Whetten R. L., Huang S.-M., Kaner R., Diederich F. and Holczer K., *Nature* 351, 632 (1991).
7. Rosseinsky M. J., Ramirez A. P., Glarum S. H., Murphy D. W., Haddon R. C., Hebard A. F., Palstra T. T. M., Kortan A. R., Zahurak S. M. and Makhija A. V., *Phys. Rev. Lett.* 66, 2830 (1991).
8. Fleming R. M., Ramirez A. P., Rosseinsky M. J., Murphy D. W., Haddon R. C., Zahurak S. M. and Makhija A. V., *Nature* 352, 787 (1991).
9. Tanigaki K., Ebbesen T. W., Saito S., Mizuki J., Tsai J. S., Kubo Y. and Kuroshima S., *Nature* 352, 222 (1991).
10. Kelty S. P., Chen C.-C. and Lieber C. M., *Nature* 352, 223 (1991).
11. Rosseinsky M. J., Murphy D. W., Fleming R. M., Tycko R., Ramirez A. P., Siegrist T., Dabbagh G. and Barrett S. E., *Nature*, 356, 416 (1992).
12. Gu C., Stepniak F., Poirier D. M., Jost M. B., Benning P. J., Chen Y., Ohno T. R., Martins J. L., Weaver J. H., Fure J. and Smalley R. E., *Phys. Rev. B* 45, 6348 (1992).
13. Fleming R. M., Rosseinsky M. J., Ramirez A. P., Murphy D. W., Tully J. C., Haddon R. C., Siegrist T., Tycko R., Glarum S. H., Marsh P., Dabbagh G., Zahurak S. M., Makhija A. V. and Hampton C., *Nature* 352, 701 (1991).
14. Zhou O., Fischer J. E., Coustel N., Kycia S., Zhu Q., McGhie A. R., Romanow W. J., McCauley Jr., J. P., Smith A. B. III and Cox D. E., *Nature* 351, 462 (1991).
15. Dresselhaus M. S. and Dresselhaus G., *Adv. Phys.* 30, 139 (1981).
16. McCauley J. P., Jr., Zhu Q., Coustel N., Zhou O., Vaughan G., Idziak S. H. J., Fischer J. E., Tozer S. W., Groski D. M., Bykovetz N., Lin C. L., McGhie A. R., Allen B. H., Romanow W. J., Denenstein A. M. and Smith A. B. III, *J. Am. Chem. Soc.* 113, 8537 (1991).
17. Yvon K., Jeitschko W. and Parthe E., University of Geneva.
18. Gabe E. J., LePage Y., Charland J. P., Lee F. L. and White P. S., *J. appl. Cryst.* 22, 384 (1989).
19. Fleming R. M., Siegrist T., Marsh P. M., Hessen B., Kortan A. R., Murphy D. W., Haddon R. C., Tycko R., Dabbagh G., Mujsce A. M., Kaplan M. L. and Zahurak S. M., *Mater. Res. Soc. Symp. Proc.* 206, 691 (1991).
20. Heiney P. A., Fischer J. E., McGhie A. R., Romanow W. J., Denenstein A. M., McCauley J. P., Smith A. B. and Cox D. E. *Phys. Rev. Lett.* 66, 2911 (1991).
21. Yannoni C. S., Johnson R. D., Meijer G., Bethune D. S. and Salem J. R., *J. Phys. Chem.* 95, 9 (1991).
22. Tycko R., Haddon R. C., Dabbagh G., Glarum S. H., Douglass D. C. and Mujsce A. M., *J. Phys. Chem.* 95, 518 (1991).
23. David W. I. F., Ibberson R. M., Matthewman J. C., Prassides K., Dennis T. J., Hare J. P., Kroto H. W., Taylor R. and Walton D. R. M., *Nature* 353, 147 (1991).
24. Liu S., Lu Y.-J., Kappes M. M. and Ibers J. A., *Science* 254, 408 (1991).
25. Tycko R., Dabbagh G., Fleming R. M., Haddon R. C., Makhija A. V. and Zahurak S. M., *Phys. Rev. Lett.* 67, 1886 (1991).
26. Kortan A. R., Kopylov N., Glarum S., Gyorgy E. M., Ramirez A., Fleming R. M., Thiel F. A. and Haddon R. C., *Nature* 355, 529 (1992).
27. Tycko R., Dabbagh G., Rosseinsky M. J., Murphy D. W., Fleming R. M., Ramirez A. P. and Tully J. C., *Science* 253, 884 (1991).
28. Rabe K. M., Vandenberg J. M. and Phillips J. C., preprint.
29. Weiss A., Sauermann G. and Thirase G., *Chem. Ber.* 116, 74 (1983).
30. Harrison M. R., Edwards P. P., Klinowski J., Thomas J. M., Johnson D. C. and Page C. J., *J. Solid State Chem.* 54, 330 (1984).

31. Zhu Q., Zhou O., Coustel N., Vaughan G. B. M., McCauley J. P., Jr., Romanow W. J., Fischer J. E. and Smith A. B. III, *Science* **254**, 545 (1991).

32. Murphy D. W., Rosseinsky M. J., Haddon R. C., Ramirez A. P., Hebard A. F., Tycko R., Fleming R. M. and Dabbagh G., *Physica C.* **185–189**, 403 (1991).

33. Dubois D., Kadish K. M., Flanagan S., Haufler R. E., Chibante L. P. F. and Wilson L. J., *J. Am. Chem. Soc.* **113**, 4364 (1991).

34. Bausch J. W., Prakash G. K. S., Olah G. A., Tse D. S., Lorents D. C., Bae Y. K. and Malhortra R., *J. Am. Chem. Soc.* **113**, 3205 (1991).

35. Wang H. H., Kini A. M., Savall B. M., Carlson K. D., Williams J. M., Lathrop M. W., Lykke K. R., Parker D. H., Wurz P., Pellin M. J., Gruen D. M., Welp U., Kwok W.-K., Fleshler S., Crabtree G. W., Schirber J. E. and Overmyer D. L., *Inorg. Chem.* **30**, 2962 (1991).

36. Wang H. H., Kini A. M., Savall B. M., Carlson K. D., Williams J. M., Lykke K. R., Wurz P., Parker D. H., Pellin M. J., Gruen D. M., Welp U., Kwok W.-K., Fleshler S. and Crabtree G. W., *Inorg. Chem.* **30**, 2838 (1991).

37. Wudl F., Hirsch A., Khemani K. C., Suzuki T., Allemand P.-M., Koch A., Eckert H., Srdanov G. and Webb H. M., *Am. Chem. Soc. Symp. Series* **481**, 161 (1992).

38. Fleming R. M., Kortan A. R., Hessen B., Siegrist T., Thiel F. A., Marsh P., Haddon R. C., Tycko R., Dabbagh R., Kaplan M. and Mujsce A. M., *Phys. Rev. B.* **44**, 888 (1991).

39. Lagrange P. and Setton R., in *Graphite Intercalation Compounds I: Structure and Dynamics* (Edited by Zabel H. and Solin S. A.), p. 283 Springer, Berlin.

40. Tycko R., Dabbagh G., Rosseinsky M. J., Murphy D. W., Ramirez A. P. and Fleming R. M., *Phys. Rev. Lett.* **68**, 1912 (1992).

41. Uemura Y. J., Keren A., Le L. P., Luke G. M., Sternlieb B. J., Wu W. D., Brewer J. H., Whetten R. L., Huang S. M., Lin S., Kaner R. B., Diederich F., Donovan S., Grüner G. and Holczer K., *Nature* **352**, 605 (1991).

42. Clem J. R. and Kogan V. G., *Jap. J. Appl. Phys. Suppl.* **26**, 1161 (1987).

43. Schirber J. E., Overmyer D. L., Wang H. H., Williams J. M., Carlson K. D., Kini A. M., Pellin M. J., Welp U. and Kwok W.-K., *Physica C* **178**, 137 (1991).

44. Sparn G., Thompson J. D., Huang S.-M., Kaner R. B., Diederich F., Whetten R. L., Grüner G. and Holczer K., *Science* **252**, 1829 (1991).

45. Sparn G., Thompson J. D., Whetten R. L., Huang S.-M., Kaner R. B., Diederich F., Grüner G. and Holczer K., *Phys. Rev. Lett.* **68**, 1228 (1992).

46. Zhou O., Vaughan G. B. M., Zhu Q., Fischer J. E., Heiney P. A., Coustel N., McCauley J. P. and Smith A. B., III, *Science* **255**, 833 (1992).

47. Haddon R. C., Hebard A. F., Rosseinsky M. J., Murphy D. W., Glarum S. H., Palstra T. T. M., Ramirez A. P., Duclos S. J., Fleming R. M., Siegrist T. and Tycko R., *Am. Chem. Soc. Symp. Series* **481**, 71 (1992).

48. Haddon R. C., Hebard A. F., Rosseinsky M. J., Murphy D. W., Duclos S. J., Lyons K. B., Miller B., Rosamilia J. M., Fleming R. M., Kortan A. R., Glarum

S. H., Makhija A. V., Muller A. J., Eick R. H., Zahurak S. M., Tycko R., Dabbagh G. and Thiel F. A., *Nature* **350**, 320 (1991).

49. Kochanski G. P., Hebard A. F., Haddon R. C. and Fiory A. T., *Science* **255**, 184 (1992).

50. Palstra T. T. M., Haddon R. C., Hebard A. F. and Zaanen J., *Phys. Rev. Lett.* **68**, 1054 (1992).

51. Wong W. H., Hanson M., Clark W. G., Grüner G., Thompson J. D., Whetten R. L., Huang S.-M., Kaner R. B., Diederich F., Petit P., Andre J.-J. and Holczer K., *Europhys. Lett.* **18**, 79 (1992).

52. Glarum S. H., Duclos S. J. and Haddon R. C., *J. Am. Chem. Soc.*, **114**, 1996 (1992).

53. Weaver J. H., Martins J. L., Komeda T., Chen Y., Ohno T. R., Kroll, G. H., Troullier N., Haufler R. E. and Smalley R. E., *Phys. Rev. Lett.* **66**, 1741 (1991).

54. Benning P. J., Martins J. L., Weaver J. H., Chibante L. B. F. and Smalley R. E., *Science* **252**, 1417 (1991).

55. Wertheim G. K., Rowe J. E., Buchanan D. N. E., Chaban E. E., Hebard A. F., Kortan A. R., Makhija A. V. and Haddon R. C., *Science* **252**, 1419 (1991).

56. Chen C. T., Tjeng L. H., Rudolf P., Meigs G., Rowe J. E., Chen J., McCauley J. P. Jr, Smith A. B. III, McGhie A. R., Romanow W. J. and Plummer E. W., *Nature* **352**, 603 (1991).

57. Haddon R. C., *Accts. Chem. Res.*, **25**, 127 (1992).

58. Slichter C. P., *Principles of Magnetic Resonance.* Springer, New York, 3rd ed. (1990).

59. McMillan W. L., *Phys. Rev.* **167**, 331 (1968).

60. Varma C. M., Zaanen J. and Raghavachari K., *Science* **254**, 989 (1991).

61. Schlüter M. A., Lannoo M., Needels M. and Baraff G. A., *Phys. Rev. Lett.* **68**, 526 (1992).

62. Schlüter M. A., Lannoo M., Needels M. and Baraff G. A., *J. Phys. Chem. Solids* **53**, 1473 (1992).

63. Zhang F. C., Ogato M. and Rice T. M., *Phys. Rev. Lett.* **67**, 3452 (1991).

64. Chakravarty S., Gelfand M. P. and Kivelson S., *Science* **254**, 970 (1991).

65. Baskaran G. and Tossati E., *Curr. Sci.* **61**, 33 (1991).

66. Anderson P. W., preprint.

67. Duclos S. J., Haddon R. C., Glarum S. H., Hebard A. F. and Lyons K. B., *Science* **254**, 1625 (1991).

68. Prassides K., Tomkinson J., Christides C., Rosseinsky M. J., Murphy D. W. and Haddon R. C., *Nature* **354**, 462 (1991).

69. Ebbesen T. W., Tsai J. S., Tanigaki K., Tabuchi J., Schimakawa Y., Kubo Y., Hirosawa I. and Mizuki J. *Nature* **355**, 620 (1992).

70. Ramirez A. P., Kortan A. R., Rosseinsky M. J., Duclos S. J., Mujsce A. M., Haddon R. C., Murphy D. W., Makhija A. V., Zahurak S. M. and Lyons K. B., *Phys. Rev. Lett.* **68**, 1058 (1992).

71. Chen C.-C. and Lieber C. M., *J. Amer. Chem. Soc.* **114**, 3141 (1992).

72. Zhang Z., Chen C.-C., Kelty S. P., Dai H. and Lieber C. M., *Nature* **353**, 333 (1991).

73. Zhang Z., Chen C.-C. and Lieber C. M., *Science* **254**, 1619 (1991).

74. Rotter L. D., Schlesinger Z., McCauley J. P., Jr., Coustel N., Fischer J. E. and Smith A. B., III, *Nature* **355**, 532 (1992).

STRUCTURE, DYNAMICS AND ORDERING TRANSITION OF SOLID C$_{60}$

Paul A. Heiney

Department of Physics and Laboratory for Research on the Structure of Matter,
University of Pennsylvania, Philadelphia, PA 19104, U.S.A.

(*Received* 26 *February* 1992)

Abstract—Solid C$_{60}$ forms a face-centered-cubic structure at room temperature, in which the molecules exhibit dynamic orientational disorder. At 255 K the molecules develop orientational order via a first order phase transition, lowering the space group symmetry to $Pa\bar{3}$ in a simple cubic lattice. Even in the ordered phase, short-range disorder persists to the lowest temperatures. The $Pa\bar{3}$ structure is stabilized by both Lennard-Jones atomic pair potentials and by Coulombic forces arising from excess charge in the double bonds. The low temperature dynamics are primarily characterized by jump rotational diffusion between equivalent orientations superimposed on small amplitude librational oscillations. The behavior at the phase transition is strongly influenced by trace impurities and solvent. Similar effects are seen in C$_{60}$O and in C$_{70}$.

Keywords: Fullerene, X-ray, neutron, phase transition, crystal structure, disorder.

1. INTRODUCTION

C$_{60}$, buckminsterfullerene, has attracted interest due to both its elegant molecular symmetry and the intriguing structural, dynamic, and electronic properties of its solid phase and doped derivatives [1–4]. The C$_{60}$ molecule, which has icosahedral symmetry, has a shape that is very close to spherical. Accordingly, the interaction between neighboring molecules is dominated by the spherically symmetric component, and at room temperature the molecules are rotationally disordered, forming a "plastic crystal". The crystal lattice is face-centered-cubic (fcc). Below 255 K, the four molecules in the conventional unit cell of an fcc lattice become orientationally inequivalent, and the system undergoes a first order transition to a simple-cubic (sc) structure with $Pa\bar{3}$ symmetry. This transition, as well as holding structural interest, has a pronounced effect on the dynamics of the crystal, and hence is observed by dynamical probes such as nuclear magnetic resonance (NMR), Raman spectroscopy, and inelastic neutron scattering.

The remainder of this article is organized as follows. In section 2 we discuss in a general way the space groups and symmetry concepts that we will need to understand the structural measurements of both the low and high temperature phases of C$_{60}$. In section 3 we discuss measurements of the room temperature properties of C$_{60}$. In section 4 we describe calorimetric and structural measurements of the phase transition to an orientationally ordered phase, and discuss in some detail the structure of the low tem-

perature phase. The properties of the low temperature phase and the ordering transition are affected in important ways by both morphological disorder and residual impurities, as described in section 5. Section 6 summarizes other techniques that have been used to probe the ordering transition. In section 7 we examine the various theoretical approaches that have been taken to understand the ordering transition and the low temperature phase. C$_{60}$ is not the only fullerene to display an orientational ordering transition; section 8 describes the current state of our knowledge of C$_{70}$ and related systems. Finally, we conclude with some speculations on future directions in section 9.

2. SYMMETRY AND DIFFRACTION

Before presenting detailed research results, it is useful to consider in general the ways in which icosahedral structures can be accommodated within a crystal lattice. Periodic translational order is incompatible with icosahedral point symmetry. Thus, there are several ways to incorporate a C$_{60}$ molecule into an ordered lattice: (i) the molecule itself may undergo a (hopefully slight) symmetry breaking to a lower symmetry structure which is consistent with a crystal symmetry; (ii) the molecule may be thermally or statically disordered so that the point symmetry is *raised*, making the molecule appear more spherical; or (iii) an icosahedral quasicrystal, rather than crystal, lattice may form.

Quasicrystals are structures which display sharp diffraction patterns with icosahedral point group symmetry, thus necessitating quasiperiodic rather than periodic translational order [5, 6]. The quasicrystals observed experimentally to date have been composed to binary or ternary metallic alloys; empirically, at least, more than one structural unit seems to be required. Although there has been one report [7] of quasicrystallinity in a solvated C_{60} sample, it is generally agreed that pure C_{60} always forms crystalline phases, and accordingly we will henceforth confine our attention to periodic crystalline symmetries.

If the C_{60} molecule is considered to be a smooth spherical ball, then one would expect that the crystalline structure would be one of two possible close packings: hexagonal-close-packed (hcp) or fcc. Both have a packing fraction of 0.74. Although minimization of a Lennard-Jones potential favors an hcp

ground state structure, it is experimentally found that all inert gas atoms except for ^{3}He and ^{4}He form fcc crystals [8]. Even if deviations from spherical symmetry are considered, the actual crystal structure should be a relatively small perturbation on the full fcc symmetry.

The possible space groups consistent with icosahedral molecular symmetry have been discussed by Fleming *et al.* [9]. The C_{60} molecule takes the form of a truncated icosahedron (Fig. 1), and has molecular point group symmetry $m\bar{3}\bar{5}$ (I_h). All carbon atoms are equivalent; pseudo-hexagonal faces are centers of three-fold rotational symmetry, pentagonal faces are centers of five-fold rotational symmetry, edges between two hexagons are centers of two-fold rotational symmetry, and the entire object is symmetric under inversion. Periodic translational symmetry is incompatible with five-fold rotational symmetry, and

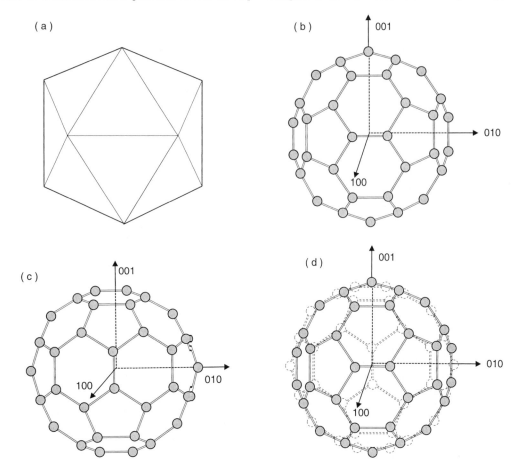

Fig. 1. Icosahedral symmetry and the C_{60} molecule (adapted from [9]). (a) A regular icosahedron. Vertices are centers of five-fold symmetry, triangular faces are centers of three-fold symmetry, and edges are centers of two-fold symmetry. The entire object has inversion symmetry. (b) Truncated icosahedron, created by flattening the vertices of a regular icosahedron. The symmetry elements remain the same. Carbon atoms are placed on the vertices. Hexagon–hexagon and hexagon–pentagon edges are inequivalent and are not in general the same length. The molecule is shown placed in one of two 'standard orientations' such that Cartesian axes in a cubic crystal pass through three orthogonal two-fold aces. (c) The other standard orientation, related to that in (b) by a rotation of $\pi/2$ about any Cartesian axis. (d) Higher-symmetry object created via 'merohedral disorder': the superposition of two molecules with the orientations in (b) and (c).

the maximal subgroup consistent with crystalline symmetry is $m\bar{3}$; when a static C_{60} molecule is placed in a cubic crystalline lattice the icosahedral symmetry is broken by the environment and there must be at least three inequivalent types of carbon atom in the molecule. However, the actual fractional distortion of bond lengths within the molecule may be unobservably small.

In the simplest possible crystalline structure, all molecules in the lattice have the same orientation. For this case, the relative orientation of a truncated icosahedron placed at the origin of a cubic lattice is shown in Fig. 1b: [100] axes pass through three orthogonal two-fold molecular axes (hexagon–hexagon edges), and the four [111] axes pass through hexagonal faces. In the remainder of the paper we will refer to this as the "standard orientation". There are actually two such standard orientations of the molecule; they are related by a $\pi/2$ rotation about any of the [100] axes, or equivalently by a reflection through a (110) plane (Fig. 1b and c). For example, consider the double bond (two-fold molecular symmetry axis) that passes through the [100] crystal axis: this bond may be directed along either the [010] or the [001] direction. In either orientation, the Cartesian [100], [010] and [001] cubic crystal axes are no longer centers of four-fold rotational symmetry, but only of two-fold symmetry. In an fcc structure, four icosahedra are placed at the origin and at the mid-points of three faces of a cubic cell, all with the same standard orientation, to give a lattice with symmetry $Fm\bar{3}$. Similarly, in a body-centered-cubic (bcc) lattice the icosahedra are placed at the origin and center of the crystal cube, with the same orientation, to form a crystal with $Im\bar{3}$ symmetry. This more loosely-packed structure is not seen in pure C_{60}, but is found in the alkali-doped compound K_6C_{60} [10].

The standard orientation is found quite commonly in nature in metallic crystals which have icosahedral local symmetry [6, 11, 12]. Both the AlLiCu and AlMnSi ternary alloys form quasicrystals under conditions of rapid quenching. The same alloys form conventional crystals when cooled more slowly, and the quasicrystalline structure of the quenched alloys is thought to be related to the icosahedral co-ordination of metal atoms in slowly cooled crystals. Crystalline Al_5CuLi_3, for example, is composed of concentric shells of metal atoms to form clusters with icosahedral symmetry, which are then organized in a bcc lattice to form a crystal with $Im\bar{3}$ symmetry. (Of course, the different icosahedral clusters share some atoms; the structure is considerably more complicated than rigid shells of 60 atoms.) α(AlMnSi) has a similar structure, although since the cluster at the body center position is slightly different from that at

the origin the lattice is primitive cubic, space group $Pm\bar{3}$. Likewise, the structural chemistry of elemental boron [13] is dominated by the regular icosahedron, although again adjacent icosahedra may share atoms, or at the very least be strongly attached via covalent bonds. There may be as many as 16 allotropes of boron. Two of the most common allotropes are the R-12 crystal, with 12 borons in a rhombohedral $Rm\bar{3}$ structure, and the R-105 crystal, in which atoms form concentric dodecahedral and icosahedral shells. The standard orientation is also seen in virus structure. One of the best examples is the tomato bushy stunt virus, which takes the form of a truncated icosahedron; it crystallizes in a bcc lattice, space group $I23$, with a unit cell edge length of 386 Å [14].

The fcc structure can also have its symmetry *raised* by incorporating statistical disorder. An X-ray diffraction experiment measures the thermodynamic average of instantaneous Fourier transforms of the charge density. If either static or thermal disorder is present in the sample, an X-ray diffraction pattern will probe the average structure within the unit cell [15]. Two plausible types of orientational disorder are "merohedral disorder" and "merohedral twinning" [9, 16]. An ordinary twinned crystal is composed of large crystallites separated by grain boundaries, within each of which the molecules have the same orientation; merohedral twinning corresponds to the case where adjacent crystallites are related by a mirror reflection. As used in the present paper, 'merohedral disorder' refers to random site-by-site population of the two different standard orientations. This means that the 'average' molecule probed has 120 carbon atom sites, each with half-occupancy, as shown in Fig. 1d. Both types of disorder raise the apparent symmetry of a nominal single crystal to $Fm\bar{3}m$, since the [100] axes now have full four-fold rotational symmetry. When analysing diffraction patterns, however, there is in principle a clear distinction between the two: in the case of merohedral twinning we add the calculated *intensities* of crystals with two different orientations, whereas in the case of merohedral disorder we add the calculated structure factor *amplitudes* of molecules with two different orientations. In contrast to the orientational order observed in K_6C_{60} [10], the C_{60} molecules in superconducting K_3C_{60} are in fact observed to display merohedral disorder [17].

The orientational disorder is further increased if one assumes that the C_{60} molecules assume completely random orientations [18]. In this case, the 60 carbon atoms at each site are replaced by a spherical shell of charge. The crystal symmetry is still $Fm\bar{3}m$, so that while X-ray diffraction peak intensities may be subtly different from the merohedral disorder case, there will be no distinction in the symmetries or systematic extinctions of diffraction peaks.

The fcc structure can also have its symmetry *reduced* in several ways. One can stretch or shrink the cubic unit cell along a single [111] direction, maintaining all icosahedra in the same relative orientation, to give a rhombohedral unit cell ($Rm\bar{3}$). Alternatively, one can make the four sites in the unit cell inequivalent, resulting in a primitive cubic lattice. At low temperature, C_{60} is observed to transform to an sc structure with approximately the same cube edge length as the high temperature fcc phase. To anticipate section 4, this comes about neither by a translation of the molecular centers of mass, nor by a distortion of the molecular structure, but by the development of inequivalent orientations of the four molecules within the cubic cell.

The possible orientationally ordered sc structures have been discussed in detail by Harris and Sachidanandam [19]. The only plausible space groups are $Pm\bar{3}$, $Pn\bar{3}$ and $Pa\bar{3}$. In $Pm\bar{3}$, the molecule at the origin is given one of the two standard orientations, and the three molecules on the faces are given the other orientation. This makes the corner and face molecules inequivalent, lowering the symmetry from fcc to sc. In both $Pn\bar{3}$ and $Pa\bar{3}$, we arrive at the simple cubic structure as follows: we start with all four molecules in one of the two standard orientations, and then each molecule is rotated by an angle Γ about a different [111] axis. (There are four molecules, and four different [111] axes.) The difference between the two structures lies in the assignment of rotation axes to molecules. Specifically, in the $Pn\bar{3}$ structure the molecules at (000), $(\frac{1}{2}0\frac{1}{2})$, $(\frac{11}{22}0)$, and $(0\frac{11}{22})$ are rotated by Γ about the [111], [$\bar{1}1\bar{1}$], [$\bar{1}\bar{1}1$], and [$1\bar{1}\bar{1}$] axes, respectively, while in the $Pa\bar{3}$ structure they are rotated about the [111], [$\bar{1}\bar{1}1$], [$1\bar{1}\bar{1}$], and [$\bar{1}1\bar{1}$] axes. (In fact, a rotation about the [111], [$\bar{1}1\bar{1}$], [$\bar{1}\bar{1}1$], and [$1\bar{1}\bar{1}$] axes is an equivalent $Pa\bar{3}$ structure.) In both the $Pa\bar{3}$ and $Pn\bar{3}$ cases, the rotation angle is not fixed by symmetry. The diffraction intensities in both cases are symmetric under cyclic permutation of Bragg indices, $I(HKL) = I(KLH) = I(LHK)$, but in general are not symmetric under exchange of indices, $I(HKL) \neq I(KHL)$. The $Pa\bar{3}$ and $Pn\bar{3}$ structures are distinct, both from a symmetry point of view and in terms of local coordinations, and can be clearly distinguished in a diffraction experiment from the systematically absent reflections. The $Pn\bar{3}$ structure can also be excluded on this basis.

3. ROOM TEMPERATURE PROPERTIES

In this section we present the current state of knowledge of the room temperature properties of C_{60}.

We will first discuss the static properties, as determined by X-ray, neutron, and electron diffraction measurements, and then the dynamical properties, as determined primarily by Raman scattering, NMR and inelastic neutron scattering.

The first measurements on crystalline C_{60} were reported in the seminal paper of Krätschmer et al. [2], which also presented the soot preparation technique that made the field of solid-state fullerene research possible. Although it is now generally accepted that the room temperature structure of pure C_{60} is fcc, Krätschmer et al. identified the lattice as hcp from X-ray powder diffraction data. (Recall that both hcp and fcc are close-packed lattices which are energetically and structurally quite similar.) There were several reasons for this misidentification. (1) The hcp and fcc lattices actually have a large number of Bragg peaks in common, so that a subset of the peaks can be indexed according to either scheme. (2) Because of systematic extinctions due to the molecular structure factor (discussed below), a number of the fcc reflections, and in particular the $H00$ peaks with H even, are missing. (3) Due to the presence of stacking faults, additional features can appear in a powder pattern. In particular, there is generally a feature, just below the fcc 111 peak, which is not an allowed fcc peak position but which coincidentally falls at the position of the hexagonal $\bar{1}10$ peak. (4) Krätschmer's sample contained higher fullerenes such as C_{70} and also undoubtedly some residual solvent. It is now understood that solvent can dramatically alter the structure, nucleating in some cases an hcp phase [20] and in other cases low symmetry monoclinic or primitive hexagonal structures [9, 16, 21]. Thus, it is quite possible that portions of the Krätschmer et al. sample were actually hcp.

Measurements by Fleming et al. [9, 16] on single crystals of C_{60}, grown by sublimation and therefore substantially free of solvent, C_{70}, and other impurities, established that the crystal structure is in fact fcc. Four equivalent molecules are contained in a unit cube with edge length $a = 14.198$ Å, at the origin and face centers. Analysis of diffraction peak intensities showed a $Fm\bar{3}m$ apparent symmetry, which could arise from either merohedral twinning or merohedral disorder of $Fm\bar{3}$ (i.e. molecules with both orientation, but with either long or short range order in the choice of standard orientation). It was also necessary to incorporate large thermal factors, indicating librational fluctuations of the molecules. The molecular radius was determined to be 3.5 Å.

A subsequent powder X-ray diffraction measurement by Fischer et al. [18] confirmed the fcc structure, but took a different approach to analysing the structure. Motivated in part by NMR measurements, discussed below, complete orientational disorder rather than merohedral disorder was assumed. In general, at scattering vector $\mathbf{Q}$ the structure factor $S(\mathbf{Q})$ is given by

$$S(\mathbf{Q}) = \sum_j f_j \exp(-i\mathbf{Q} \cdot \mathbf{r}_j), \qquad (1)$$

where f_j and $\mathbf{r}_j$ are the atomic form factor and position of the jth atom within the unit cell. (We use the convention $\mathbf{Q} = \mathbf{k}_i - \mathbf{k}_f$, with magnitude $Q = (4\pi \sin \theta)/\lambda = 2\pi/d$, where λ is the X-ray wavelength and d is the plane spacing probed. At the Bragg peak with indices HKL, $\mathbf{Q} = \mathbf{G}_{HKL}$. The X-ray peak intensity $I(\mathbf{G})$ is proportional to $|S(\mathbf{G})|^2$.) If we assume that the

C$_{60}$ molecule consists of 60 carbon atoms (each with atomic factor f_c) spread "smoothly" over a sphere of radius R, then the structure factor for each molecule becomes

$$S(\mathbf{Q}) = \frac{60f_c}{4\pi} \int_0^{2\pi} d\phi \int_{-1}^1 d(\cos \theta) e^{-iQR\cos\theta}$$
$$= \frac{60f_c \sin(QR)}{QR} \equiv 60f_c j_0(QR), \quad (2)$$

where j_0 is the zeroth order spherical Bessel function. This simple model, with only one adjustable parameter, provides excellent agreement with the measured X-ray peak intensities once geometrical terms such as the Lorentz-polarization factor and the peak multiplicity are included [18, 22], as shown in Fig. 2b and c. (The improved data of [22] give $R = 3.52 \pm 0.01$ Å and $a = 14.17 \pm 0.01$ Å at 300 K.) In particular, the model explains why even-order $H00$ peaks are not observed in powder diffraction experiments, even though they are allowed by the fcc selection rules: they fall very close to zeros of the Bessel function. By coincidence, even-order $H00$ peaks fall at positions $Q = G_{2n,0,0}$ with

$$(G_{2n,0,0})(R) = \left(\frac{2\pi(2n)}{14.17 \text{ Å}}\right)(3.52 \text{ Å}) = 0.99n\pi \quad (3)$$

so that $\sin(G_{2n,0,0}R) \simeq 0$. This fact also makes the $H00$ reflections useful probes of deviations from the pure atmospheric pressure C$_{60}$ structure: if the lattice contracts under pressure, or if foreign atoms are introduced into the lattice, the 200 peak, for example, may acquire measurable intensity [24].

Actually, powder diffraction measurements to date really cannot distinguish between the merohedral disorder and spherical shell models; both give crystallographic goodness-of-fit R-factors of 8–10%. Since dynamical measurements such as NMR indicate a high degree of rotational disorder at room temperature, the spherical shell model is preferred on physical grounds, but the crystal lattice surely imposes some degree of preferential orientation on the molecules. To the extent that it will ever be possible to establish the degree of orientational order, high-statistics data at very large momentum transfer G will be required, preferably on non-twinned single crystals.

The conclusions of X-ray diffraction measurements are generally supported by electron microscopy studies [20, 25–30], which shed additional light on the types of disorder present in crystalline C$_{60}$. A C$_{60}$/C$_{70}$ powder extracted from solvent forms a heavily defected hcp phase [20], which irreversibly transforms to fcc under electron irradiation; presumably the primary effect of the electron beam in this case is to evaporate the residual solvent. As first observed by

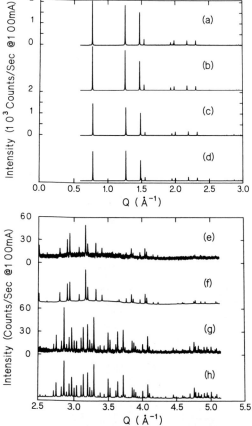

Fig. 2. Synchrotron X-ray powder diffraction patterns from C$_{60}$ [23]. (a)–(d) show the lower scattering-angle portion of the data, and (e)–(f) the higher-angle portion. X-ray intensities are calibrated in photons per second, normalized to a synchrotron ring current of 100 mA. Data collection times were typically 2–10 s per point with the ring current between 100 and 200 mA. (a) and (e): Measured diffraction pattern at 300 K. (b) and (f): Pattern calculated by least-squares refinement of the 300 K data, using the spherical disorder model outlined in eqn (2). The background was smoothly interpolated from data points far from Bragg peaks. (c) and (g): Measured diffraction pattern at 15 K. (d) and (h): Pattern calculated by least-squares refinement of the 15 K data. As discussed in section 5, this fit used the $Pa\bar{3}$ model with 85% of the molecules at the 24° orientation and 15% at the 84° orientation.

and can be disregarded in the remainder of our discussion. I am grateful to R. Moret for bringing this correction to my attention.

53. Michel K. H., Copley J. R. D. and Neumann D. A., *Phys. Rev. Lett.* **68**, 2929 (1992).
54. See the article by O. Zhou and D. E. Cox in this issue. [*J. Phys. Chem. Solids* **53**, 1373 (1992).]
55. Heiney P. A., Fischer J. E., McGhie A. R., Romanow W. J., Denenstein A. M., McCauley J. P. Jr, Smith A. B. III and Cox D. E., *Phys. Rev. Lett.* **67**, 1468 (1991).
56. Liu S., Lu Y. J., Kappes M. M. and Ibers J. A., *Science* **254**, 408 (1991).
57. David W. I. F., Ibberson R. M., Matthewman J. C., Prassides K., Dennis T. J., Hare J. P., Kroto H. W., Taylor R. and Walton D. R. M., *Nature* **353**, 147 (1991).
58. Guo Y., Karasawa N. and Goddard W. A. III, *Nature* **351**, 464 (1991).
59. Lu J. P., Li X.-P. and Martin R. M., unpublished.
60. Heiney P. A., Vaughan G. B. M., Fischer J. E., Coustel N., Cox D. E. Copley J. R. D., Neumann D. A., Kamitakahara W. A., Creegan K. M., Cox D. M., McCauley J. P. Jr and Smith A. B. III, *Phys. Rev.* **B45**, 4544 (1992).
61. David W. I. F., Ibberson R. M., Dennis T. J. S., Hare J. P. and Prassides K., *Europhys. Lett.* **18**, 219 (1992).
62. Figure 4 supplied by A. Cheng and M. E. Klein, University of Pennsylvania.
63. Buerger M. J., *Crystal Structure Analysis*, pp. 53–74. Wiley, New York (1960).
64. Hu R., Egami T., Li F. and Lannin J. S., *Phys. Rev.* **B45**, 9517 (1992).
65. See, e.g. Toby B. H., Egami T., Jorgensen J. G. and Subramanian M. A., *Phys. Rev. Lett.* **64**, 2414 (1990), and references therein.
66. See, e.g. Zemansky M. W., *Heat and Thermodynamics*, Fifth Edition, pp. 607–626. McGraw-Hill, New York (1968).
67. McGhie A. R., University of Pennsylvania, unpublished.
68. Milliken J., Keller T. M., Baronavski A. P., McElvany S. W., Callahan J. H. and Nelson H. H., *Chemistry of Materials* **3**, 386 (1991).
69. Atake T., Takana T., Kawaji H., Kikuchi K., Saito K., Suzuki S., Ikemoto I. and Achira Y., *Physica C* **185–189**, 427 (1991)
70. Gensterblum G. *et al.*, unpublished HREELS results, and Lucas A. A., private communication; see also the article by A. A. Lucas and J. H. Weaver in this issue. [*J. Phys. Chem. Solids* **53**, xx (1992).]
71. Vaughan G. B. M., Heiney P. A., Fisher J. E., Luzzi D. E., Ricketts-Foot D. A., McGhie A. R., Hui Y. W., Smith A. L., Cox D. E., Romanow W. J., Allen B. H., Coustel N., McCauley J. P. Jr and Smith A. B. III, *Science* **254**, 1350 (1991).
72. Failey M. P., Anderson D. L., Zoller W. H., Gordon G. E. and Lindstrom R. M., *Anal. Chem.* **51**, 2209 (1979).
73. Kiefl R. F., Schneider J. W., McFarlane A., Chow K., Duty T. L., Kreitzman S. R., Estle T. L., Hitti B., Lichti R. L., Ansaldo E. J., Schwab C., Percival P. W., Wei G., Wlodek S., Kojima K., Romanow W. J., McCauley J. P.
74. Jr, Coustel N., Fischer J. E. and Smith A. B. III, *Phys. Rev. Lett.* **68**, 1347 (1992).
74. Neumann D. A., Copley J. R. D., Kamitakahara W. A., Rush J. J., Cappelletti R. L., Coustel N., McCauley J. P. Jr, Fischer J. E., Smith A. B. III, Creegan K. M. and Cox D. M., *J. Chem. Phys.*, in press.
75. Tolbert S. H., Alvisatos A. P., Lorenzana H. E., Kruger M. B. and Jeanloz R., *Chem. Phys. Lett.* **188**, 163 (1992).
76. Hsuang Y., Gilson D. F. R. and Butler I. S., *J. Phys. Chem.* **95**, 5723 (1991).
77. Samara G. A., Schirber J. E., Morosin B., Hansen L. V., Loy D. and Sylwester A. P., *Phys. Rev. Lett.* **67**, 3136 (1991).
78. Yu R. C., Tea N., Salamon M. B., Lorents D. and Malotra R., *Phys. Rev. Lett.* **68**, 2050 (1992).
79. Another topic that has excited considerable interest in the theoretical community, the origin of superconductivity observed in alkali-doped C_{60}, is discussed in the article by M. Schlüter in this issue. [*J. Phys. Chem. Solids* **53**, 1473 (1992).]
80. Zhang Q.-M., Yi J.-Y. and Bernholc J., *Phys. Rev. Lett.* **66**, 2633 (1991).
81. Cheng A. and Klein M. L., *J. Phys. Chem.* **95**, 6750 (1991).
82. Cheng A. and Klein M. L., *Phys. Rev.* **B45**, 1889 (1992).
83. Gunnarsson O., Satpathy S., Jepsen O. and Andersen O. K., *Phys. Rev. Lett.* **67**, 3002 (1991).
84. Lu J. P., Li X.-P. and Martin R. M., *Phys. Rev. Lett.* **68**, 1551 (1992).
85. Sprik M., Cheng A. and Klein M. L., *J. Phys. Chem.* **96**, 2027 (1992).
86. Cullen J. R., Mukamel D., Shtrikman S., Levitt L. C. and Callen E., *Solid State Commun.* **10**, 195 (1972).
87. James H. M. and Keenan T. A., *J. Chem. Phys.* **31**, 12 (1959).
88. Hawkins J. M., Meyer A., Lewis T. A., Loren S. and Hollander F. J., *Science* **252**, 312 (1991).
89. Fagan P. J., Calabrese J. C. and Malone B., *Science* **252**, 1160 (1991).
90. Creegan K. M., Robbins J. L., Robbins W. K., Millar J. M., Sherwood R. D., Tindall P. J., Cox D. M., McCauley J. P. Jr, Jones D. R., Gallagher R. T. and Smith A. B. III, *J. Am. Chem. Soc.* **114**, 1103 (1992).
91. Vaughan G. B. M., Heiney P. A., Cox D. E., McGhie A. R., Jones D. R., Strongin R. M., Cichy M. A. and Smith A. B. IV, submitted to *Chem. Phys.* (1992).
92. Gelfand M. P. and Lu J. P., *Phys. Rev. Lett.* **68**, 1050 (1992).
93. Diederich F., Ettl R., Rubin Y., Whetten R. L., Beck R., Alvarez M., Anz S., Sensharma D., Wudl F., Khemani K. C. and Koch A., *Science* **252**, 548 (1991).
94. Ettl R., Chao I., Diederich F. and Whetten R. L., *Nature* **353**, 149 (1991).
95. Diederich F., Whetten R. L., Thilgen C., Ettl R., Chao I. and Alvarez M. A., *Science* **254**, 1768 (1991).
96. Chai Y., Guo T., Jin C., Haufler R. E., Chibante L. P. F., Fure J., Wang L., Alford J. M. and Smalley R. E., *J. Phys. Chem.* **95**, 7564 (1992).

NEUTRON SCATTERING STUDIES OF C_{60} AND ITS COMPOUNDS

J. R. D. COPLEY,† D. A. NEUMANN,† R. L. CAPPELLETTI‡ and
W. A. KAMITAKAHARA†

†Materials Science and Engineering Laboratory, National Institute of Standards and Technology,
Technology Administration, U.S. Department of Commerce, Gaithersburg, MD 20899, U.S.A.

‡Department of Physics and Astronomy and Condensed Matter and Surface Sciences Program,
Ohio University, Athens, OH 45701, U.S.A.

(*Received* 10 *February* 1992)

Abstract—We describe neutron scattering studies of the structure and dynamics of C_{60}-fullerite. Above the order–disorder phase transition temperature, T_C, diffraction and quasielastic scattering measurements indicate that the molecules are constantly reorienting while their centers remain on a face-centered cubic lattice. Below T_C the molecules are orientationally ordered with four molecules per unit cell. Recent diffraction results suggest that any given molecule occupies one of two possible orientations with unequal probability, and inelastic scattering measurements show that the molecules librate about their equilibrium orientations. The librational amplitude becomes large as the phase transition is approached. Pulsed source measurements yield direct information regarding the structure of the C_{60} molecule and also support the low temperature two-orientation model. High energy inelastic studies of undoped and alkali metal-doped buckminsterfullerene provide detailed information about intramolecular vibrational frequencies and point to the important role of specific modes in promoting superconductivity at low temperatures.

Keywords: Neutron diffraction, neutron scattering, C_{60}, orientational disorder, order–disorder phase transition, rotational diffusion, libration, intramolecular vibrations.

1. INTRODUCTION

Over the past year neutron scattering investigations have significantly added to our overall understanding of the structural and dynamical properties of buckminsterfullerene, C_{60}. Despite some difficulty in obtaining sufficiently large samples, the broad range of capabilities of the neutron scattering technique has been well demonstrated in a variety of studies of this remarkable molecule. Comparable results for compounds of C_{60}, such as the superconducting materials K_3C_{60} and Rb_3C_{60}, have started to appear, and it is likely that much more will be achieved in the reasonably near future.

Thermal neutrons are a powerful microscopic probe of condensed matter, because their wavelengths are comparable with interatomic distances, and at the same time their energies are of the same order as typical solid state excitational energies. Thus neutrons are used to investigate atomic and molecular structures in analogy with X-rays, and they are also used to probe dynamical phenomena such as molecular vibrations and rotations; in this respect they complement spectroscopic methods such as Raman scattering and infrared absorption. Furthermore typical wavelengths and energies of thermal neutrons, *in combination*, allow one to determine the

spatial character of an excitation in a manner unmatched by other spectroscopic probes. Since neutrons interact primarily with nuclei rather than with atomic electrons, and because the form of the interaction is very simple, the interpretation of experimental measurements is generally quite straightforward. For example the scattering length b, which is the neutron equivalent of the X-ray atomic scattering factor, is independent of the scattering wave vector. The neutron–nucleus interaction is weak so that, with a few notable exceptions, neutrons probe bulk material. By the same token they readily penetrate the walls of sample containers so that experiments at low or high temperature (for example) are not very much more difficult than experiments at ambient temperature. In addition, the nature of the neutron's interaction is such that all the vibrational modes of a molecule are visible whereas only some of these modes can be observed in optical spectroscopy because of selection rules.

In Section 2 we summarize aspects of the neutron scattering technique which are relevant to our later discussion of experiments on C_{60} and its compounds. We also describe neutron scattering instruments and a method to determine the hydrogen content of candidate samples for neutron scattering experiments. In the following section we describe

Figure 6 shows a portion of the room temperature diffraction pattern for C_{60}, as measured at NIST [21]. The Bragg peaks and the diffuse scattering are clearly apparent. The complete pattern is well described by a very simple model which consists of C_{60} molecules with centers on a face-centered cubic (fcc) lattice, randomly oriented, with no correlation between the orientations of different molecules. In this case the powder diffracted intensity may be written as

$$I(Q) = I_B(Q) + I_D(Q), \qquad (8)$$

where the Bragg intensity is

$$I_B(Q) = \overline{|\langle F(\vec{Q})\rangle|^2} \left| \sum_{m=1}^{N} e^{i\vec{Q}\cdot\vec{R}_m} \right|^2 \qquad (9)$$

and the diffuse intensity is

$$I_D(Q) = N \left[\overline{\langle |F(\vec{Q})|^2\rangle} - \overline{|\langle F(\vec{Q})\rangle|^2} \right]. \quad (10)$$

Here $\vec{R}_m$ is the position of molecule m, angle brackets denote an average over all possible molecular orientations with respect to the lattice, an overbar represents a powder average, and the sum in eqn (9)

extends over all N molecules in the system. Furthermore the molecular structure factor

$$F(\vec{Q}) = b \sum_{n=1}^{60} e^{i\vec{Q}\cdot\vec{r}_n}, \qquad (11)$$

where $\vec{r}_n$ is the vector joining atom n to its molecular center. The solid line in Fig. 6 represents the sum of $I(Q)$, scaled and convoluted into the known resolution of the instrument, and a flat background. Note that the relative intensity of the Bragg peaks and the diffuse scattering is determined by the model and is therefore not an adjustable parameter. In Section 4.1 we shall see that the diffuse scattering is quasielastic; there is no evidence of purely elastic diffuse scattering. The rotational disorder is therefore dynamic, with molecules constantly tumbling into different orientations.

Below the phase transition the situation is more interesting and more complicated. Both the increased density of Bragg peaks and the reduced effects of thermal motion on Bragg peak intensities make it hard to separate out the diffuse scattering. In principle the solution is to use very good resolution but this generally implies a restricted Q range. It is also important to minimize the instrumental background.

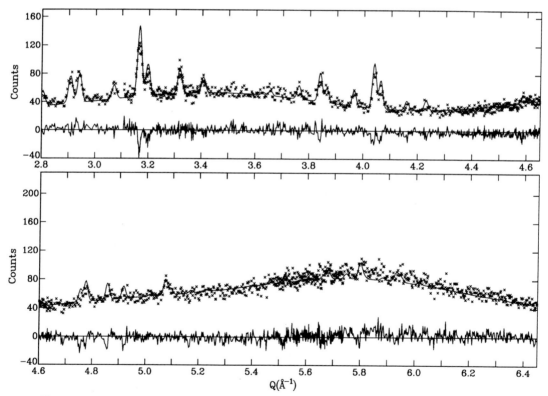

Fig. 6. A portion of the room temperature diffraction pattern of C_{60}. Points represent measurements at NIST [21], normalized to a fixed monitor count. Upper lines were calculated as described in Section 3.2, assuming $d_1 = 1.45$ Å, $d_2 = 1.40$ Å and $a_0 = 14.17$ Å. Lower lines show the difference between measurement and calculation.

To date the highest resolution neutron diffraction measurements on C_{60} are those of David *et al.* [14]. At 5 K these data show definite evidence of diffuse scattering maxima at d-spacings of ~ 1.8 and 1.05–1.1 Å, corresponding to wave vectors of ~ 3.5 and ~ 5.8 Å^{-1}, respectively. The high background in these measurements precludes quantitative conclusions about the size and shape of the diffuse scattering. The measurements of Copley *et al.* [21], though of more modest resolution, have a much lower background level (almost exclusively due to the vanadium sample container [22]), and also suggest that at low temperatures there is diffuse scattering which is considerably reduced in strength but has roughly the same shape as at room temperature (Fig. 7).

In order to model the diffuse scattering below T_C we have adopted the two-state model of David *et al.* [18], allowing molecules to librate isotropically with respect to their equilibrium orientations. Thus the diffuse intensity depends on p (defined in Section 3.1), and on σ, the root mean square (rms) amplitude of libration of the molecules, which is assumed to be the same for both of the equilibrium orientations of a given molecule. The parameters p and σ affect both

the Bragg peak intensities and the diffuse scattering intensity, and the relative intensity of the two types of scattering is (again) not adjustable. It is also assumed that a_0, Γ, d_1 and d_2 are known. Inelastic scattering studies, which are discussed in Section 4.2, place important constraints on the value of σ at a given temperature.

The low temperature Bragg diffracted intensity differs from eqn (9) because each of the four molecules per unit cell has its own limited distribution of orientations. The sum over N molecules, in eqn (9), becomes a sum over $N/4$ unit cells and a sum over the four molecules in a unit cell. Also the separation into a product of powder averages, in eqn (9), is no longer possible. We therefore obtain

$$I_B(Q) = \overline{\left| \sum_{k=1}^{4} e^{i\vec{Q} \cdot \vec{R}_k} \langle F_k(\vec{Q}) \rangle \right|^2 \left| \sum_{m=1}^{N/4} e^{i\vec{Q} \cdot \vec{R}_m} \right|^2}, \quad (12)$$

where $\vec{R}_m$ now labels one of the $N/4$ simple cubic lattice vectors in the system, $\vec{R}_k$ locates the kth molecule within a unit cell, and the molecular structure factor has become a function of k. The angle brackets now represent an average over a restricted distribution of orientations of the molecule with

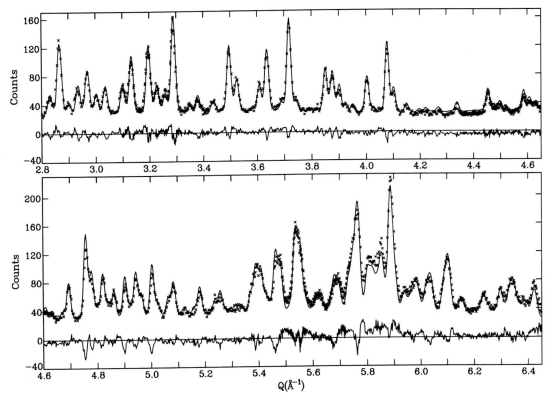

Fig. 7. A portion of the 14 K diffraction pattern of C_{60}. Points represent measurements at NIST [21], normalized to the same fixed monitor count as the data shown in Fig. 6. Upper lines were calculated as described in Section 3.2, assuming $d_1 = 1.45$ Å, $d_2 = 1.40$ Å, $a_0 = 14.04$ Å, $\Gamma = 24°$, $p = 0.833$ and $\sigma = 1.9°$; a constant background is also included. Lower lines show the difference between measurement and calculation.

190   J. R. D. COPLEY *et al.*

respect to the lattice. With this modification the expression for $I_D(Q)$, eqn (10), is unchanged.

To compute the orientational averages in eqns (10) and (12) a Monte Carlo method has been employed (J. R. D. Copley, unpublished). For a given molecule the starting orientation is orientation 1 (cf. Section 3.1) with probability p, orientation 2 with probability $(1 - p)$. A rotation axis is selected from an isotropic distribution, and a rotation angle is then chosen from a Gaussian distribution with standard deviation σ. In this way representative sets of configurations are generated. Calculations of $I_B(Q)$ and $I_D(Q)$ are straightforward, though tedious. In Fig. 7 we show the results of a typical calculation, with $p = 0.833$ (cf. [18]) and $\sigma = 1.9°$ (an appropriate value at low temperature, see Section 4.2). The agreement with experiment is good. On the other hand, calculations which only include static disorder (of the type proposed in [18]), i.e. calculations which ignore librational disorder, do not agree well with experiment.

3.3. *Radial distribution function studies*

Information about the structure and orientation of C_{60} molecules in the solid state can be obtained by measuring the set of lengths between atoms and comparing the result with model calculations. There are 23 distinct interatomic distances within a truncated icosahedral molecule when two different bond

lengths (d_1 and d_2) are allowed. This number increases rapidly with departures from icosahedral symmetry.

The set of distinct lengths between atoms is described by the pair distribution function $g(r)$ which is obtained from the Fourier transform of the measured scattering function $S(Q)$ [2, 11, 12]

$$g(r) = 1 + \frac{1}{2\pi^2 \rho_0 r} \int_0^\infty [S(Q) - 1]\sin(Qr)Q \, dQ, \quad (13)$$

where ρ_0 is the number density of atoms in the sample. In performing this transform it is sometimes necessary to correct for effects due to the finite maximum attainable value of Q.

Total scattering measurements to large Q have been performed using the special environment powder diffractometer SEPD at the IPNS [23–25]. $S(Q)$ was measured from $Q = 0.7 \, Å^{-1}$ to $45 \, Å^{-1}$ on a 0.65 g powder sample of purified C_{60} at room temperature and at 10 K. The results are shown in Fig. 8a and b (300 K and 10 K, respectively). The corresponding reduced radial distribution functions

$$G(r) = 4\pi r \rho_0 [g(r) - 1], \quad (14)$$

calculated from $S(Q)$ using a Lorch modification function to correct for the finite cutoff in Q, are shown in Fig. 9a and b. (In [24] $G(r)$ is variously

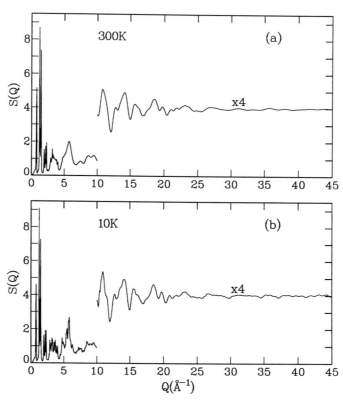

Fig. 8. The total scattering function $S(Q)$ for C_{60}, (a) at 300 K and (b) at 10 K, measured using the SEPD instrument at IPNS (after Ref. 24).

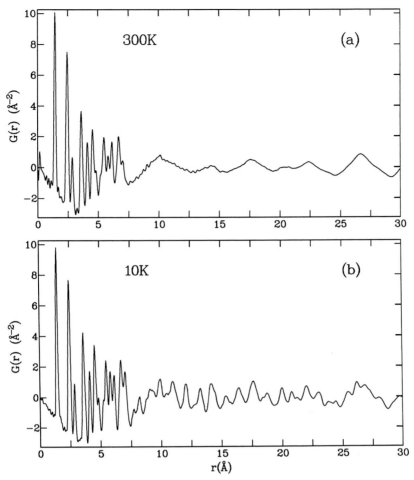

Fig. 9. The real space function $G(r)$, which is defined in Section 3.3, for C$_{60}$, (a) at 300 K and (b) at 10 K, derived from the measurements shown in Fig. 8 (after Ref. 24).

called the pair distribution or pair correlation function.)

The standard truncated icosahedron model of C$_{60}$, with adjustable bond lengths, accounts very well for all of the short distance ($r < 8$ Å) correlations exhibited in Fig. 9a. Li *et al.* [23] analysed selected peak positions from these data, obtaining inter- and intra-pentagon distances $d_2 = 1.39$ Å and $d_1 = 1.46$ Å, respectively, in good agreement with more indirect experimental estimates [26, 27]. The function shown in Fig. 8a also agrees quite well with a room temperature calculation of $I_D(Q)$ (eqn (10)).

A more refined analysis by Hu *et al.* [25] of the low temperature results shown in Fig. 9b yields bond lengths of 1.40 ± 0.01 Å and 1.45 ± 0.01 Å. It also provides insights into the nature of interfullerene ordering, insights which are in substantial agreement with the structural model proposed by David *et al.* [18]. The analysis suggests localization of the minority ($\Gamma_1 + 60°$) orientation and a substantially higher concentration of this phase ($\sim 40\%$), compared with

the analysis of David *et al.* [18] which gives $\sim 16.5\%$ at this temperature.

In summary, the truncated icosahedral model of C$_{60}$ has been confirmed by direct measurements of the lengths between atoms within a molecule. This work also provides additional experimental evidence in support of the two-state model for the low temperature structure of C$_{60}$-fullerite [18].

4. ROTATIONAL DYNAMICS

The diffraction measurements reported in the previous section establish that there is considerable disorder in C$_{60}$-fullerite both above and below the first-order phase transition temperature T_C. Furthermore the characteristic but highly unusual Q-dependence of the diffuse scattering intensity shows that orientational disorder predominates: there is no reason to expect that any other type of disorder would result in this type of scattering pattern. Although time-dependent aspects of orientational disorder in C$_{60}$ can be studied by techniques such as

NMR, neutron inelastic scattering provides unique insights for the reasons discussed in Section 1.

Because of the near-spherical shape of the C_{60} molecule, the scattering associated with orientational dynamics is expected to be extremely small at small wave-vectors. Indeed scattering needs to be measured to large values of Q in order to distinguish between different models of the dynamics. In addition the large moment of inertia of the molecule means that the energy scale of the rotational motions is small and that good energy resolution is required. In what follows we shall see that Q values as high as 6 Å^{-1} and resolutions of better than 1 meV are required. [1 meV is approximately 8.1 cm^{-1}, or 0.24 THz.] Below T_C the Q resolution must also be good because of the high density of Bragg peaks. These requirements, and the small size of available samples, constitute a major challenge to the experimentalist.

To date the only neutron scattering studies of the rotational dynamics of C_{60} have been those performed at NIST [21, 28, 29].† In brief, this work has shown that below T_C molecules librate about their equilibrium positions with energies of order 2–3 meV (frequencies of order 0.5 THz) and that the librational amplitude increases with increasing temperature such that close to T_C it is a substantial fraction of the angular separation between nearest neighbor atoms in a molecule. Above T_C the scattering results are consistent with a model in which there are no intermolecular correlations, and each molecule reorients completely randomly, as opposed to jump rotations among a restricted set of sites.

4.1. *Rotational motion above the transition*

Figure 10a shows a representative inelastic neutron scattering spectrum obtained at $Q = 5.65$ Å^{-1} and 260 K [28]. These and all other data have been corrected for detailed balance ('symmetrized') by multiplying by $\exp(-\hbar\omega/2kT)$, and for variations in instrumental resolution. The single quasielastic line, which is much broader than the resolution (1.2 meV FWHM), is characteristic of diffusive motion, directly demonstrating that orientational disorder above the transition is dynamic. Given that molecular centers lie on a fcc lattice, it is clear that the diffusive motion involves angular displacements of the molecules about their centers.

The reorientational character of diffusive motion in C_{60} is reflected in the Q dependence of the quasielastic scattering. Integrated symmetrized intensities for two temperatures above the transition are shown in Fig. 11 as square symbols. Sears [30] has developed a model in which the rotations of adjacent molecules

---

†See the Note added in proof (p. 1369).

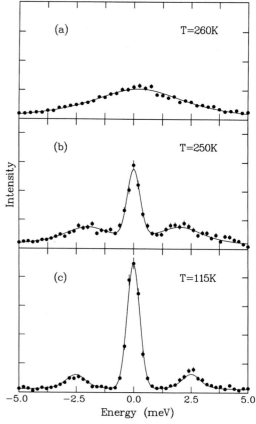

Fig. 10. Inelastic neutron scattering spectra taken at $Q = 5.65$ Å^{-1} at 260 K, 250 K and 115 K. Above T_C the spectrum consists of a single Lorentzian centered at zero energy transfer, characteristic of diffusion. Below T_C the spectra show peaks at non-zero energy transfers, corresponding to librations of C_{60} molecules about their equilibrium positions.

are uncorrelated and individual molecules undergo rotational diffusion about their fixed centers. In a classical treatment the angular motion of each molecule satisfies the differential equation

$$D_r \nabla_\Omega^2 p(\Omega, \Omega_0, t) = \frac{\partial}{\partial t} p(\Omega, \Omega_0, t), \qquad (15)$$

where D_r is the rotational diffusion constant, ∇_Ω^2 is the Laplacian operator in the space of Euler angles Ω, and $p(\Omega, \Omega_0, t)$ is the probability that a molecule has orientation Ω at time t, having had orientation Ω_0 at time 0. This model implies that the molecule tumbles through a continuum of orientational angles and does not, for example, undergo rotational jumps between some set of discrete orientations. The powder-averaged rotational component of the coherent neutron scattering function, $S_r(Q, \omega)$, can then be expressed as the sum of Lorentzians

$$S_r(Q, \omega) = \sum_{\ell=1}^\infty a_\ell (2\ell + 1) j_\ell^2(QR) \frac{1}{\pi} \frac{\tau_\ell}{1 + \omega^2 \tau_\ell^2} \qquad (16)$$

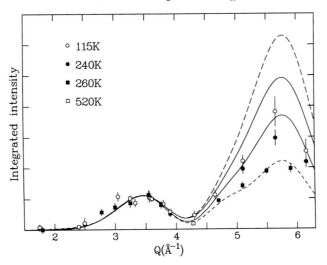

Fig. 11. Q dependence of the total integrated rotational scattering. Points represent data at various temperatures. The lower dashed line corresponds to complete rotational disorder (or extremely large amplitude librations). The upper dashed line corresponds to vanishingly small amplitude librations. Upper and lower solid lines correspond to librations with amplitudes of 4.0° and 6.7°, respectively. Calculations and data are arbitrarily scaled to the $\ell = 10$ peak at $Q \approx 3.5$ Å^{-1}.

where

$$\tau_\ell = \frac{1}{\ell(\ell + 1)D_r}, \tag{17}$$

j_ℓ is a spherical Bessel function, and R is the radius of the C$_{60}$ molecule. The coefficients a_ℓ are in this case given by

$$a_\ell = \sum_{n,n'=1}^{60} P_\ell(\cos\theta_{nn'}), \tag{18}$$

where P_ℓ is a Legendre polynomial and $\theta_{nn'}$ is the angle between the position vectors joining the molecular center to atoms n and n' within a single molecule. The expression for a_ℓ differs from that obtained for incoherent scattering [5] in that the sum over the Legendre polynomials for such a case only contains terms with $n = n'$ and therefore reduces to the constant value of 60 since $P_\ell(1) = 1$ for all ℓ. Thus this sum completely accounts for the fact that motions between atoms within a single molecule are correlated, and it therefore reflects the molecular geometry. Due to the high degree of symmetry of C$_{60}$, one finds that all odd-ℓ and many even-ℓ terms are identically zero. In fact, the only terms that significantly contribute to the scattering in the Q-range of these experiments, are those with $\ell = 6, 10, 12, 16, 18$ and 20, for which a_ℓ takes the values 6.3, 224, 31, 123, 496 and 90, respectively. Thus most of the scattering is contained in the $\ell = 10$ term, which is responsible for the peak in the integrated intensity at $Q \approx 3.5$ Å^{-1}, and in the $\ell = 18$ term which dominates contributions to the peak at $Q \approx 5.75$ Å^{-1}.

To compare with the data of Fig. 11, we can calculate the total rotational scattering function $S_r(Q) = \int S_r(Q, \omega)\,d\omega$. Using a standard identity we find that $S_r(Q)$ is identical to $I_D(Q)/Nb^2$ with $I_D(Q)$ given by eqn (10) for the diffuse scattering above the transition. The lower dashed line in Fig. 11 is a plot of $S_r(Q)$ multiplied by an arbitrary constant. Within this model, which is in good agreement with the data, we find that all the diffuse scattering is dynamic, and that there is no elastic component. Further confirmation of this model is found in the Q-dependence of the widths of the quasielastic peaks (corrected for instrumental resolution) (Fig. 12). Symbols are data taken at 260 K [28] while the solid line is obtained from the model treating D_r as an adjustable parameter. We obtain $D_r = (1.4 \pm 0.4) \times 10^{10}\,\mathrm{s}^{-1}$ at 260 K. NMR measurements of T_1 have also been used to determine the 'correlation time' τ_{NMR} of the rotational dynamics. The values of $\tau_{NMR} = 12$ ps obtained at 300 K [31] and 9 ps at 283 K [32] are consistent with the results reported here since for rotational diffusion, $D_r = 1/(6\tau_{NMR})$ [32]. However, the NMR measurements reported to date are essentially $Q = 0$ measurements and therefore contain little if any spatial information about the diffusion mechanism; for example, they do not distinguish between reorientational jumps and rotational diffusion.

4.2. Rotational motion below the transition

Although diffuse scattering persists below the transition, its origin and nature are quite different, reflecting a dramatic change in the rotational dynamics of the molecules. Figure 10b and c show symmetrized spectra for $Q = 5.65$ Å^{-1} at 250 K and 115 K,

194 J. R. D. COPLEY *et al.*

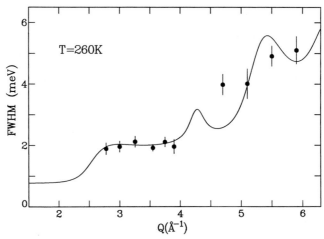

Fig. 12. Q-dependence of the width of the quasielastic scattering at 260 K. Points represent data. The solid line is calculated for isotropic rotational diffusion with $D_r = 1.4 \times 10^{10}\,\mathrm{s}^{-1}$.

respectively [29]. At 115 K, well-defined peaks are observed which are due to librational modes of the C_{60} molecules. Considering that the sample was polycrystalline, the observed peaks are remarkably sharp, suggesting that the orientational potential is very symmetric, i.e. it depends little on the molecular axis about which the angular displacement occurs. However, at this and at all other temperatures the widths of the peaks are not resolution-limited, an effect which could arise in several ways. The frequencies of the librations about different axes may indeed be slightly different, the interactions between molecules may result in significant dispersion, and/or significant damping may exist even at 20 K. Studies using single crystals of C_{60} are necessary to determine which of these effects dominates the low-temperature line width. As the temperature is increased to 250 K, the inelastic peak softens, broadens, and becomes more intense (Fig. 13). Above T_C the inelastic scattering collapses to a single broad quasielastic peak characteristic of diffusion, as we have seen above. From Fig. 10, we also note that the intensity of the elastic component decreases with increasing temperature. The most probable explanation is that the density of Bragg peaks in the simple cubic phase is so large that, despite efforts to avoid them, their tails contaminate the elastic position.

Figure 13a and b show that the librational energy softens by $\sim 35\%$ from 20 K to 250 K and that the peaks broaden by more than a factor of 6 over this same temperature range. The excitations become highly damped, but not overdamped (i.e. the librational peaks remain at non-zero energy), as the temperature approaches T_C. The peak shift of a damped harmonic oscillator is insufficient to account for the observed softening; using the measured width at 250 K this model yields a peak shift of only 10%

of the low temperature frequency. Most of the librational frequency decrease probably arises from a softening of the orientational potential itself as the increasing amplitude of libration disrupts the nesting condition. Figure 13c shows the temperature variation of the integrated intensity of the symmetrized librational peaks. The solid line is proportional to

$$\frac{1}{\omega_0(T)}\frac{e^{\hbar\omega_0(T)/2kT}}{e^{\hbar\omega_0(T)/kT}-1},\qquad(19)$$

which is itself proportional to the intensity expected for harmonic librations of frequency $\omega_0(T)$. For $\omega_0(T)$ we have taken the experimentally determined frequency at temperature T.

Assuming that the excitations are completely harmonic, the rms librational amplitude σ can also be estimated. It is found that σ increases from less than $2°$ at low temperature to more than $7°$ just below the transition. This should be compared with a typical nearest neighbor interatomic angle of about $23°$. These large amplitude librations are significant both in understanding the transition and in the analysis of the scattering data.

The assignment of the observed inelastic scattering to librations is based primarily on the Q dependence of the energy-integrated intensities. Assuming that librations of individual molecules are neither coupled to one another nor to other types of excitation, and that angular displacements from equilibrium are equally likely to occur about any axis, the powder-averaged function $S_r(Q,\omega)$, uncorrected for an angular Debye–Waller factor, is given by

$$S_r(Q,\omega)=\frac{4}{3}\sum_{\ell=0}^{\infty}a_\ell(2\ell+1)[\ell(\ell+1)]j_\ell^2(QR)$$

$$\times\left[\frac{1}{2\pi\hbar}\int_{-\infty}^{\infty}\langle\phi^*(0)\phi(t)\rangle e^{-i\omega t}\,dt\right],\quad(20)$$

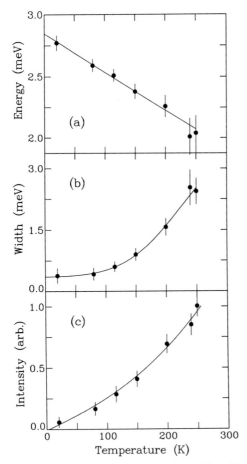

Fig. 13. Temperature dependence of (a) the position, (b) the width, and (c) the intensity of the peaks due to librational excitations below T_C. Solid lines in (a) and (b) are guides to the eye, whereas the solid line in (c) shows the expected temperature dependence of the intensity for a simple harmonic oscillator. For these measurements $Q = 5.65\,\text{Å}^{-1}$.

where the term in angle brackets is a thermally averaged angular correlation function (R. L. Cappelletti, unpublished). The energy integral of this expression, $S_r(Q)$, resembles the result for rotational diffusion except that the sum over ℓ contains an additional factor $\ell(\ell + 1)$ which lends extra weight to the higher ℓ terms.

A plot of $S_r(Q)$ derived from this model, and normalized to the $\ell = 10$ peak at $Q \approx 3.5\,\text{Å}^{-1}$, is shown as the upper dashed line in Fig. 11. The data at 115 K and 240 K, which have also been arbitrarily normalized to the $\ell = 10$ peak, are shown as circles. A clear trend toward increased weighting of the predominantly $\ell = 18$ peak at $Q \approx 5.75\,\text{Å}^{-1}$ is seen in the data, but the trend is not as strong as predicted by eqn (20). The difference arises from a lack of inclusion of the angular Debye–Waller factor, which is given by an unwieldy analytical expression. Notwithstanding the arbitrary normalization which was used to create Fig. 11, it is of course the case that

at all Q the diffuse scattering decreases with decreasing temperature.

An alternative approach to the calculation of $S_r(Q)$ is to use the Monte Carlo technique described in Section 3.2. For extremely small values of σ the Monte Carlo calculation reproduces the small libration scattering result, which is readily derived from eqn (20), whereas for very large σ it gives the result for rotational diffusion, eqn (10). The solid lines in Fig. 11 represent integrated intensities calculated for rms angular displacements of 4.0° and 6.7°, values that (within the harmonic approximation) roughly correspond to temperatures of 115 K and 240 K, respectively. This comparison with experiment shows the pronounced effect of the angular Debye–Waller factor on the Q-dependence of the rotational scattering intensity, and emphasizes the importance of including it in all calculations of the scattering expected from C$_{60}$, including the Bragg peaks. It also suggests that the rms angular displacements are somewhat underestimated in the harmonic approximation.

4.3. Comparison with molecular dynamics simulations; summary remarks

The neutron scattering results in both phases can be directly compared with molecular dynamics simulations, providing a stringent test of the interatomic potentials used in these calculations. At 100 K simulations using only the intermolecular van der Waals atom–atom interaction [33] predict that librations occur in a band of width ~ 0.8 meV which is centered at ~ 1.3 meV. This implies that the orientational potential felt by the molecules in this simulation is about a factor of 4 too small and that it is not as rotationally symmetric as is indicated by the measurements. A more recent simulation by Sprik et al. [34] includes explicit interaction sites on the short, electron-rich intramolecular bonds and an electrostatic contribution to the intermolecular potential in addition to the van der Waals contribution. At 100 K these authors predict a somewhat higher energy librational band centered at ~ 1.9 meV, but the predicted width of the band has also increased to ~ 1.1 meV, again implying that the model orientational potential is not as symmetric as the experimental results indicate. Molecular dynamics simulations above T_C predict that the molecules undergo rotational diffusion with $D_r = 2.4 \times 10^{10}\,\text{s}^{-1}$ at 260 K. Though the simulations are not in quantitative agreement with experiment, the apparent qualitative similarities are gratifying.

The neutron scattering work described in this section has given us a fairly complete picture of the orientational dynamics of C$_{60}$ in the solid phase. Below T_C the molecules librate about their equilibrium positions. These excitations become quite anharmonic and their amplitude increases to about 1/3 of the near neighbor interatomic angles within a buckyball as the temperature is increased to just below the orientational order–disorder transition. One may therefore regard the transition as the rotational analogue of displacive melting which, in the Lindemann picture, occurs when atomic displacements are $\sim 1/5$ of interatomic distances [35]. Furthermore the 'molten' state can in both cases be characterized as one in which diffusion occurs: translational diffusion in the case of a liquid and rotational diffusion for C_{60} in the 'orientational liquid' state. In addition, these measurements yield stringent tests of the intermolecular potentials used to describe the dynamics and phase transitions in solid C_{60}. They should therefore lead to improved mathematical descriptions of the intermolecular forces and consequently more accurate microscopic theories of these phenomena.

5. INTRAMOLECULAR VIBRATIONAL MODES

We turn now to a discussion of the internal (intramolecular) modes of vibration of C_{60} molecules in C_{60} and its compounds. Of the 174 intramolecular modes of the C_{60} molecule in undoped C_{60}, corresponding to 46 distinct frequencies, only four are active with respect to infrared absorption, while 10 are Raman-active. In neutron scattering there are no selection rules such as those which apply to i.r. or Raman scattering, so that one can observe all vibrational modes with appropriate methods. Furthermore, relatively straightforward data reduction procedures permit the determination of a function which is very nearly the vibrational density of states, a quantity which is readily compared with theoretical predictions. This type of treatment has not been given to the data reported in this section but the spectra, as shown, nonetheless represent good approximations to densities of states except at the lowest energy transfers. The advantages of neutron scattering relative to optical spectroscopies are to some extent offset by its poorer energy resolution and the need for large samples.

Because carbon–carbon bonds are among the strongest found in nature, and because the mass of the C atom is small, the vibrational modes of C_{60} extend to rather high energies, of order 200 meV. Special neutron scattering techniques are needed to access such energies. At reactor sources a variant of the triple-axis method is often used, in which energy analysis of the scattered neutrons is accomplished by a low-pass polycrystalline filter (Fig. 2c), rather than by Bragg reflection from a crystal (Fig. 2b). The

neutron distribution emerging from an accelerator-based spallation source is richer at the high energy end of the spectrum (roughly 100 meV to a few eV) than the distribution from a steady-state reactor source. In one type of instrument at a pulsed spallation source (Fig. 3b), scattered neutrons are energy analysed by a crystal while the incident neutron energy and hence the energy transfer are defined by the arrival time at the detector. Rather similar spectra are obtained from a reactor-based filter–detector spectrometer and a pulsed-source crystal analyser spectrometer. The latter has the advantage at energy transfers above about 100 meV because of the source spectrum, while the former performs particularly well in the intermediate energy range between roughly 50 and 100 meV.

5.1. *Pure* C_{60}

In the first experiment reported on neutron spectroscopy of C_{60}, that of Cappelletti *et al.* [36], the energy transfer range between 25 and 215 meV was examined with the filter–detector spectrometer at the NIST research reactor. This energy range encompasses all the expected intramolecular modes of C_{60}. Two filter detectors were used, a Be filter which gave lower overall resolution (5–18 meV FWHM), but permitted the observation of the highest energy modes, and a Be/graphite filter which gave better resolution (2–4 meV FWHM) at the lower energy transfers, $25\,\text{meV} < \hbar\omega < 80\,\text{meV}$. A highly structured spectrum was observed, as shown in Fig. 14, which was compared in detail with existing theoretical predictions [37–39], and with Raman and i.r. experiments [40]. Good qualitative agreement between experiment and various theoretical predictions was evident, especially considering that most of the calculations proceed from a fairly fundamental quantum mechanical basis. The higher resolution data of Fig. 15 permit additional mode assignments in the lower energy range, using the theoretical predictions as a guide. (Figure 15 shows more recent (unpublished) measurements which are somewhat improved in resolution and were made using a larger sample than the sample used in the measurements reported in [36].)

Two subsequent measurements [41–43] of intramolecular modes in C_{60} have been reported, both of which were carried out at ISIS using the TFXA (time-focused crystal analyser) spectrometer, an instrument of the type shown in Fig. 3b. The measurements are in principle similar, but the later data of Coulombeau *et al.* [42, 43], using a larger sample, have much better counting statistics. These data are shown in Figs 16 and 17. For energy transfers above about 80 meV, the pulsed source data have significantly better energy resolution than the data obtained

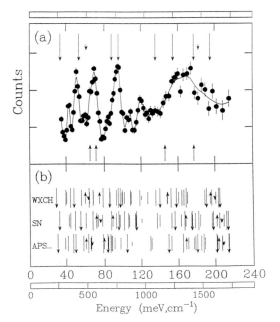

Fig. 14. (a) Vibrational spectrum for C$_{60}$, measured on a filter–detector spectrometer at the NIST research reactor [36], using a Be filter. The ordinate (counts) is proportional to the amplitude-weighted vibrational density of states, except that a smooth background of multi-phonon scattering is present, which rises at higher energies. Upward- and downward-pointing arrows represent observed [40] i.r.-active and Raman-active modes, respectively, with the lengths of the arrows proportional to mode degeneracies. (b) Calculated frequencies from Refs 37 to 39. Arrows and their lengths have the same meanings as in (a).

by Cappelletti *et al.* [36] at a reactor source. Coulombeau *et al.* [43] were thus able to make a number of additional mode assignments, identifying all the modes observed by optical spectroscopy, and 23 of the 46 frequencies overall.

All three experiments on pure C$_{60}$ are in excellent agreement between 25 and 80 meV. (We shall briefly discuss the energy transfer range below 25 meV in Section 6.) Table 1 shows the observed frequencies, together with the assignments made by Prassides *et al.* [41], which we favor over those of Coulombeau *et al.* [43]. Above 80 meV, the theoretical calculations which have so far been reported do not correspond closely enough to experiment to guide assignments of modes. Hence the optically observed modes are the only ones which can be identified and assigned with confidence for $E > 80$ meV. Improved calculations of the modes would be very helpful in this regard.

Above energy transfers of 40 meV or so, the vibrational spectrum of C$_{60}$ resembles that of graphite (see [44]), with major peaks for C$_{60}$ lying a few per cent lower in energy. No doubt the resemblance arises from the similar local bonding and coordination in the two materials. The lowest-energy intramolecular mode of C$_{60}$, basically a 'squashing' mode of the ball,

occurs at 33 meV. There is a gap between this energy and the top of the band of rigid-molecule modes. No such gap exists for graphite, since isolated-layer modes can have frequencies down to zero energy and mix with the rigid-layer modes, which extend from 0 to 16 meV.

5.2. *Doped* C$_{60}$

Reaction of C$_{60}$ with alkali metals has recently led to the exciting discovery of a new class of superconducting materials M$_3$C$_{60}$ with surprisingly high transition temperatures, 19 K for M = K and 29 K for M = Rb. While the precise mechanism leading to superconductivity in these materials has yet to be determined, most attention has focused on the standard mechanism of the electron–phonon interaction, with special properties of the electronic states and vibrational modes (in particular the intramolecular modes in some theories) coming into play.

Two experiments on the vibrational modes of M$_3$C$_{60}$ have already been reported, both of which made use of the TFXA at ISIS. Prassides *et al.* [45] noted a number of changes in the spectrum of K$_3$C$_{60}$ relative to pure C$_{60}$. In particular, the high energy region between 130 and 200 meV appears to show substantially more scattering in the case of K$_3$C$_{60}$. Changes in some of the lower energy modes (around

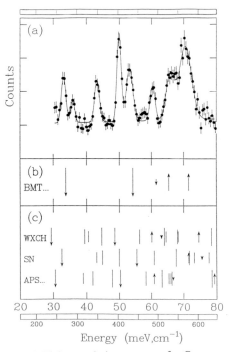

Fig. 15. (a) Higher-resolution spectrum for C$_{60}$, measured on the NIST filter-detector spectrometer with a Be/graphite filter. The feature at 36 meV is a phonon peak from the sample can. (b) Observed i.r.-active and Raman-active mode frequencies (Ref. 40). (c) Calculated frequencies, as in Fig. 14b. Upward- and downward-pointing arrows in (b) and (c) have the same meanings as in Fig. 14.

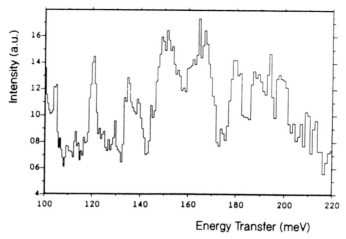

Fig. 16. Neutron scattering spectrum for C_{60} above 100 meV, measured on a crystal-analyser spectrometer at the ISIS pulsed neutron source (after Ref. 43).

50 meV) were also noted, which can be attributed to a non-uniform broadening of the peaks of C_{60}; a mode of H_g symmetry at 54 meV in C_{60} appears to have broadened so much in K_3C_{60} as to disappear into the background. Prassides *et al.* [45] remark that their results favor the theoretical picture of Schlüter *et al.* [46], which emphasizes coupling of electrons to vibrational modes of H_g symmetry.

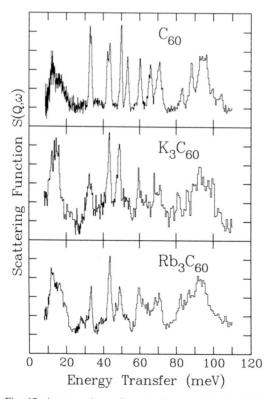

Fig. 17. A comparison of spectra for pure C_{60} (at 25 K) (Ref. 43), K_3C_{60} (summed data at 5 K and 30 K) (Ref. 45), and Rb_3C_{60} (summed data at 22 K and 35 K) (Ref. 47). The peaks at 54 and 66 meV in C_{60} are essentially absent in K_3C_{60} and Rb_3C_{60}.

Similar measurements were carried out on Rb_3C_{60} by White *et al.* [47], who measured the scattering above and below the superconducting transition temperature (29 K). The scattering at high energy transfers is rather weak, but a difference spectrum, obtained by subtracting 35 K data from 22 K data, appears to show changes near 135 and 180 meV. More striking is the disappearance of peaks observed in C_{60} at 54 meV and 66 meV, suggesting strong broadening of these modes, presumably due to the electron–phonon interaction. Similar changes are also present in the data obtained by Prassides *et al.* [45] for K_3C_{60}. Their data and the White *et al.* data [47] for Rb_3C_{60} are shown as the middle and lower curves in Fig. 17. Selected H_g modes are also observed to be highly broadened in Raman spectra of these materials [48–50].

The data which have been reported on the vibrational modes of pure C_{60} already provide a basis for improved theoretical understanding of the dynamics of the molecule. The experiments on the superconducting compounds have yielded tantalizing results, and further experiments are clearly desirable.

6. CONCLUDING REMARKS

Since the first measurement, in December 1990, much has been learned from neutron scattering experiments about the microscopic properties of C_{60}-fullerite. Crystal structure, rotational disorder and molecular structure have all been studied, and unique insights into the high and low temperature rotational dynamics and intermolecular potential, and into the intramolecular vibrational spectrum, have been forthcoming. Much less has been reported about compounds of C_{60}, but the situation shows signs of changing dramatically. Thus far little is known about intermolecular translational modes in C_{60} and its

Table 1. Observed peak positions, in meV, for vibrational modes below 80 meV in pure C_{60} (1 meV = 8.066 cm^{-1}). Assignments are those of Ref. 41. R and IR denote Raman and infrared-active modes, respectively. Neutron peaks assigned to T_{1u} symmetry very probably include contributions from other modes

Assignment	IR/Raman (Ref. 40)	Neutron (Fig. 15)	Neutron (Ref. 41)	Neutron (Ref. 43)
H_g	33.8 (R)	33.3	33	33.2
T_{2u}, G_u	—	43.4	43, 44	42.7, 44.0
H_u	—	50.1	50	50.1
H_g	54.2 (R)	53.4	54	53.6
G_g, A_g	61.5 (R)	60.6	60	60.5
T_{1u}	65.3 (IR)	66.0	66	65.8
T_{1u}	71.5 (IR)	70.8	71	70.6

compounds. The intense scattering observed in the TFXA measurements between 10 and 20 meV (cf. Fig. 17) has been attributed to lattice modes [41–43, 45, 47]. This seems unlikely given the size and mass of the molecule, its bulk modulus [51] and Debye temperature [52], and the predictions of simulations [33] and calculations [53]; these results and considerations suggest that the highest intermolecular mode energy in pure C_{60} should be less than 10 meV. Possible alternative explanations include (hydrogenous) impurity scattering and/or some unintended instrumental effect.

At the time of writing we know of several experimental studies at ISIS, by the Sussex, AT&T Bell Labs and RAL groups, which have not yet been submitted for publication. These include a very high Q diffraction experiment on C_{60}, crystal structure measurements on C_{70} [54] and on $K_3 C_{60}$ and $Rb_3 C_{60}$ [55], and vibrational spectroscopy studies of C_{70} and $Rb_6 C_{60}$ [56]. Low energy inelastic scattering measurements of $K_3 C_{60}$ and $Rb_6 C_{60}$ have also been completed at NIST.

Most of the measurements already reported have not been easy, in part due to the small size of available samples. Given significantly larger samples (ideally several tens of grams) it will be possible to improve instrumental resolution sufficiently to look at the structure and dynamics in greater detail. For example, better Q resolution will enable clearer separation of the Bragg and diffuse intensities in diffraction experiments. Better energy resolution will facilitate the assignment of peaks in the high energy vibrational spectrum and will also allow us to search for scattering associated with long time-scale dynamical processes in the low temperature phase, such as jump-diffusion between the two orientational states proposed in the model of David et al. [18].

Additional exciting opportunities will present themselves if and when large single crystals become available. For example a careful study of the diffuse scattering will yield new information about orientational probability distributions. It should also be possible to probe the phonon and librational dispersion relations of C_{60}, as well as the nature and extent of the coupling between these excitations. The results of such experiments will yield detailed information on the orientational potential and on the role that intermolecular phonons play in the orientational phase transition.

Similar remarks apply to future studies of doped C_{60} materials such as the alkali metal compounds. Larger samples will make possible more detailed studies of the intramolecular modes and the rotational dynamics, and the availability of single crystals will further enhance our understanding of these systems. Molecular derivatives of C_{60}, in which various atoms or molecular groups are covalently attached to C_{60} molecules, are presently being synthesized in several laboratories. Neutron studies of such materials, which have not yet been attempted for lack of sufficiently large and verifiably pure samples, represent another promising future direction, especially the vibrational spectroscopy and orientational dynamics of products which contain hydrogenous groups. We look forward to exciting new results within the next year or two.

Note added in proof

Neutron scattering studies of C_{60} and its compounds have been vigorously pursued during the 6 months since this paper was completed. In this Note we list recent developments; cited references should be consulted for additional information. We have attempted to include all known neutron scattering work on C_{60} and its compounds and we apologize if a particular piece of research is inadvertently omitted.

Very high Q time-of-flight diffraction measurements on C_{60} have been reported using the ISIS small angle neutron diffractometer for liquids and amorphous samples (SANDALS) [57]. Recent high resolution measurements using HRPD are presently being analysed [58].

New low and high energy inelastic studies of C_{60}, with improved resolution, have been made at ISIS using the high resolution quasielastic spectrometer IRIS and the multi-angle rotor instrument MARI, respectively [59].

In the first experiment on single crystals (which had a total volume of 5 mm^3), dispersion curves for intermolecular translational and librational modes in C_{60} are reported [60]. These excitations extend to energies of 6–7 meV, confirming our expectation (Section 6) that the highest energy be < 10 meV. In addition the low energy phonon density of

STRUCTURES OF C_{60} INTERCALATION COMPOUNDS

Otto Zhou† and David E. Cox‡

†Department of Materials Science and Engineering and Laboratory for Research on the Structure of Matter, University of Pennsylvania, Philadelphia, PA 19104-6272, U.S.A.
‡Physics Department, Brookhaven National Laboratory, Upton, NY 11973, U.S.A.

(Received 30 June 1992)

Abstract—In the past two years, many intercalation compounds of C_{60} have been prepared and characterized by X-ray powder diffraction techniques. The M–C_{60} systems with M = Li, Na, K, Rb, Cs and Ca have been studied, and the structures of several compounds have been determined in considerable detail, including the face-centered cubic superconductors K_3C_{60} and Rb_3C_{60}, body-centered tetragonal K_4C_{60}, body-centered cubic M_6C_{60} with M = K, Rb and Cs, simple hexagonal $C_{60}I_4$, and the monoclinic molecular ferromagnet tetrakis-dimethylamino-ethylene (TDAE) $C_2N_4(CH_3)_8C_{60}$. The Rietveld profile technique for structure refinement coupled with high resolution synchrotron X-ray powder data has played an important role in providing accurate structural data for many of these compounds.

In this article, a detailed review of the literature dealing with structural studies on the intercalated compounds of C_{60} will be given.

Keywords: Fullerene intercalation compounds, X-ray powder diffraction, structure refinement.

1. INTRODUCTION

Since the discovery by Krätschmer, Huffman and colleagues of an efficient laboratory method [1] for producing "macroscopic" quantities of buckminsterfullerene C_{60}, knowledge of the molecular and solid state chemistry of C_{60} has advanced rapidly. Many fullerene-derived solid phases have been synthesized with novel properties and potential applications. These compounds can generally be divided into three categories: (1) intercalation compounds, in which the fullerene host retains its molecular identity and foreign atoms occupy the large interstices in the lattice, such as alkali metal-doped C_{60} compounds [2–6]; (2) exohedral solids, formed from fullerenes to which foreign atoms or molecules are covalently bonded on the outside of the carbon cage, such as fluorinated fullerides [7]; and (3) endohedral cluster solids derived from non-carbon atoms encapsulated by fullerene, $La@C_{82}$ being the first confirmed such molecule; where the @ indicates the La is inside the cage [8, 9]. The crystal structures of many of these fullerene derivatives have now been determined by X-ray powder and single crystal diffraction techniques. In particular, the Rietveld refinement method [10] applied to synchrotron X-ray powder data has played an important role in the structural characterization of the intercalation compounds, since for these materials the application of single crystal X-ray and neutron powder diffraction techniques has been limited by the difficulty of synthesizing solvent-free single crystals or gram quantities of high quality polycrystalline samples.

Among the various intercalation compounds, the synthesis, composition, and phase equilibria of alkali metal-doped fullerides have drawn most of the attention, mainly because of the occurrence of superconductivity in these systems [11–13]. The structures of many of these compounds have now been determined, and a schematic temperature–composition phase diagram of Rb_xC_{60} has been established experimentally [6]. Very recently, iodine has been successfully intercalated into the fullerene lattice to form a compound $C_{60}I_4$, and the structure of the latter has been characterized [14]. Another interesting intercalated derivative of C_{60} is the tetrakis-dimethylamino-ethylene (TDAE) compound $C_2N_4(CH_3)_8C_{60}$, which is ferromagnetic with a Curie temperature of 16 K, the highest known for an organic ferromagnet [15, 16]. Among the exohedral type of compounds, some quite complex structures have been determined by X-ray single crystal diffraction, including Os [17] and Pt [18] derivatives of C_{60} and the compound $C_{60}Br_{24}$ [19]. However, as yet no crystal structure determination of an endohedral solid has been reported, due to difficulties associated with sample preparation and purification.

In the present article, we shall be concerned with the C_{60} intercalation compounds, with main emphasis on the alkali metal-doped systems.

2. STRUCTURAL CHARACTERIZATION BY POWDER DIFFRACTION TECHNIQUES

Characterization of sample purity by X-ray powder diffraction techniques plays a vital role in the preparation of high quality specimens. Laboratory data are usually adequate for the identification of major impurities or to show if phase segregation has occurred, and also for the determination of unit cell parameters and basic structural features in simple cases, and some examples of this type will be described in the following sections. However, they often do not permit a detailed structure refinement to be undertaken. In the case of fullerenes and derivative compounds, the difficulties are compounded by the fact that carbon is a weak scatterer of X-rays, and also because many of the intercalated compounds are highly reactive towards air and moisture, and must be handled in sealed capillaries rather than in the flat-plate holders typically used with laboratory diffractometers. On the other hand, synchrotron X-ray powder diffraction offers several advantages, such as high resolution with either capillary or flat-plate samples, excellent peak-to-background discrimination, and ease of operation at both low and high temperatures. With perfect crystal-analyser geometry, the instrumental angular resolution (full-width at half-maximum, FWHM) is about 0.01–$0.03°$ over an extended range of scattering angle, 2θ, or around 0.05–$0.07°$ if narrow receiving-slit geometry is used [20, 21]. Because of the absence of systematic errors due to sample displacement effects, the low-angle peak positions can be determined with very high accuracy, which is a great advantage in phase-transition studies and in the determination of unknown unit cells by auto-indexing techniques. In the many cases where only a few milligrams of well-crystallized material are available because of the difficult and tedious nature of the purification process for fullerenes, synchrotron X-ray powder data have usually proved to be adequate for structure determination and refinement. High resolution neutron powder diffraction, which is also a powerful tool for structural characterization, is not in general an option because, except for C_{60} [22], the gram quantities of high quality material needed for such studies are not available at the present time.

Traditional methods of structure refinement based on integrated intensity data from a limited number of resolved powder diffraction peaks have now been largely superseded by the much more powerful Rietveld profile technique in which the entire observed pattern is fitted with a calculated pattern composed of individual peak profiles generated from a structural model and a peak shape function [10, 23]. Such a profile can contain several hundred possible reflections, and allows all the data, including heavily overlapped regions, to be used in the refinement. In this way, the maximum amount of structural information can be obtained from the pattern, including unit cell parameters, peak widths, atomic positions, Debye–Waller temperature factors and site occupancies. However, a word of caution is necessary, in that the weights used in the least-squares fitting procedure are based solely upon the counting statistics, which are seldom the limiting source of error in powder diffraction experiments. Thus the estimated errors quoted for the refined parameters represent a lower limit, and may not reflect the real accuracy of the refinement within a factor of 2–3, probably more in the case of thermal factors and site occupancies, which tend to be highly correlated. Furthermore, the commonly-quoted profile R-factors, R_p and R_{wp} (for definitions see [23]), cannot in general be used for standard crystallographic significance tests [24] to determine whether agreement is improved by the addition of parameters to the structural model, and this therefore becomes more a matter of subjective judgment based on a decrease in the integrated intensity R-factor, R_I. An alternative approach for simple patterns which works especially well for cubic structures is to use one of the standard pattern decomposition techniques for extracting integrated intensities [21], and then to use these for structure refinement as described by Will and coworkers [25]. This method usually yields more realistic estimated errors, and the results of significance tests based on the weighted integrated intensity R-factors, R_{WI}, can be regarded with more confidence.

Some caution must be exercised in the interpretation of apparently large thermal factors (e.g. $B > 4$ Å^2), which may in fact result from large static displacements of atoms from their assigned positions rather than thermal fluctuations, and can be checked by carrying out data collection at low temperatures. A high value of B could also be an artifact caused by less-than-complete occupancy of the site in question. In principle this can be checked by introducing a variable occupancy factor, but in practice the reliability of the result may be questionable because the two parameters are usually highly correlated.

In devising a suitable strategy for structure refinement of C_{60} and its intercalates, one needs to consider the extent to which the C_{60} molecules may be orientationally ordered, e.g. complete spherical disorder, as in the room temperature structure of C_{60} [26, 27]; partial disorder over two or more symmetry-equivalent orientations, as in the $Fm\overline{3}m$ structure of K_3C_{60} [3]; or fully ordered with one orientation, as in the $Im\overline{3}$ structure of K_6C_{60} [5], and also the constraints to be placed on the atomic positions of the C_{60} atoms in a molecule. An unconstrained description of the latter requires three inequivalent C atoms and eight positional parameters in the case of point symmetry $m\bar{3}$, increasing to 10 inequivalent atoms and 30 positional parameters for $\bar{3}$ point symmetry. Even with high resolution synchrotron data our experience has been that an unconstrained refinement of this type does not yield reliable results; however, if the symmetry is constrained to that of a rigid truncated icosahedron with two C–C distances, very satisfactory results are obtained in most cases. The two distances correspond to C–C bonds shared by a pentagon and a hexagon (D_1), or by two hexagons (D_2), respectively (typically 1.45 and 1.39 Å). The atomic coordinates in Å can be generated from the expressions given in Table 1 for one of the standard orientations with the $\langle 100 \rangle$ axes passing through three orthogonal two-fold molecular axes (hexagon–hexagon edges) [27]. These can be transformed easily to the corresponding fractional coordinates for any given unit cell parameters. In the case of complete orientational disorder, D_1 and D_2 are irrelevant, and the only variable positional parameter is the radius of the spherical shell (R_0), typically about 3.55 Å.

In the following sections we shall describe the structures obtained for intercalated C_{60} compounds, mainly from Rietveld refinement of synchrotron X-ray powder data, but in a few cases with integrated intensities from laboratory X-ray patterns.

3. SOLID C_{60}

Structural work on C_{60} is reviewed in detail in the article by Heiney in this issue [27], and here we give only a brief description of the room temperature structure in order to provide a starting point for the subsequent discussion of intercalated structures.

Solid C_{60} is a van der Waals bonded molecular crystal [28] shown by Fleming and coworkers in an X-ray single crystal diffraction study to have a closed-packed face-centered cubic (fcc) crystal structure at 300 K, with four fullerene molecules centered on the fcc Bravais lattice sites and a lattice parameter a of 14.198 Å [29]. A slightly smaller value of 14.17 Å was

obtained in a subsequent study by Heiney et al. [26]. The nearest-neighbor molecular center-to-center distance is 10.02 Å, implying a van der Waals separation of 2.9 Å for a calculated C_{60} diameter of 7.1 Å [30]. Disorder is intrinsic to the fcc solid C_{60}, due to the incompatibility of the icosahedral symmetry of the fullerene molecule and the fcc lattice. The solid state ^{13}C nuclear magnetic resonance (NMR) spectrum of C_{60} shows a sharp single line at 300 K, indicating that the C_{60} molecules are tumbling isotropically about a fixed point in the lattice at a frequency greater than $10^5\,\text{s}^{-1}$ [31, 32]. Since the NMR result cannot distinguish between random spinning and "ratcheting" motions of the fullerene molecule, the 300 K crystal structure can be described in at least two ways, as discussed briefly below.

The X-ray single crystal diffraction data of Fleming et al. showed the 300 K structure to be closed-packed fcc with four fullerene molecules per unit cell [29]. The structure was described in space group $Fm\bar{3}$ in terms of a merohedral twinning model, obtained by the superposition of 90° rotations of the C_{60} molecules about the cubic [100] direction. The effect of this is to impose an apparent four-fold symmetry axis along the [100] direction and hence to raise the symmetry to $Fm\bar{3}m$. The three non-equivalent carbon atoms necessary to generate the molecule were assigned the following positions: C1 in 0.0, 0.252(2), 0.042(3); C2 in 0.218(3), 0.125(3), 0.083(5); and C3 in 0.139(3), 0.213(3), 0.044(5), where the numbers in parentheses are estimated errors. However, the corresponding C–C bond lengths and the large Debye–Waller factors were indicative of considerable residual disorder.

An alternative approach was taken by Heiney et al., by assuming that the C_{60} molecules are completely orientationally disordered [26, 27]. The ensemble-averaged electron density of each fullerene molecule is then a uniform spherical shell with the diameter of the carbon skeleton. The molecular form factor is simply the zero order Bessel function j_0, where $j_0(Q) = \sin(QR_0)/QR_0$, where $Q = 4\pi \sin\theta/\lambda$ is the magnitude of the scattering vector, and R_0 is the radius of the C_{60} molecule. Least-squares fits based on this model with four fullerene molecules per fcc unit cell yielded good agreement with the X-ray powder diffraction data, with $R_0 = 3.52$ Å and an essentially zero overall Debye–Waller factor, B [26]. The small value of the latter indicates that there is very little residual disorder. A recent synchrotron X-ray single crystal study by Chow and coworkers has revealed that there are small deviations from a uniform probability distribution corresponding to a significant degree of orientational order at room temperature [33].

Table 1. Coordinates in Å for a truncated C_{60} icosahedron in one of the standard orientations with $m\bar{3}$ symmetry. $C_1 = (\sqrt{5}+1)/4$, $C_2 = (\sqrt{5}-1)/4$, and D_1 and D_2 are typically 1.45 and 1.39 Å

	x	y	z
C(1)	$D_2/2$	0	$C_1(2D_1 + D_2)$
C(2)	$(D_1 + D_2)/2$	$C_1 D_2$	$C_1(2D_1 + D_2) - C_2 D_1$
C(3)	$D_1/2$	$C_1(D_1 + D_2)$	$C_2 D_1 + (2D_1 + D_2)/2$

4. ALKALI METAL-DOPED PHASES

The early reports of superconductivity at 19 K [11] and 29 K [12, 13] in potassium- and rubidium-doped C_{60} have stimulated an intense amount of research directed towards determining the compositions and phase equilibria of this new class of synthetic metals. The 300 K crystal structures of M_xC_{60} (M = K, Rb) are now well-established. Four distinct crystalline phases have been identified, namely fcc(I) for the parent phase with $x = 0$ [29, 26], a superconducting fcc (II) phase with $x = 3$ [3, 34, 35, 6, 36], a body-centered tetragonal (bct) phase with $x = 4$ [4, 35, 6, 36], and a body-centered cubic (bcc) phase with $x = 6$ [5, 6, 36], which represents saturation doping. The binary phase diagram for Rb_xC_{60}, established experimentally by Zhu and coworkers [6], is shown in Fig. 1. The temperature scale in this diagram is defined approximately by the 300 K label (lower left). While the superconducting phase and the bct phase have been shown to be line compounds, the fcc (I) and bcc phases are stable over a finite range of x at 300 K. The crystalline stoichiometric RbC_{60} phase recently observed above room temperature [37, 38] and the low temperature simple cubic C_{60} phase [26] are not shown in this phase diagram.

The Cs–C_{60} system has not been characterized in such detail, except for the saturation-doped Cs_6C_{60} phase [5]. Cs_4C_{60} has been identified as having a bct structure in a two-phase sample at the nominal composition Cs_3C_{60} [4]. Superconductivity at 30 K has been reported in samples synthesized by reaction of C_{60} with CsM_2 (M = Hg, Tl, Bi) alloys at a nominal composition of Cs_3C_{60} [39]. Recent X-ray diffraction work by Messaoudi et al. has shown that the super-

conducting Cs_3C_{60} phase can be indexed as a bcc structure with a lattice parameter similar to that of the Cs_6C_{60} phase [40]. A compound with the nominal concentration CsC_{60} has been found to be fcc with a lattice parameter $a = 14.09$ Å, smaller than the value for C_{60} [40], suggesting a structure similar to that of RbC_{60} [37, 38].

Intercalation of C_{60} with lithium has been studied by Chabre and co-workers [41] by the technique of solid-state electrochemical doping. Metallic lithium was used as the negative electrode and polyethylene oxide lithium perchlorate $(P(EO)_8LiClO_4)$ polymer film as the electrolyte. Cyclic voltammetry results indicated the formation of stoichiometric phases of Li_xC_{60} at $x = 0.5, 2, 3, 4$ and 12. The maximum doping concentration of 12 Li ions per C_{60} molecule would be twice that obtained from vapor doping of C_{60} with K, Rb and Cs [5]. However, the $x = 6$ phase formed by the reaction of heavy alkali metals with C_{60} was not observed in the Li_xC_{60} system in the electrochemical study. Photoemission studies also indicate the existence of a stable stoichiometric M_2C_{60} phases with M = Li and Na [42].

C_{60} doped with Na has been reported recently by Rosseinsky et al. [43], and shown to have very different phase relationships, compared to the systems doped with the heavier alkali metals. Three stoichiometric Na_xC_{60} phases ($x = 2, 3$ and 6) have been characterized, with a maximum dopant concentration of six Na atoms per C_{60} molecule in an fcc structure [43]. The Na_2C_{60} compound has a simple cubic structure at room temperature, and Na_3C_{60} is isostructural with K_3C_{60} and Rb_3C_{60}, but does not superconduct down to 2 K.

A series of ternary alkali metal-doped C_{60} compounds has been synthesized with the general formula $M_{3-x}M'_xC_{60}$, all of which are isostructural with K_3C_{60} [34, 44, 43]. Chemical ordering in which the larger alkali metal atoms occupy the octahedral sites and smaller atoms occupy the tetrahedral sites, has been reported in Na_2MC_{60} (M = Rb, Cs) [43, 44] and M_2CsC_{60} (M = Na, K and Rb) [45], where the difference in ionic radii is large. Most of these fcc binary and ternary $M_{3-x}M'_xC_{60}$ compounds are superconducting with T_cs increasing with increasing fcc lattice parameter [34, 46]; however, exceptions are the Na-doped compounds. Na_3C_{60} is not superconducting, although the above-mentioned correlation between T_c and a would predict a T_c of ~ 16 K, as discussed in more detail later. Na_2KC_{60} and Na_2RbC_{60} have been studied by two groups with different results. In one study [44], these materials were found to superconduct at 2.5 K with very low shielding fractions, but in the other [43], no superconductivity was observed down to 2 K.

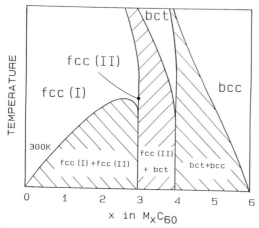

Fig. 1. Provisional binary phase diagram for Rb-doped C_{60}. Shaded regions denote two-phase coexistence. The (linear) temperature scale is indicated approximately by the 300 K label in the lower left corner. The RbC_{60} and the low temperature simple cubic C_{60} phases are not shown in this figure. Reprinted with permission from the authors and *Science* **254**, 545 (1991). Copyright 1992 by the AAAS.

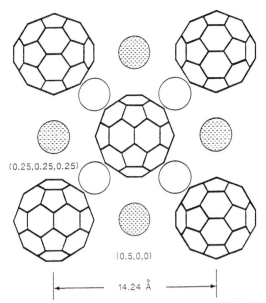

Fig. 2. The structure of K_3C_{60}. The open and hatched spheres represent the potassium at the tetrahedral and octahedral sites, respectively. Reprinted with permission from the authors and *Nature* **351**, 633 (1991). Copyright (C) 1992, Macmillan Magazines Ltd.

The crystal structures of the various intercalated M_xC_{60} phases will now be described in more detail, both under ambient conditions and also as a function of temperature and pressure when known.

(A) Superconducting fcc M_3C_{60}

(1) *Ambient conditions*. It is now well-established that there is a single stoichiometric superconducting phase in the alkali metal-doped M–C_{60} system with a composition of three alkali metal ions per C_{60} molecule. The crystal structure of the 19 K superconducting K_3C_{60} phase was first determined by Stephens and colleagues by Rietveld refinement of synchrotron X-ray powder diffraction data [3].

The structure of K_3C_{60} is illustrated in Fig. 2, shown as the (100) projection of the K_3C_{60} unit cell. As in pristine solid C_{60}, fullerene molecules are centered at the fcc (0,0,0) and translationally equivalent lattice sites. K ions occupy all the available tetrahedral and octahedral vacancies of the host lattice, shown by open and hatched spheres respect-

ively in the figure. Intercalation of K expands the unit cell dimension from 14.17 Å in pure C_{60} to 14.24 Å. The orientation of the C_{60} molecules in this doped phase is of particular interest. Free rotation of the C_{60} molecules was ruled out on steric grounds, since the radius of the tetrahedral site would be only 1.12 Å in this case, much smaller than the 1.33 Å radius of the K^+ ion. The absence of measurable intensity for the (642) peak allowed a choice to be made in favor of $Fm\overline{3}m$ symmetry, in which the molecules are distributed between two equally-populated orientations (merohedral disorder). They are oriented such that eight of their 20 hexagonal faces are perpendicular to the cubic $\langle 111 \rangle$ directions, the two orientations being related by a 90° degree rotation along the [001] direction, as shown in Fig. 2. The corresponding atomic positions are listed in Table 2. The average C–C radius is 3.54 Å, and there is little residual disorder of the C_{60} lattice, as indicated by the low value of the carbon Debye–Waller factor.

Rb_3C_{60} has been found to be isostructural with K_3C_{60}, with a lattice parameter of 14.42 Å [6, 36]. Some recent synchrotron X-ray powder data collected from Rb_3C_{60} at 300 K and a wavelength of 1.2984 Å are shown in Fig. 3. The sample contains a minority phase whose strongest Bragg peak lies between the fcc Rb_3C_{60} (311) and (222) reflections. Superposed on the same figure is the fit resulting from Rietveld refinement in space group $Fm\overline{3}m$ with two orientations for the C_{60} molecules. All the carbon–carbon bond lengths were constrained to be equal, the refined value being 1.42 Å. As shown by the intensity difference plot, a very respectable fit is obtained, with most of the discrepancies arising from the intensities of the minority peaks. The atomic coordinates and the thermal factors of the C and Rb atoms are listed in Table 3. The R-factor was not improved significantly if two different carbon bond lengths were refined. The result is very similar to that reported by Stephens et al. [36]. Refinement based on models with fullerene molecules fixed in one orientation (space group $Fm\overline{3}$) or freely rotating did not yield such good agreement with the experimental data.

Table 2. Atomic coordinates, fractional occupancies, N, and thermal factors, B, for K_3C_{60} obtained from Rietveld refinement. Space group $Fm\overline{3}m$, $a = 14.24$ Å. Estimated errors are given in parentheses referred to the least significant digit(s). Reprinted with permission from the authors and *Nature* **351**, 632 (1991). Copyright (C) 1992, Macmillan Magazines Ltd

	Site	x	y	z	N	B (Å^2)
C1	96j	0.0	0.046	0.245	0.5	1.2 (0.5)
C2	192i	0.213	0.084	0.096	0.5	1.2 (0.5)
C3	192i	0.184	0.160	0.051	0.5	1.2 (0.5)
K1	8c	0.25	0.25	0.25	1.0 (0.2)	6.5 (2.0)
K2	4b	0.5	0.5	0.5	1.0 (0.2)	16 (6)

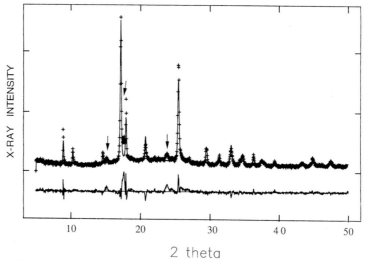

Fig. 3. The 300 K synchrotron X-ray powder diffraction pattern of Rb_3C_{60} collected at $\lambda = 1.2984$ Å, together with the fit obtained from Rietveld refinement in space group $Fm\bar{3}m$ ($R_I = 0.097$ and $R_{wp} = 0.13$). The peaks indicated by the arrows are from the RbC_{60} phase and were not excluded in the refinement.

Table 3. Atomic coordinates, fractional site occupancies, N, and thermal factors, B, for Rb_3C_{60}. Space group $Fm\bar{3}m$, $a = 14.42$ Å

	Site	x	y	z	N	B (Å^2)
C1	96j	0.0	0.0493	0.2391	0.5	0.36
C2	192i	0.2087	0.0797	0.0984	0.5	0.36
C3	192i	0.1781	0.1593	0.0493	0.5	0.36
Rb1	8c	0.25	0.25	0.25	1.0	3.3
Rb2	4b	0.5	0.5	0.5	1.0	17

The small B values of 1.2 and 0.4 Å^2, obtained for C in the structure refinements for K_3C_{60} and Rb_3C_{60}, respectively, indicate that there is little residual disorder of the C_{60} molecule. In contrast, the much larger values for K and Rb, especially in the octahedral sites, reveal the presence of substantial thermal fluctuations, or perhaps more likely, static displacements away from the ideal positions towards one of the surrounding C_{60} molecules. However, a refinement of the Rb_3C_{60} structure, in which the Rb^+ ions were allowed to move away from the ideal tetrahedral and octahedral sites to $(1/4 + \delta, 1/4 + \delta, 1/4 + \delta)$ and $(1/2 + \delta, 0, 0)$ did not improve the result significantly. Additional work at low temperature is needed to resolve this point.

Table 4 summarizes some of the structural parameters of K_3C_{60} and Rb_3C_{60} derived from the refinements. The nearest neighbor center-to-center

Table 4. Some structural parameters for the fcc C_{60} and M_3C_{60} phases. C–M is the closest carbon–metal distance, and ρ the crystallographic density

	a(Å)	C_{60}–C_{60}(Å)	C–M(Å)	ρ(g cm^{-3})
C_{60}	14.17	10.02	—	1.70
K_3C_{60}	14.24	10.06	3.27	1.93
Rb_3C_{60}	14.43	10.20	3.33	2.16

distance between two C_{60} molecules is found to increase from 10.02 Å in pristine C_{60} to 10.07 and 10.20 Å in K- and Rb-doped C_{60}, respectively, with corresponding closest C–M distances of 3.27 Å and 3.33 Å. The crystallographic densities of K_3C_{60} and Rb_3C_{60} are 1.93 and 2.16 g cm^{-3}, both larger than the 1.68 g cm^{-3} value of undoped C_{60}.

Structure refinements based on laboratory X-ray powder diffraction data have shown that in the ternary compounds of the type Na_2MC_{60} with M = Rb and Cs, there is ordering of Rb and Cs on the octahedral sites, as would be expected on the basis of the respective ionic radii [43, 44].

(2) *Temperature dependence studies.* At 300 K, a solid state ^{13}C NMR study of K_3C_{60} by Tycko *et al.* [47] showed a single sharp line with a frequency of 186 ppm with respect to tetramethylsilane and a width of 15 ppm, indicating that the C_{60} molecules are dynamically disordered on the NMR time scale ($\sim 10^{-5}$ s), such that the chemical shift anisotropy of the ^{13}C is averaged out. In the light of the X-ray refinement results, which show little thermal disorder, the dynamics of the C_{60} molecules at 300 K were interpreted in terms of jumping between the symmetry-equivalent orientations. The NMR line-width becomes much broader at 220 K, implying a freezing

out of this jumping motion. However, no structural transitions have been observed in either K_3C_{60} or Rb_3C_{60} down to 10 K in X-ray diffraction experiments. This can be attributed to the quite different time scale these two techniques are capable of probing, that of 10 keV X-rays being around 10^{-19} s.

The fcc Rb_3C_{60} phase is stable in an inert gas atmosphere up to about 400°C, and the lattice parameter is found to increase linearly with temperature. The isobaric thermal expansion coefficient $d \ln a/dT$ of $3.1 \times 10^{-5} \, K^{-1}$ [28] is substantially larger than the averaged value of $2.1 \times 10^{-5} \, K^{-1}$ for C_{60} [48] (Table 5). No phase transition is observed in differential scanning calorimetry (DSC) measurements up to 350°C.

The crystallinity of the minority phase present in the sample of nominal composition Rb_3C_{60} previously discussed (Fig. 3) increases with temperature up to about 70°C. This phase has an fcc structure above room temperature, with all the allowed reflections except the (222) visible [38]. A recent structure refinement indicates that it has the stoichiometry RbC_{60} with the C_{60} molecules orientationally disordered and the Rb atoms occupying the octahedral sites only [37].

(3) *Pressure dependence studies.* The structural behavior of the superconducting compounds under hydrostatic pressure is of considerable interest for understanding the mechanism of superconductivity. In common with most of the organic superconductors, T_c for the M_3C_{60} compounds decreases with applied hydrostatic pressure. For K_3C_{60}, an initial slope of -0.6 K kbar^{-1} was reported by Schirber et al. [49], and a somewhat higher slope of -0.8 K kbar^{-1} by Sparn et al. [50], while for Rb_3C_{60} a slope of -1.0 K kbar^{-1} was observed [51]. Furthermore, as noted by Fleming et al. [34], at atmospheric pressure, the T_cs of the isostructural $M_{3-x}M'_xC_{60}$ superconductors, with M and M′ = K, Rb or Cs, increase monotonically with the intermolecular distance, $a/\sqrt{2}$. These results suggest that the superconducting transition temperature is correlated with the overlap between the adjacent C_{60} molecules, and is relatively insensitive to the nature of the metal intercalates.

Synchrotron X-ray powder diffraction data collected on samples loaded in a diamond anvil cell have shown that both K_3C_{60} and Rb_3C_{60} are stable up to 30 kbar of hydrostatic pressure, no Bragg

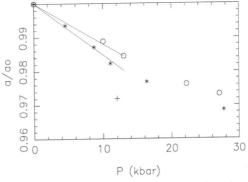

Fig. 4. Fractional reduction of lattice parameter a/a_o vs pressure P; open circle and asterisk symbols represent data for K_3C_{60} and Rb_3C_{60}, while the + symbol is for pure C_{60}, taken from [28]. The experimental errors in a and P are ± 0.002 Å and ± 0.3 kbar, respectively. Solid lines are linear fits to the data points up to 13 kbar.

reflections other than those from the fcc phase having been observed [46]. However, these results are not sufficient to rule out the possibility of an orientational ordering transition under pressure, in which the C_{60} molecules settle in one orientation rather than being distributed between the two symmetry equivalent positions. This in part results from the weak diffracted intensities from the M_3C_{60} powder samples in the diamond anvil cell, due to the small sample diameter (about 0.25 mm), especially at high diffraction angles. However, such a pressure-induced transition is highly plausible, from the viewpoint of the energetics of the system. For example, DSC measurements on C_{60} have shown that the orientational ordering temperature increases with pressure at the rate of 10 K kbar^{-1} [52], as discussed in more detail in the article by Heiney in this issue [27].

Figure 4 shows the change of the cubic lattice parameter versus applied hydrostatic pressure for K_3C_{60} and Rb_3C_{60} up to 30 kbar. For comparison, a data point for undoped C_{60} measured by Fischer et al. [28] is superposed on the same graph. The contraction of the lattice constants is linear up to ~ 13 kbar within experimental error, corresponding to average linear compressibilities $d \ln a/dP$ of $1.2 \pm 0.1 \times 10^{-3}/$kbar and $1.5 \pm 0.1 \times 10^{-3}/$kbar for K_3C_{60} and Rb_3C_{60}, respectively. These values are both less than the values of $2.3 \pm 0.2 \times 10^{-3}/$kbar and $1.8 \pm 0.2 \times 10^{-3}/$kbar for C_{60} reported in [28] and [53], respectively. These results are compared to similar data for KC_8, RbC_8 and graphite in Table 6.

By scaling the $a(P)$ data point-by-point with the $T_c(P)$ data for K_3C_{60} [50] and Rb_3C_{60} [51], and then combining with the atmospheric pressure data of T_c vs a for $M_{3-x}M'_xC_{60}$ [34], we may extend the monotonic variation noted in [34] to a first-order "universal relationship" between T_c and the lattice

Table 5. Isobaric thermal expansion coefficients of Rb_xC_{60} ($x = 0, 1, 3$ and 6), from temperature-dependence X-ray powder diffraction measurements [38]. Units are $10^{-5}/K$

	C_{60}	RbC_{60}	Rb_3C_{60}	Rb_6C_{60}
$\partial \ln a/\partial T$	2.1†	2.1	3.1	2.4

† Ref. 48.

Table 6. The compressibilities of K_3C_{60}, Rb_3C_{60} [46], and pure C_{60} [28, 53]. Values for analogous graphite intercalation compounds and pure graphite are included for comparison [67]. The first row gives the average linear compressibilities at low P. Values in the second row are the leading terms in fits to the equations-of-state over a wide pressure range. Units are $10^{-12}\,cm^2\,dyne^{-1}$

	K_3C_{60}	Rb_3C_{60}	C_{60}	KC_8	RbC_8	Graphite
d ln a/dP, d ln c/dP	1.2	1.5	2.3 (1.8)	2.06	2.07	2.74
1/K_o	4.8	5.7	5.5			

parameter a, over a fairly broad range, i.e. $13.9\,\text{Å} < a < 14.5\,\text{Å}$ and $5\,K < T_c < 30\,K$, as shown in Fig. 5a. The implication of this empirical correlation between T_c and a is that, to first order, the main role of the alkali metal ions in the superconducting mechanism is as electron donors. However, closer inspection of the results reveals that there is a subtle difference in the pressure dependence of T_c for K_3C_{60} and Rb_3C_{60}, corresponding to $\partial T_c/\partial a$ values of 33 ± 2 and $45 \pm 1\,K/\text{Å}$, respectively, as shown in Fig. 5b. The difference is significantly larger than the experimental error, and indicates that the nature of the alkali metal intercalates does in fact have a second-order effect on T_c [46].

The $T_c(a)$ data of a recently reported series of ternary compounds $M_2M'C_{60}$ [44] are superimposed on Fig. 5a. The measured T_c values of Na_2CsC_{60} and Li_2CsC_{60} are consistent with the values predicted from the above-mentioned correlation between T_c and a. However, the T_c values of $2.5\,K$ reported for Na_2KC_{60} and Na_2RbC_{60} in [44] are lower than the values predicted.

(B) M_6C_{60}

C_{60} doped to saturation with K and Cs was found to form non-superconducting stoichiometric compounds with six alkali metal atoms per fullerene molecule in a body-centered cubic lattice [5], the first direct structural evidence to show that C_{60} can be intercalated. Figure 6 shows the results of a Rietveld refinement of the 300 K X-ray powder diffraction data from Cs_6C_{60} (shown as dots), in the form of a profile fit (top) and a difference plot (below). All reflections can be indexed on a bcc lattice with a unit cell parameter of 11.79 Å, the corresponding value for K_6C_{60} being 11.39 Å. The calculated fit, based on the space group $Im\bar{3}$ with the C_{60} molecules orientationally ordered, is shown as a solid line in the same

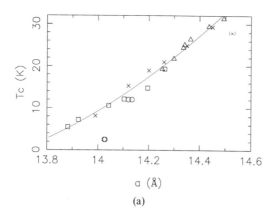

(a)

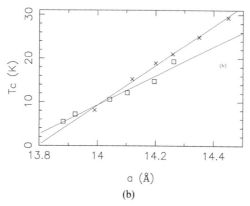

(b)

Fig. 5. Superconducting transition temperature versus lattice parameter. (a) Open square and cross symbols represent K_3C_{60} and Rb_3C_{60} data obtained from pressure measurements [46], and crosses are 1 bar $T_c(a)$ data from [34]. The solid line is a quadratic fit to the composite data. Recent atmospheric pressure T_c data from [44] are superposed on the graph as open circles. (b) Open square and cross symbols represent K_3C_{60} and Rb_3C_{60}, respectively. The solid lines are linear fits to the two sets of data. The slopes are 33 ± 2 and $45 \pm 1\,K\,\text{Å}^{-1}$, respectively.

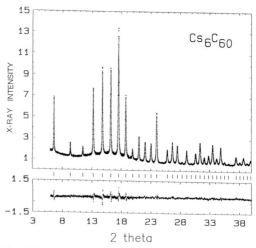

Fig. 6. 300 K powder diffraction pattern of Cs_6C_{60}. Intensity is plotted on a linear scale. The data were recorded with a 2θ step of $0.01°$ at a wavelength of 0.9617 Å. The solid line represents a fit from Rietveld refinement with $Im\bar{3}$ symmetry ($R_I = 0.043$).

Table 7. Atomic coordinates, fractional site occupancies, N, and thermal factors, B, in K$_6$C$_{60}$.
Space group $Im\bar{3}$, $a = 11.39$ Å

	Site	x	y	z	N	B (Å^2)
C1	24g	0.0633	0.0	0.3075	1.0	1.04
C2	48h	0.1266	0.1025	0.2683	1.0	1.04
C3	48h	0.0633	0.2049	0.2292	1.0	1.04
K	12e	0.0	0.5	0.2804	1.0	3.25

figure, for which an integrated intensity R factor of 0.043 was achieved. The two near neighbor C–C distances within the fullerene molecule were constrained to be equal, the refined value being 1.44 Å. With this constraint, there are only two variable positional parameters, one each for C and Cs. The R factor was not improved significantly by allowing two different bond lengths in the hexagons, nor by unconstrained refinement of all eight C positional parameters. Rb$_6$C$_{60}$ has also been shown to have this type of bcc structure with $a = 11.54$ Å [6, 36]. The atomic coordinates and Debye–Waller temperature factors of C and M (M = K, Rb and Cs) obtained from the refinements are listed in Tables 7–9. The B values of the alkali metal atoms are relatively small compared to the values in the M$_3$C$_{60}$ phases.

In contrast to the M$_3$C$_{60}$ phases, the C$_{60}$ molecules are orientationally ordered in the saturation-doped phases. The best fit to the Cs$_6$C$_{60}$ data with randomly-oriented C$_{60}$ molecules yielded an integrated intensity R factor of 0.109, much worse than the value of 0.043 obtained from the orientationally-ordered model. Similar results were also obtained for Rb$_6$C$_{60}$ and K$_6$C$_{60}$ [54]. The solid state ^{13}C NMR study by Tycko et al. [47] also indicates orientational ordering in K$_6$C$_{60}$, and this doping-induced ordering could be a consequence of the electrostatic field associated with electron transfer from M to C$_{60}$, or it could be a signature of orbital hybridization (partial covalent bonding).

Figure 7 illustrates the atomic arrangement in the bcc unit cell. Two equivalent C$_{60}$ molecules per cell are centered at (0,0,0) and (1/2, 1/2, 1/2) oriented with their two-fold axes along the cube edges. Twelve Cs atoms per cell are located in the 12(e) positions at (0, 0.5, 0.28), which can be visualized as four-atom motifs centered at the (1/2, 1/2, 0) and equivalent positions. In Fig. 7 a typical motif is shown in the (001) plane with atoms displaced ~ 0.28 along x and ~ 0.22 along y from the face center. This distortion has been shown to be a lower energy configuration by an ab initio molecular dynamics simulation [55]. A slight dilation of the fullerene molecule is also observed both in the experiment and in the simulation. Each C$_{60}$ is surrounded by 24 Cs atoms, and each Cs is in a distorted tetrahedral environment of four C$'_{60}$s.

Table 10 summarizes some of the structural results for the three bcc M$_6$C$_{60}$ compounds. The nearest-neighbor distance between C$_{60}$ centers is found to be 9.86, 9.99, and 10.21 Å for K$_6$C$_{60}$, Rb$_6$C$_{60}$, and Cs$_6$C$_{60}$, respectively, compared to the intermolecular distance of 10.02 Å in pure C$_{60}$. The nearest-neighbor C–M distances lie in the range of 3.20–3.38 Å, larger than the sum of the van der Waals C radius and the respective alkali metal ionic radii. All the M–M near-neighbor distances are considerably larger than the respective ionic diameters.

Rb$_6$C$_{60}$ is stable in an inert gas atmosphere up to 673 K (the highest temperature reached in the experiment), and the bcc lattice parameter a was found to increase linearly with temperature [38]. The thermal expansion coefficient (d ln a/dT), 2.4×10^{-5}/K, is larger than the value for C$_{60}$ [48] (Table 5). Structural refinement with the 520 K X-ray powder diffraction

Table 8. Atomic coordinates, fractional site occupancies, N, and thermal factors, B, in Rb$_6$C$_{60}$.
Space group $Im\bar{3}$, $a = 11.54$ Å

	Site	x	y	z	N	B (Å^2)
C1	24g	0.0624	0.0	0.3030	1.0	1.73
C2	48h	0.1247	0.1010	0.2644	1.0	1.73
C3	48h	0.0624	0.2019	0.2258	1.0	1.73
Rb	12e	0.0	0.5	0.2822	1.0	2.97

Table 9. Atomic coordinates, fractional site occupancies, N, and thermal factors, B, in Cs$_6$C$_{60}$.
Space group $Im\bar{3}$, $a = 11.79$ Å

	Site	x	y	z	N	B (Å^2)
C1	24g	0.0611	0.0	0.2968	1.0	1.43
C2	48h	0.1222	0.0989	0.2590	1.0	1.43
C3	48h	0.0611	0.1978	0.2212	1.0	1.43
Cs	12e	0.0	0.5	0.2781	1.0	3.11

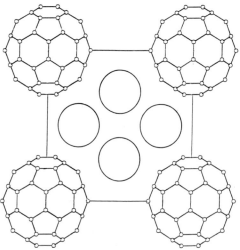

Fig. 7. Part of the Cs_6C_{60} structure projected on the cube face normal to [001], derived from Rietveld refinement of the profile shown in Fig. 6. Large circles represent Cs^+ on the same scale as the cube edge. Small circles are C atoms, not to scale. An equivalent C_{60} molecule is centered at (1/2,1/2,1/2) (not shown). Cs coordinates (clockwise from top) are (0.5,0.72,0), (0.78,0.5,0), (0.5,0.28,0) and (0.22,0.5,0) with respect to an origin at the bottom left corner. Faces normal to x or y may be visualized by rotating the diagram $\pm 90°$. This is a consequence of the molecular orientation with the two-fold axes along the cube edges. Reprinted with permission from the authors and *Nature* **351**, 463 (1991). Copyright (C) 1992, Macmillan Magazines Ltd.

data indicates that the C_{60} molecules are still orientationally ordered with no significant change in the Rb position. Although the X-ray diffraction pattern collected at 673 K did not cover a sufficiently extended range for Rietveld refinement, the close resemblance of the 520 K and the 673 K patterns suggests that orientational ordering of the fullerene molecules is retained, consistent with the observed linear expansion of the lattice parameter.

(C) M_4C_{60}

In addition to the cubic M_3C_{60} and M_6C_{60} phases discussed above, a body-centered tetragonal (bct) phase with four alkali metal atoms per fullerene molecule was also shown to exist by Fleming *et al.* [4] and subsequently confirmed in two other studies [6, 36]. Powder diffraction patterns of the M_4C_{60} compounds (M = K, Rb and Cs) are consistent with the space group $I4/mmm$. The lack of four-fold

symmetry in the C_{60} molecule indicates orientational disorder in this bct structure, which can be described as either completely orientationally-disordered, as in the room temperature phase of C_{60}, or merohedrally-disordered, analogous to the model used by Stephens *et al.* for the K_3C_{60} phase [3]. In this model, the fullerene molecules are equally distributed among two symmetry-equivalent orientations, both of which have their two-fold axes aligned along the unit cell $\langle 100 \rangle$ directions and are related by a 90° rotation about the [001] axis (the four-fold symmetry direction). For $I4/mmm$ symmetry, nine inequivalent C atoms and one K atom are necessary to generate the structure. Fleming *et al.* carried out a refinement of the M_4C_{60} structure based on the first model, with a uniform shell of electron density centered at (0,0,0) and alkali metal atoms at (x, 1/2, 0), and obtained a best fit with $x = 0.22$ and a C_{60} radius of 3.55 Å [4].

We have compared the above two models by carrying out Rietveld refinements with X-ray powder diffraction data collected from K_4C_{60} at 300 K with Cu Kα radiation on a laboratory diffractometer equipped with a flat graphite (002) monochromator and an Inel curved position-sensitive detector spanning a range of 120°. Refinement based on the second model was carried out with the constraint that all the C–C bond lengths were equal, previously shown to be a reasonable approximation for the M_6C_{60} phases [5]. Figures 8a and 8b show the experimental data and the profile fits obtained from the refinements based on the orientationally-disordered and merohedrally-disordered models, respectively. The latter clearly provides a better fit to the experimental data in the higher angle between 30 and 40°, as shown both in the figure and also by the improvement in the integrated intensity R factor. The atomic coordinates and temperature factors obtained from the merohedrally-disordered model are listed in Table 11. The K ion is displaced from the ideal (0.25, 0.5, 0) position to (0.21, 0.5, 0), similar to the value reported in [4]. The single C–C bond distance is 1.44 Å, and the closest K–C distance is 3.33 Å, comparable to the values found in the K_3C_{60} and K_6C_{60} compounds. The B factors correspond to mean-square displacements of 0.010 and 0.059 Å^2 for carbon and potassium, respectively.

Table 10. Some structural parameters for the bcc M_6C_{60} phases obtained from Rietveld refinement. C–M and M–M denote the near-neighbor carbon–metal and metal–metal distances, respectively. Values of ionic radii are taken from [68]

	a (Å)	C_{60}–C_{60} (Å)	C–M (Å)	M–M (Å)	r_{M+} (Å)	ρ (g cm^{-3})
K	11.38	9.86	3.20	4.06	1.33	2.21
Rb	11.54	9.99	3.25	4.11	1.48	2.66
Cs	11.79	10.21	3.38	4.19	1.67	3.08

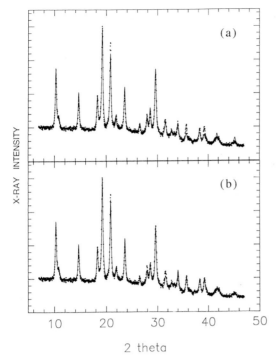

X-RAY INTENSITY

(a)

(b)

10   20   30   40   50

2 theta

Fig. 8. Fits to laboratory X-ray powder diffraction pattern collected from K_4C_{60} at $\lambda = 1.5406$ Å based on Rietveld refinements with: (1) a completely orientationally-disordered model ($R_I = 0.13$ and $R_{wp} = 0.146$) and (b) a merohedrally-disordered model ($R_I = 0.121$ and $R_{wp} = 0.122$). Note the improved fit for the latter in the region between 35 and 40°.

The bct K_4C_{60} phase was found to be stable down to ~ 50 K from synchrotron X-ray powder diffraction experiment, and no evidence for the charge-density-wave (CDW) distortion, suggested by a recent local-density-approximation (LDA) calculation [56], was observed.

(D) Rb_xC_{60}

The temperature-composition phase diagram of Rb_xC_{60} compounds has been studied in detail [6, 38, 37]. In addition to the stoichiometric doped phases described above, direct experimental evidence

has been obtained for a high temperature fcc RbC_{60} phase, and a sub-stoichiometric bcc phase with $x \sim 5.6$ [54]. In contrast, the fcc phase with $x = 3$ and the bct phase with $x = 4$ appear to be line compounds with essentially zero solubility for vacancies or Rb interstitials.

The structure of the RbC_{60} phase has been studied recently by Zhu et al. A Rietveld refinement based on X-ray powder diffraction data collected at 200°C has shown that the C_{60} molecules are orientationally disordered and the Rb atoms occupy the octahedral sites only [37]. The room temperature structure of this phase has not yet been determined. The thermal expansion coefficient $d\ln a/dT$ of this compound is listed in Table 6, along with those of the $x = 3$ and $x = 6$ materials.

An early study [6] of Rb_xC_{60} samples prepared with nominal x values of 2.9 and 3.3 has shown both samples to phase separate into a dominant fcc phase with $x = 3$ and a lattice parameter of 14.42 ± 0.02 Å, together with small amounts of different minority phases. The minority component in the $x = 2.9$ sample was found to be the above-mentioned RbC_{60}. ^{13}C NMR spectra indicate phase separation in K_xC_{60} samples with nominal values of $x = 1.5$ and 2.0 [47], consistent with the present results. The minority phase in the $x = 3.3$ sample was found to be bct Rb_4C_{60}, with lattice parameters of $a = 11.96$ Å and $c = 10.98$ Å, essentially the same as the values of $a = 11.962$ Å and $c = 11.022$ Å measured from a single phase bct Rb_4C_{60} sample [4]. This suggests that the bct phase is also a line compound at 300 K. The miscibility gap between the fcc phase with $x = 3$ and the bct phase should extend all the way to the melting point due to the different crystal symmetries, but the range of unstable compositions is very likely to be temperature dependent for entropy reasons.

Figure 9 shows the X-ray profile of a sample prepared with a nominal Rb concentration $x = 5$ (top) and a saturation-doped sample with nominal $x = 6$ (bottom). Both are dominated by bcc-type

Table 11. Atomic coordinates, fractional site occupancies, N, and thermal factors, B in K_4C_{60}, obtained from Rietveld refinement with a model in which the fullerene molecules are equally distributed between the two symmetry-equivalent orientations. Space group $I4/mmm$, $a = 11.84$ Å, $c = 10.75$ Å

	Site	x	y	z	N	B (Å^2)
C11	16n	0.0000	0.0589	0.3243	0.5	0.76
C12	16n	0.0589	0.2944	0.0000	0.5	0.76
C13	16n	0.2944	0.0000	0.0649	0.5	0.76
C21	32o	0.9950	0.1205	0.2824	0.5	0.76
C22	32o	0.1205	0.2564	0.1096	0.5	0.76
C23	32o	0.2564	0.0995	0.1327	0.5	0.76
C31	32o	0.0615	0.2200	0.2147	0.5	0.76
C32	32o	0.2200	0.1949	0.0677	0.5	0.76
C33	32o	0.1949	0.0615	0.2423	0.5	0.76
K	8j	0.2103	0.500	0.0000	1.0	4.7

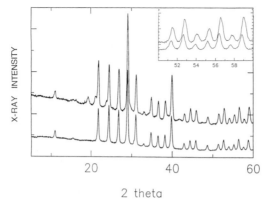

X-RAY INTENSITY

20   40   60

2 theta

Fig. 9. Diffraction patterns from a sample with nominal composition Rb_5C_{60} (top curve) and saturation-doped material Rb_6C_{60} (bottom curve) measured on a laboratory diffractometer with Cu characteristic radiation. The inset shows the majority bcc phase has a different lattice parameter in the two samples. Reprinted with permission from the authors and *Science* **254**, 545 (1991). Copyright 1992 by the AAAS.

reflections. The $x = 5$ sample contains a small amount of the bct phase, implying that the bcc component has a composition with an x value significantly less than 6. Structural refinement based upon the M_6C_{60} unit cell (space group $Im\bar{3}$) but with variable Rb occupancy converged to an x value of 5.6 ± 0.1 with an integrated intensity R factor of 0.12. The lattice parameter of the $x = 5.6$ bcc phase is 11.518 Å, slightly smaller than the value of 11.54 Å found for the saturation-doped $x = 6$ phase, but otherwise the structural features are essentially identical. A recent X-ray powder diffraction study of the same sample, carried out as a function of temperature, showed the volume fraction of the bct phase to decrease with increasing temperature, no bct peaks being detectable above 250°C [38]. This result confirms that the solubility of interstitial vacancies increases with increasing temperature in the bcc phase. The bct–bcc transition must be first order for symmetry reasons, and the miscibility gap should therefore extend up to the melting point.

The X-ray results indicate that in the Rb_xC_{60} phase diagram, the upper boundaries separating mixed-phase regions from the high-T single phase regions are asymmetric for $0 < x < 3$ and $4 < x < 6$. This is a consequence of the finite interstitial or vacancy densities in fcc(I) or bcc, respectively, and the corresponding zero densities in bct and fcc(II). The asymmetry for $4 < x < 6$ is reminiscent of a model proposed for graphite intercalation compounds which takes into account elastic dilation of the host lattice [57]; the smaller the concentration the lower the strain, and hence the more stable the phase mixture against an entropy-driven order–disorder transition. The corresponding strain in these systems

could be associated with shear distortions from fcc(II) to bct and from bct to bcc. In this context it is interesting to note that the volume per C_{60} molecule is anomalously large in the bct phase, despite the monotonic increase in M concentration. Based upon the above lattice parameters, the excess volume per molecule for Rb_xC_{60} relative to C_{60} is 38, 74, 53 and 57 Å^3 at $x = 3$, 4, 5.6 and 6, respectively. The asymmetry of the $0 < x < 3$ region is reversed; stoichiometric fcc(II) stabilizes the phase mixture while dilute fcc(I) destabilizes it. This could be due to the higher configurational entropy associated with random site occupancy in fcc(I), and/or the absence of shear distortions between fcc(I) and fcc(II). The mixed fcc(II) + bct regime can be tentatively regarded as having greater stability than the other two, as suggested by observations that bct is still present in samples with $x \sim 2.2$ even after extended annealing at relatively low temperatures.

The proposed M_xC_{60} phase diagram resembles that of other well-studied guest–host intercalation compounds, i.e. graphite intercalation compounds [57] and doped conjugated polymers. While the attractive and repulsive elastic–dipole interactions between guest species are responsible for ordering (staging) of intercalants in layer intercalation compounds, a qualitatively different competition can be envisaged in M_xC_{60} that would enhance the stability of high- and low-concentration phases. The electrostatic energies associated with various Rb_xC_{60} phases have been calculated by assuming point charges for Rb^+ and C_{60}^{x-} [4, 47]. The results indeed show that the three observed stoichiometric phases have the lowest electrostatic energies.

(E) Na_xC_{60}

The structural properties of the Na_xC_{60} system have been reported recently by Rosseinsky and colleagues [43], and are described in detail in the article by Murphy *et al.* in this issue [58]. Since they differ in several respects from those of the heavier alkali metal systems discussed in the previous sections, a brief summary is given here.

Although as previously noted Na_3C_{60} is not superconducting, it was found to be isostructural with K_3C_{60} and Rb_3C_{60}, with $a = 14.191$ Å and Na in octahedral and tetrahedral sites. At 250 K there is evidence in the diffraction pattern of separation into two fcc phases. However, unlike the other systems, a distinct phase was observed for $x = 2$, with $a = 14.189$ Å. Structural refinement based on laboratory X-ray powder data revealed the Na to reside mainly in tetrahedral sites. An interesting feature was the appearance of weak reflections consistent with simple cubic rather than fcc symmetry.

A new type of structure was found for Na$_6$C$_{60}$ quite unlike that of the K and Rb compounds, with fcc rather than bcc symmetry ($a = 14.380$ Å). In addition to Na occupying the tetrahedral sites, there are four Na atoms arranged in a tetrahedral cluster around the center of each octahedral position, with a Na–Na distance of 2.8 Å, opening up the possibility of covalent interactions which might reduce the charge transfer.

The existence of an fcc structure across the whole system allows the possibility of significant regions of solid solution, and thus the $x = 3$ composition is not necessarily a line phase as in the other systems.

5. ALKALINE-EARTH METAL-DOPED PHASES

The Ca$_x$C$_{60}$ system has been studied by Kortan and colleagues over the range $x = 1.5$–8 [59]. The compounds have an fcc structure for $x < 5$, with $a \approx 14.0$ Å, but at $x = 5$, extra peaks characteristic of a simple cubic structure are observed, and this material becomes superconducting below 8.4 K. The proposed structural model for the compositions with $x > 3$ involves multiple occupancy of the octahedral sites by Ca^{2+} ions, analogous to the arrangement of Na ions in Na$_6$C$_{60}$. The correlation between T_c and a illustrated in Fig. 5a is also satisfied for Ca$_5$C$_{60}$.

6. IODINE-DOPED C$_{60}$

The alkali metal-doped C$_{60}$ compounds discussed thus far are donor-type intercalation compounds, which involve electron transfer from metal dopants to the C$_{60}$ molecules. The large C$_{60}$ molecular ionization potential [60] seems to rule out the possibility of forming acceptor-type compounds. The recently reported stoichiometric iodine-doped compound C$_{60}$I$_4$ is the first example of a fullerene intercalation compound with no electron transfer between C$_{60}$ and the intercalate [14]. The crystalline phase can be obtained easily by vapor phase reaction of iodine with pure C$_{60}$ at 250°C for several days in evacuated pyrex tubes. It is non-metallic with a resistivity at 300 K that exceeds 10^9 Ω cm, and no superconductivity is observed down to 4 K.

X-ray diffraction data for this new phase are shown in Fig. 10 (top pattern). The particle size estimated from the width of the strongest peak exceeds 1000 Å, and the diffraction profile remained unchanged after exposing the sample to air in the laboratory for two days. Thermogravimetric analysis gave a C$_{60}$:I ratio of 3.7 ± 0.7. The data were indexed in terms of a simple hexagonal cell with $a = 9.96$ Å and $c = 9.98$ Å, corresponding to a unit cell volume of 857 Å^3 per molecule, compared to 711 Å^3 in fcc C$_{60}$.

There are no systematically absent reflections, and the symmetry is therefore $P6/mmm$ or one of its hexagonal or trigonal subgroups. Of the latter, only $P\bar{3}$ is consistent with the point symmetry of the C$_{60}$ molecule if complete orientational order is assumed. It is interesting to note that the diffraction data can be equally well indexed on the basis of a large cubic unit cell, but this was ruled out because of the implausible number of missing reflections.

In this simple hexagonal cell, there are two interstices large enough to accommodate an iodine atom, one with trigonal prismatic coordination around 2d sites at (1/3, 2/3, 1/2) and one with square coordination around 3f sites at (1/2, 0, 1/2). These are located 7.62 Å and 7.05 Å, respectively from the origin, compared to the van der Waals radii of about 5 Å and 2 Å for C$_{60}$ and I.

Various models of orientational order and disorder were investigated in the analysis of the data. The first series of Rietveld refinements was carried out in space group $P\bar{3}$. For an orientationally-ordered structure, this requires a total of 10 C atoms in the 6(g) general positions at x, y, z. The coordinates of the latter were generated from an ideal truncated icosahedron with C–C distances of 1.39 Å for the bonds shared by two hexagons, and 1.45 Å for the bonds shared by a hexagon and a pentagon. The directions of the hexagonal [100], [010] and [001] axes were defined as [10$\bar{1}$], [1$\bar{1}$0] and [111] of an fcc cell with the three mutually orthogonal twofold C$_{60}$ axes aligned along the cubic $\langle 100 \rangle$ directions. I(1) and I(2) atoms were placed in the 2d sites at (1/3, 2/3, 1/2) and the 3f sites at (1/2, 0, 1/2). A common temperature factor B was assigned to all C atoms, and individual Bs and occupancy factors to the I(1) and I(2) atoms. This resulted in a mediocre fit with a surplus of I(1) ($\simeq 2.4$ atoms) and about 50% occupancy of the I(2) sites ($\simeq 1.5$ atoms), and extremely large thermal factors. The latter are usually indicative of large static displacements, and models of this type were therefore tried next. A very satisfactory fit was eventually obtained with a model in which the I(1) and I(2) atoms were allowed to relax to positions $x, 2x, z$ with a common thermal factor. This gave approximately two atoms of I on each site, corresponding to an ideal formula C$_{60}$I$_4$. A slight improvement in the fit was obtained for a model in which the C$_{60}$ molecule was rotated 2.3° about the hexagonal [001] axis from the [100] axis towards the [010] axis. The results of this refinement are summarized in Table 12.

A variety of increasingly disordered models were next tried, based upon higher symmetry space groups $P\bar{3}m1$, $P\bar{3}/m$, $P6/m$ and $P6/mmm$, corresponding to different types of twinning disorder, but without any noticeable improvement in the fit. Finally, a model of

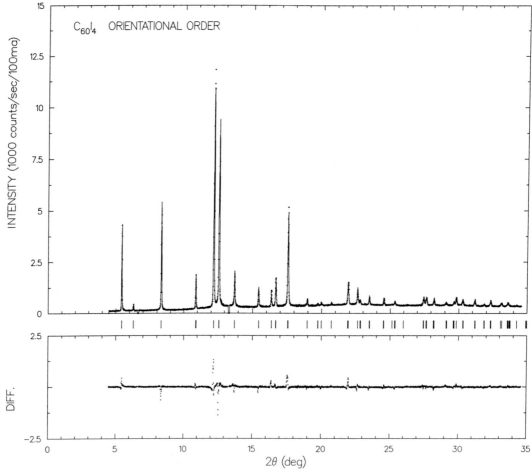

Fig. 10. X-ray powder diffraction profile of the product resulting from vapor phase reaction of a 3/2:1 molar ratio of molecular iodine I_2 with pure solid C_{60} at 250°C. Dots are experimental data, the solid curve is a fit based on Rietveld refinement in space group $P\bar{3}$ ($a = 9.962$ Å, $c = 9.984$ Å). The difference plot is shown underneath. The short vertical markers represent allowed reflections.

Table 12. Atomic coordinates, fractional site occupancies, N, and thermal factors, B, in $C_{60}I_4$ obtained from Rietveld refinement. Space group $P\bar{3}$, $a = 9.962$ Å, $c = 9.984$ Å. C–C distances fixed at 1.45 and 1.39 Å. I11 and I12 constrained to $(x_1, 2x_1, z_1)$ and $(x_1, 2x_1, -z_1)$ with equal occupancies, I2 constrained to $(x_2, 2x_2, 0.5)$. Numbers in parentheses are estimated errors referred to the least significant digit(s). The corresponding values obtained for I11, I12 and I2 in an integrated intensity refinement are listed underneath

	Site	x	y	z	N	B (Å^2)
C11	6g	0.0868	0.2930	0.2409	1.0	2.8 (4)
C12	6g	0.2183	0.3634	0.1605	1.0	2.8 (4)
C21	6g	0.0579	0.1624	0.3248	1.0	2.8 (4)
C22	6g	0.2077	0.4093	0.0248	1.0	2.8 (4)
C23	6g	0.3264	0.0608	0.1891	1.0	2.8 (4)
C24	6g	0.3265	0.3062	0.1605	1.0	2.8 (4)
C31	6g	0.1617	0.1076	0.3248	1.0	2.8 (4)
C32	6g	0.3804	0.0709	0.0590	1.0	2.8 (4)
C33	6g	0.2988	0.1810	0.2409	1.0	2.8 (4)
C34	6g	0.3829	0.3167	0.0248	1.0	2.8 (4)
I11	6g	0.6423 (9)	0.2846	0.4463 (6)	0.164 (1)	4.7 (3)
		0.640 (3)	0.280	0.448 (3)	0.164 (6)	4.7 (1.3)
I12	6g	0.6423 (9)	0.2846	0.5537 (6)	0.164 (1)	4.7 (3)
		0.640 (3)	0.280	0.552 (3)	0.164 (6)	4.7 (1.3)
I2	6g	0.5373 (4)	0.0746	0.5	0.3101 (2)	4.7 (3)
		0.538 (2)	0.076	0.5	0.325 (12)	4.7 (3)

complete orientational disorder was tried, analogous to the room temperature structure of C_{60} itself [26]. In this case the C_{60} molecular structure factor is the zero-order Bessel function $\sin(QR_o)/QR_o$. Rather surprisingly, with $R_o = 3.554$ Å this model gave a somewhat better profile fit. In order to check the significance of this result, a refinement based on integrated intensities obtained by pattern decomposition was carried out. In this case, a distinctly better fit was obtained for the orientationally-ordered model ($R_{WI} = 0.131$ against 0.141). The results obtained in this refinement are in good agreement with those from the profile fit (Table 12), but it is interesting to note that the estimated errors are typically a factor of 5 higher, as discussed earlier in section 2. Attempts to refine the C–C distances did not improve the fit significantly, most likely because of the limited range of the data.

The atomic coordinates listed in Table 12 for the orientationally-ordered model with $P\bar{3}$ symmetry correspond to displacements of the I(1) and I(2) atoms from the ideal sites by 0.68 and 0.64 Å, respectively. These displacements are very likely to be correlated, for example, an I(2) atom situated at 0.537, 0.075, 0.5 will have an I(1) neighbor at one of the four equivalent sites 0.642, 0.358, ±0.446 and 0.715, 0.358, ±0.446, at a distance of 2.53 Å. This is fairly close to the intramolecular distance 2.72 Å reported for elemental I_2 [61]. One can then construct a model having two such I_2 molecules per unit cell, with intermolecular separations of 3.78 Å or 4.91 Å but no long-range correlations between the displacement from the ideal sites. Again, these are comparable to the intermolecular distances of 3.50, 3.97 and 4.27 Å observed in solid I_2 [61]. The I(1) and I(2) atoms are located 7.12 and 7.08 Å, respectively from the center of the C_{60} molecules, with C–I near-neighbor distances between 3.6 and 4.0 Å.

The structure has many satisfying aspects; it allows dumbbell-shaped I_2 molecules to be accommodated in a primitive hexagonal C_{60} lattice completely consistent with the respective van der Waals distances, and also retains many of the features of the I_2 elemental structure.

One remaining question is whether the C_{60} molecules are orientationally ordered or not. One might predict an order–disorder transition analogous to that observed at 255 K in C_{60}, but most likely significantly above room temperature because of the extra steric hindrance from the I_2 molecules. Orientational order is also suggested by the 9.97 Å average intermolecular separation; in pure C_{60} this separation contracts from 10.02 Å to 9.99 Å below the ordering transition temperature at 255 K [26, 62]. Unfortunately the precision of the present structure determination is not good enough to resolve this issue. A further point is the possibility that a structural transition may occur below room temperature due to ordering of the I_2 molecules. Any such transition would most likely be accompanied by a fairly drastic change of symmetry.

The structure of $C_{60}I_4$ is shown in Fig. 11, and is very similar to that of stage-1 graphite intercalation compounds. The {111} planes of fcc C_{60} become the {100} planes, analogous to the covalently-bonded C planes in graphite, and the guest iodine layers reside in the van der Waals galleries between the host C_{60} {100} planes. The c-axis stacking sequence is transformed from ABCABC ... in C_{60} to A/A/A/ ... in $C_{60}I_4$, where capital letters denote different host layers

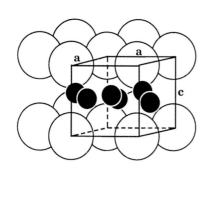

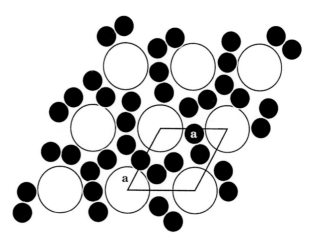

Fig. 11. Schematic crystal structure of $C_{60}I_4$. Large open spheres are C_{60} molecules and small filled spheres represent iodine atoms. (a) Three-dimensional view. (b) Projection in basal plane, showing the fractionally-occupied $6g$ sites.

and "/" represents a guest layer. This process presumably nucleates from stacking faults in solid C_{60} [63] which "heal" in the process since we find no evidence for similar faulting in the compound.

The volume per C_{60} molecule derived from the cell parameters is 857 Å^3, compared to 711 Å^3 in fcc C_{60}. If the fullerene van der Waals radius is taken as 5 Å, this means that 39% of the trigonal cell volume is available to accommodate guest species compared to only 26% in the close-packed fcc structure. Intercalation of C_{60} with K, which has an ionic radius of 1.33 Å, to form K_3C_{60} results in only a small dilation of the fcc cell. Since the atomic volume of iodine is much smaller, 45 Å^3 (van der Waals radius of ~ 2.2 Å), there is ample room in the trigonal cell for four iodine equivalents even with the slightly compressed intermolecular spacing.

An intercalation compound of C_{60} with iodine has been described independently by Akahama et al. [64] prepared under fairly similar conditions but with a final composition reported to be $C_{60}I_{1.6}$ [65]. A diffraction pattern obtained with Mo Kα radiation was interpreted in terms of a C-centered orthorhombic cell with $a = 17.33$, $b = 9.99$, $c = 9.97$ Å, containing two formula units. Iodine atoms were placed in sites slightly shifted from the centers of the trigonal prisms, with near-neighbor and next near-neighbor distances of 3.90 and 7.21 Å, respectively.

It appears that since a is almost exactly $\sqrt{3}b$, this orthorhombic cell can equally well be indexed in terms of a simple hexagonal cell with $a = 9.99$ and $c = 9.97$ Å, i.e. essentially identical to that for $C_{60}I_4$ described above, but with I atoms rather than I_2 molecules in the trigonal prismatic sites based on the near-neighbor separation of 3.90 Å, which is close to the van der Waals distance.

7. TDAE-C_{60}

The final example is a rather different type of intercalation compound containing a relatively large organic molecule, tetrakis-dimethylamino-ethylene, TDAE, which has the formula $C_2N_4(CH_3)_8$. This compound is ferromagnetic with a T_c of 16 K, the highest value known for an organic molecular ferromagnet [15]. In a synchrotron X-ray powder study, the material was found to have the stoichiometric composition TDAE-C_{60}, with a C-centered monoclinic unit cell, $a = 15.807$, $b = 12.785$, $c = 9.859$ Å, $\beta = 94.02°$, and two formula units per cell [16]. Since the TDAE molecule lacks an inversion center, the logical choice of space group is C2, with the C_{60} molecules centered at the 2a sites at (0, 0, 0) and (1/2, 1/2, 0), and the organic molecules in the

large interstices centered around the 2d sites at (1/2, 0, 1/2) and (0, 1/2, 1/2). Rietveld refinement in this group with an assumed rigid structure for the molecule generated from a molecular mechanics model gave a reasonable fit with the long axis of the molecule directed along the c-axis, but the quality of the data was not good enough to permit a detailed determination of the structure.

The separation between C_{60} molecules is 9.98 Å along the c-axis and 10.15 Å between corner and face-centered molecules. The packing resembles that found in the simple hexagonal structure of $C_{60}I_4$, since it can be derived by opening up the 120° angle in the hexagonal basal plane to 129° followed by a small tilt of the [001] axis away from the normal to the plane, yielding large, rather elongated interstices centered at the (1/2, 0, 1/2) and equivalent positions. If the van der Waals radius for C_{60} is taken as 5.0 Å, then 47% of the cell volume is available to accommodate the guest TDAE molecule, even larger than in the simple hexagonal structure of $C_{60}I_4$.

8. SUMMARY

Since the first reports of doping C_{60} with foreign species in 1991, tremendous progress has been made in understanding the composition and phase equilibria of this new class of three dimensional-intercalation compounds. The results so far have shown that the phase behavior of the alkali metal-doped M_xC_{60} compounds is very sensitive to the ionic radii of the metal dopants. In K- and Rb-doped systems, four stoichiometric crystalline phases have been found. These are the high temperature fcc MC_{60} phase, the superconducting fcc M_3C_{60} phase, the bct M_4C_{60} phase, and the saturation-doped bcc M_6C_{60} phase. A crystalline fcc RbC_{60} phase has been determined above room temperature. The orientational order of the C_{60} molecules in these compounds varies with alkali metal concentration, changing from completely-disordered at 300 K in C_{60} and above room temperature in RbC_{60}, to merohedrally-disordered in M_3C_{60} (M = K, Rb) and K_4C_{60}, and then to ordered in M_6C_{60} (M = K, Rb and Cs). Multiple-occupancy of the C_{60} octahedral interstitial sites has been observed in the Na- and Ca-doped systems. In fcc Na_6C_{60}, there is a Na_4 cluster in every octahedral site, and recent preliminary results indicate that as many as eight Na atoms can be accommodated in this site [66]. The Cs–C_{60} system has not been studied in such detail, except for the bcc Cs_6C_{60} phase. There has been X-ray evidence suggesting the formation of a bcc superconducting Cs_3C_{60} phase [40], which, if confirmed, would be the only known fullerene superconductor with this type of structure.

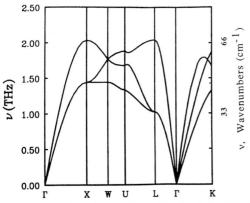

Fig. 3. Phonon dispersion relations $v(k)$ of bulk fcc fullerite from Ref. 16.

orbitals is partly responsible for the softening of the radial A_g mode frequency in C$_{60}$ (to 492 cm^{-1}) [29]. For an isolated C$_{60}$ molecule, two of the 10 Raman-active modes are non-degenerate A_g modes, and the remaining eight are five-fold degenerate H_g symmetry modes, each of which are expected to split into a 3-fold (T_g) and 2-fold (E_g) mode in a cubic crystal field, when the balls crystallize into solid C$_{60}$ [30]. As we shall see, the sharpness and the simplicity of the observed Raman spectrum indicates the molecular nature of solid C$_{60}$. Significant solid interactions between the balls would produce detectable $H_g \rightarrow T_g + E_g$ line splittings in the molecular modes, introduce new lines and initiate line broadening in the cubic field of the solid. Since C$_{60}$ is a molecular solid, knowledge of the properties of the isolated molecules is therefore important for understanding the behavior of the solid and doped solid.

In view of the charge transfer between alkali metal atoms around C$_{60}$, it is rather remarkable that the addition of interstitial alkali metal dopants in C$_{60}$ perturbs the Raman spectra of C$_{60}$ only slightly [26, 27]. Optical data, for example, are consistent with the complete charge transfer of 1e per M atom to the C$_{60}$ ball. Thus the M$_6$C$_{60}$ compounds should be insulating with the six electrons transferred from the M atoms filling the t_{1u}-derived band. We can, in fact, identify many of the lines in the M$_6$C$_{60}$ Raman spectra with those of pristine C$_{60}$, and very little change in the spectra is found from one alkali metal dopant to another [27].

Also in the graphite intercalation compounds, the lattice modes associated with the carbon atoms are only slightly perturbed upon the intercalation of charged species. For example, in the alkali metal intercalation into graphite, electrons are transferred to the graphene layers, resulting in a positively charged K-layer, and causing both a small intralayer C–C bond expansion and a concomitant intralayer

mode softening [29]. A similar situation is observed for the tangential modes in C$_{60}$, which systematically soften by ~ 6 cm^{-1} per transferred electron to the C$_{60}$ molecule on doping with alkali metals [26, 27].

Another common feature between the Raman spectra in stage 1 alkali metal graphite intercalation compounds and metallic M$_3$C$_{60}$ is the observed Breit–Wigner–Fano lineshapes [31]. The Fano lineshape of the lowest Raman line in K$_3$C$_{60}$ at 271 cm^{-1} is similar to that observed for the lines at 566 cm^{-1} in the first stage compound C$_8$K. For K$_3$C$_{60}$, the coupling implied by the Fano lineshape is to a continuum of low-lying states [31], either associated with electronic states or with a multiphonon continuum. Furthermore, the H_g-derived modes in M$_3$C$_{60}$ are quite broad [31], possibly indicating a strong electron–phonon coupling which might be significant in formulating a model for superconductivity in these compounds. This has been addressed in the recent theoretical calculations [32].

Because of the weak interaction of the balls with each other and with the alkali metal dopants, solid M$_x$C$_{60}$ can be viewed as a molecular solid having energy bands with little dispersion, thereby giving rise to a very high density of states near the Fermi level. This observation is also important in understanding the high temperature superconductivity in M$_3$C$_{60}$. Our optical reflection/transmission studies reveal that the widths of these bands are ~ 0.5 eV [33]. Also favoring superconductivity is the enhanced electron–phonon interaction arising from the curvature of the buckyballs relative to a graphene sheet [32]. This curvature on the C$_{60}$ shell mixes carbon σ and π orbitals and enhances the coupling of electrons to low energy phonons [32].

2. EXPERIMENTAL DETAILS

2.1. Synthesis of pristine and doped solid C$_{60}$ films

An a.c. discharge (20 V rms, 150 A) between graphite electrodes in 200 torr of He gas was used to produce a carbon soot containing approximately 15% fullerenes, a method similar to that first reported by Krätschmer and coworkers [3]. Soxhlet extraction with toluene was carried out to remove a mixture of fullerenes C$_n$ from the soot. Separation of C$_{60}$ from C$_{70}$ and higher fullerenes was accomplished using a gel permeation liquid chromatography (GPLC) technique developed by Meier and Selegue [34] using pure toluene as the solvent. The identity of the high performance liquid chromatography (HPLC) separated fractions was verified by comparison of the u.v.–vis spectra with published results.

Table 1. Character table for point group I_h, where $\tau = (1 + \sqrt{5})/2$. Note: C_5 and C_5^{-1} are in different classes, labeled $12C_5$ and $12C_5^2$ in the character table. Then $iC_5 = S_{10}^{-1}$ and $iC_5^{-1} = S_{10}$ are in the classes labeled $12S_{10}^3$ and $12S_{10}$, respectively. Also $iC_2 = \sigma$

I_h	E	$12C_5$	$12C_5^2$	$20C_3$	$15C_2$	i	$12S_{10}^3$	$12S_{10}$	$20S_3$	$15\sigma_v$	Basis functions
A_g	+1	+1	+1	+1	+1	+1	+1	+1	+1	+1	$x^2 + y^2 + z^2$
F_{1g}	+3	$+\tau$	$1-\tau$	0	-1	+3	τ	$1-\tau$	0	-1	(R_x, R_y, R_z)
F_{2g}	+3	$1-\tau$	$+\tau$	0	-1	+3	$1-\tau$	τ	0	-1	
G_g	+4	-1	-1	+1	0	+4	-1	-1	+1	0	
H_g	+5	0	0	-1	+1	+5	0	0	-1	+1	$\begin{cases} 2z^2 - x^2 - y^2 \\ x^2 - y^2 \\ xy \\ xz \\ yz \end{cases}$
A_u	+1	+1	+1	+1	+1	-1	-1	-1	-1	-1	
F_{1u}	+3	$+\tau$	$1-\tau$	0	-1	-3	$-\tau$	$\tau-1$	0	+1	(x,y,z)
F_{2u}	+3	$1-\tau$	$+\tau$	0	-1	-3	$\tau-1$	$-\tau$	0	+1	(x^3, y^3, z^3)
G_u	+4	-1	-1	+1	0	-4	+1	+1	-1	0	$\begin{cases} x(z^2 - y^2) \\ y(x^2 - z^2) \\ z(y^2 - x^2) \\ xyz \end{cases}$
H_u	+5	0	0	-1	+1	-5	0	0	+1	-1	

group T_h^5 or $Im3$ has been identified experimentally [39].

3.1. Group theory for molecular C_{60}

The character table for the full icosahedral group I_h which describes the symmetry of an isolated C_{60} molecule is shown in Table 1. To calculate the symmetries of the vibrational modes (or of the electronic orbitals) one must know the symmetries associated with linear combinations of the atomic sites of the molecule or the unit cell of the lattice. The site symmetries for an isolated C_{60} molecule are summarized in Table 2 where the characters for the equivalence transformation of the atom sites $\chi^{a.s.}$ are given for various symmetric placements of atoms in positions of high symmetry within the I_h point group: X atoms (e.g. guest species) at the 12 vertices of the regular icosahedron, carbon atoms on the centers of the 20 hexagonal faces, the centers of the 30 edges and at the 60 vertices of the regular truncated icosahedron. The irreducible representations of the point group I_h corresponding to each of these configurations are also listed in Table 2. The I_h symmetry of the C_{60} molecule is the highest possible point group symmetry and is the starting point for all the applications of group theory in doped C_{60}. For example, a hypothetical X_1C_{60} molecule, where the single X

atom is placed in the center of the icosahedron, maintains I_h symmetry and the characters for the transformation of the atoms on the X_1 sites is $\chi^{a.s.}(X_1) = A_g$. A hypothetical $X_{12}C_{60}$ molecule, where the 12 X atoms are at the vertices of the regular icosahedron, also exhibits I_h symmetry. Following Table 2, the equivalence transformation for the 12 X atoms in $X_{12}C_{60}$ contains the following irreducible representations of the I_h point group

$$\chi^{a.s.}(X_{12}) = A_g + F_{1u} + F_{2u} + H_g$$

which occur in addition to those for the 60 carbon atoms,

$$\chi^{a.s.}(C_{60}) = A_g + F_{1g} + F_{2g} + 2G_g \\ + 3H_g + 2F_{1u} + 2F_{2u} + 2G_u + 2H_u.$$

Although there are 180 degrees of freedom (3×60) for each C_{60} molecule, the icosahedral symmetry gives rise to a number of degenerate modes, so that only 46 distinct mode frequencies are expected for the C_{60} molecule.

The symmetries of the vibrational modes of a molecule with I_h symmetry are found by taking the direct product of the characters for the atom sites $\chi^{a.s.}$ in Table 2 with the characters for a vector in I_h symmetry which corresponds to the irreducible

Table 2. Characters for the equivalent transformation $\Gamma^{\text{atom sites}}$ in I_h symmetry

I_h	E	$12C_5$	$12C_5^2$	$20C_3$	$15C_2$	i	$12S_{10}^3$	$12S_{10}$	$20S_3$	$15\sigma_v$		$\chi^{a.s.}$
X_{12}	12	2	2	0	0	0	0	0	0	4	$\Rightarrow$	$A_g + H_g$ $+ F_{1u} + F_{2u}$
C_{20}	20	0	0	2	0	0	0	0	0	4	$\Rightarrow$	$A_g + G_g + H_g$ $+ F_{1u} + F_{2u} + G_u$
C_{30}	30	0	0	0	2	0	0	0	0	4	$\Rightarrow$	$A_g + G_g + 2H_g$ $+ F_{1u} + F_{2u} + G_u + H_u$
C_{60}	60	0	0	0	0	0	0	0	0	4	$\Rightarrow$	$A_g + F_{1g} + F_{2g} + 2G_g + 3H_g$ $+ 2F_{1u} + 2F_{2u} + 2G_u + 2H_u$

Table 3. Vibrational modes in molecules with I_h symmetry

Molecule	A_g†	F_{1g}	F_{2g}	G_g	H_g‡	A_u	F_{1u}§	F_{2u}	G_u	H_u
X_{12}	1			1	2		1	1	1	1
X_{20}	1		1	2	3		1	2	2	2
X_{30}	1	1	2	3	4		2	3	3	3
C_{60}	2	3	4	6	8	1	4	5	6	7
C_{60} (radial)	1	1	1	2	3		2	2	2	2
C_{60} (tangential)	1	2	3	4	5	1	2	3	4	5
$X_{12}C_{60}$	3	4	4	7	10	1	6	6	7	8
$X_{12}C_{60}$ (radial)	2	1	1	2	4		3	3	2	2
$X_{12}C_{60}$ (tangential)	1	3	3	5	6	1	3	3	5	6

†Raman-active mode is seen only in $\parallel,\parallel$ polarization.
‡Raman-active mode is seen in both $\parallel,\parallel$ and $\parallel,\perp$ polarizations.
§Infrared-active mode.

representation F_{1u}. The resulting symmetries for the normal modes of the icosahedral molecules are listed in Table 3 according to their symmetry types and multiplicities. The Raman-active modes have A_g and H_g symmetry and the i.r. active modes have F_{1u} symmetry. One can see from the basis functions listed in Table 1 that the symmetry of the Raman tensor allows $(\parallel,\parallel)$ scattering for A_g modes, and both $(\parallel,\parallel)$ and $(\parallel,\perp)$ scattering for H_g modes. The remaining modes are silent unless some symmetry-lowering perturbation (e.g. the crystal field associated with the condensed phase) turns them on. Such symmetry-lowering perturbations would also modify the Raman scattering selection rules. In the extreme case where the point group symmetry is lowered to C_1, there would be $180 - 6 = 174$ vibrational modes and each mode would be Raman active and would be seen in both the $(\parallel,\parallel)$ and $\parallel,\perp)$ scattering geometries.

In listing the vibrational modes for the molecular units C_{60} and $X_{12}C_{60}$ in Table 3, the six degrees of freedom associated with the center of mass (translations and rotations) have been omitted. The vibrational modes in the solid state arising from these subtracted molecular modes are treated specially in our discussion below for the lattice modes in the space group.

3.2. Symmetry of solid C_{60} and doped C_{60}

Below the orientational ordering temperature of ~ 249 K, it has recently been determined by X-ray diffraction [22, 38] that pristine C_{60} crystallizes in

a simple cubic structure with space group T_h^6 [5] where there are four C_{60} molecules (or 240 carbon atoms) per unit cell. The character table for the T_h point group is given in Table 4 along with the basis functions for each irreducible representation. The basis functions indicate that the A_g, E_g and T_g modes are Raman-active, with scattering from the A_g and E_g modes appearing only for the $(\parallel,\parallel)$ geometry and from the T_g modes only for the $(\parallel,\perp)$ geometry.

We model the room temperature structure of single crystal C_{60} as a simple cubic lattice with 4 molecules per unit cell having T_h^6 space group symmetry, and each C_{60} molecule is in a site with T_h symmetry (see Table 4). The phonon dispersion relations for C_{60} can then be modeled in terms of weakly coupled, isotropically oriented, freely rotating (above 249 K) [40, 41]) icosahedral C_{60} molecules located at fcc T_h^6 site positions ('a' sites), shown in Table 5. A dopant atom (e.g. an alkali metal atom) can be placed in an octahedral 'b' site giving the crystalline stoichiometry M_1C_{60}, or in a tetrahedral 'c' site giving the stoichiometry M_2C_{60}, or if both tetrahedral and octahedral sites are fully occupied then the composition M_3C_{60} is achieved [42]. Finally, occupation of the general 'd' sites with T_h^6 symmetry in Table 5 would result in a stoichiometry M_6C_{60}. The space group T_h^6 has been identified for the doped M_xC_{60} material for $x \leq 3$ [42], while the fully doped material M_6C_{60} has been reported to crystallize in the bcc structure [43], consistent with the bcc space group T_h^5 with T_h site symmetry for the two C_{60} molecules per non-primitive

Table 4. Character table for T_h

T_h	E	$3C_2$	$4C_3$	$4C_3'$	i	$3\sigma_v$	$4S_3$	$4S_3'$	Basis functions
A_g	1	1	1	1	1	1	1	1	$x^2+y^2+z^2$
E_g	$\begin{cases}1\\1\end{cases}$	1 1	ω ω^2	ω^2 ω	1 1	1 1	ω ω^2	ω^2 ω	$x^2+\omega y^2+\omega^2 z^2$ $x^2+\omega^2 y^2+\omega z^2$
T_g	3	-1	0	0	3	-1	0	0	$(R_x, R_y, R_z), (yz, zx, xy)$
A_u	1	1	1	1	-1	-1	-1	-1	xyz
E_u	$\begin{cases}1\\1\end{cases}$	1 1	ω ω^2	ω^2 ω	-1 -1	-1 -1	$-\omega$ $-\omega^2$	$-\omega^2$ $-\omega$	$x^3+\omega y^3+\omega^2 z^3$ $x^3+\omega^2 y^3+\omega z^3$
T_u	3	-1	0	0	-3	1	0	0	(x,y,z)

Table 5. Equivalent sites of space group T_h^6

Sites	Notation	Site symmetry	Site coordinates	$\chi^{a.s.}$
4	a	$\bar{3}$	$0,0,0;\ 0,\tfrac{1}{2},\tfrac{1}{2};\ \tfrac{1}{2},0,\tfrac{1}{2};\ \tfrac{1}{2},\tfrac{1}{2},0$	$A_g + T_g$
4	b	$\bar{3}$	$\tfrac{1}{2},\tfrac{1}{2},\tfrac{1}{2};\ \tfrac{1}{2},0,0;\ 0,\tfrac{1}{2},0;\ 0,0,\tfrac{1}{2}$	$A_g + T_g$
8	c	3	$\begin{cases} x,x,x;\ \tfrac{1}{2}+x,\tfrac{1}{2}-x,\bar{x};\ \bar{x},\tfrac{1}{2}+x,\tfrac{1}{2}-x;\ \tfrac{1}{2}-x,\bar{x},\tfrac{1}{2}+x \\ \bar{x},\bar{x},\bar{x};\ \tfrac{1}{2}-x,\tfrac{1}{2}+x,x;\ x,\tfrac{1}{2}-x,\tfrac{1}{2}+x;\ \tfrac{1}{2}+x,x,\tfrac{1}{2}-x \end{cases}$	$A_g + T_g + A_u + T_u$
24	d	1	$\begin{cases} x,y,z;\ \tfrac{1}{2}+x,\tfrac{1}{2}-y,\bar{z};\ \bar{x},\tfrac{1}{2}+y,\tfrac{1}{2}-z;\ \tfrac{1}{2}-x,\bar{y},\tfrac{1}{2}+z \\ z,x,y;\ \tfrac{1}{2}+z,\tfrac{1}{2}-x,\bar{y};\ \bar{z},\tfrac{1}{2}+x,\tfrac{1}{2}-y;\ \tfrac{1}{2}-z,\bar{x},\tfrac{1}{2}+y \\ y,z,x;\ \tfrac{1}{2}+y,\tfrac{1}{2}-z,\bar{x};\ \bar{y},\tfrac{1}{2}+z,\tfrac{1}{2}-x;\ \tfrac{1}{2}-y,\bar{z},\tfrac{1}{2}+x \\ \bar{x},\bar{y},\bar{z};\ \tfrac{1}{2}-x,\tfrac{1}{2}+y,z;\ x,\tfrac{1}{2}-y,\tfrac{1}{2}+z;\ \tfrac{1}{2}+x,y,\tfrac{1}{2}-z \\ \bar{z},\bar{x},\bar{y};\ \tfrac{1}{2}-z,\tfrac{1}{2}+x,y;\ z,\tfrac{1}{2}-x,\tfrac{1}{2}+y;\ \tfrac{1}{2}+z,x,\tfrac{1}{2}-y \\ \bar{y},\bar{z},\bar{x};\ \tfrac{1}{2}-y,\tfrac{1}{2}+z,x;\ y,\tfrac{1}{2}-z,\tfrac{1}{2}+x;\ \tfrac{1}{2}+y,z,\tfrac{1}{2}-x \end{cases}$	$\begin{cases} A_g + E_g + 3T_g \\ + A_u + E_u + 3T_u \end{cases}$

unit cell. In addition, a body centered tetragonal (bct) phase has been reported for $x = 4$ [4, 39] for this lower symmetry space group, the C_{60} molecules would occupy sites with C_{2v} symmetry. The symmetry for the T_h^5 bcc space group given in Table 6 is based on two molecular units per non-primitive unit cell at the indicated sites. If a dopant is at an 'a' site the stoichiometry would be X_1C_{60} while occupation of the 'b' sites would yield an X_3C_{60} stoichiometry, and occupation of the 'c' sites would yield a X_4C_{60} composition. (The X_6C_{60} stoichiometry can be achieved by occupation of either 'd' or 'e' sites, but no definitive experiment has yet been done to distinguish between these two options.) Since the space groups T_h^6 and T_h^5 are probably the highest symmetries that can be found for C_{60}-related materials, it is useful to consider the T_h^6 and T_h^5 space groups as an approximate symmetry for the doped materials (T_h^6 for weakly doped, and T_h^5 for nearly fully doped materials) and to consider the lower symmetry space groups, when they are identified experimentally in the future, as resulting from perturbations of these high symmetry groups.

3.3. Phonon modes of pristine and doped C_{60} in space groups T_h^6 and T_h^5

The number of allowed Raman-modes for C_{60} in the T_h^6 structure is very large and includes 29 one-dimensional A_g modes, 29 two-dimensional E_g modes and 87 three-dimensional T_g modes (see Table 7). However, the simplicity of the observed room temperature Raman spectra suggests the perturbative treatment outlined in Table 7. Starting with the $(3 \times 60) - 6$ vibrational degrees of freedom for the free molecule with I_h symmetry (first column of Table 7), the lower T_h point symmetry in the crystal gives rise to the mode splittings indicated in the second column of Table 7. When the four C_{60} molecules are placed in the T_h^6 space group sites, the appropriate mode symmetries are found (see third column of the table) by taking the direct product of the characters for the molecular site symmetries of the four balls ($A_g + T_g$) (see last column of Table 5) with the characters for the vibrational modes for the C_{60} molecule in T_h symmetry (second column of Table 7).

In addition to the high frequency modes listed in Table 7, for the T_h^6, phase there are 24 low frequency intermolecular modes ($\omega < 200$ cm^{-1}) associated with the molecular transitional (F_{1u} or T_u) and rotational (F_{1g} or T_g) degrees of freedom of the C_{60} molecules in the T_h^6 phase

$$(T_g + T_u) \otimes (A_g + T_g) \rightarrow A_g + E_g + 3T_g + A_u + E_u + 3T_u,$$

where $\otimes$ denotes the direct product (see footnote (§) to Table 7). Of these, the acoustic mode is of

Table 6. Equivalent sites of space group T_h^5

Sites	Notation	Site symmetry	Site coordinates: $(0,0,0);\ (\tfrac{1}{2},\tfrac{1}{2},\tfrac{1}{2}) +$	$\chi^{a.s.}$
2	a	m^3	$0,0,0$	A_g
6	b	mmm	$0,\tfrac{1}{2},\tfrac{1}{2};\ \tfrac{1}{2},0,\tfrac{1}{2};\ \tfrac{1}{2},\tfrac{1}{2},0$	$A_g + E_g$
8	c	$\bar{3}$	$\tfrac{1}{4},\tfrac{1}{4},\tfrac{1}{4};\ \tfrac{1}{4},\tfrac{3}{4},\tfrac{3}{4};\ \tfrac{3}{4},\tfrac{1}{4},\tfrac{3}{4};\ \tfrac{3}{4},\tfrac{3}{4},\tfrac{1}{4}$	$A_g + T_g$
12	d	mm	$\begin{cases} x,0,0;\ 0,x,0;\ 0,0,x \\ \bar{x},0,0;\ 0,\bar{x},0;\ 0,0,\bar{x} \end{cases}$	$\begin{cases} A_g + E_g \\ + T_u \end{cases}$
12	e	mm	$\begin{cases} x,0,\tfrac{1}{2};\ \tfrac{1}{2},x,0;\ 0,\tfrac{1}{2},x \\ \bar{x},0,\tfrac{1}{2};\ \tfrac{1}{2},\bar{x},0;\ 0,\tfrac{1}{2},\bar{x} \end{cases}$	$\begin{cases} A_g + E_g \\ + T_u \end{cases}$
16	f	3	$\begin{cases} x,x,x;\ x,\bar{x},\bar{x};\ \bar{x},x,\bar{x};\ \bar{x},\bar{x},x \\ \bar{x},\bar{x},\bar{x};\ \bar{x},x,x;\ x,\bar{x},x;\ x,x,\bar{x} \end{cases}$	$\begin{cases} A_g + T_g \\ + A_u + T_u \end{cases}$
24	g	m	$\begin{cases} 0,y,z;\ z,0,y;\ y,z,0;\ 0,\bar{y},\bar{z};\ \bar{z},0,\bar{y};\ \bar{y},\bar{z},0 \\ 0,y,\bar{z};\ \bar{z},0,y;\ y,\bar{z},0;\ 0,\bar{y},z;\ z,0,\bar{y};\ \bar{y},z,0 \end{cases}$	$\begin{cases} A_g + E_g \\ + T_g + 2T_u \end{cases}$
48	h	1	$\begin{cases} x,y,z;\ z,x,y;\ y,z,x;\ \bar{x},\bar{y},\bar{z};\ \bar{z},\bar{x},\bar{y};\ \bar{y},\bar{z},\bar{x} \\ x,\bar{y},\bar{z};\ z,\bar{x},\bar{y};\ y,\bar{z},\bar{x};\ \bar{x},y,z;\ \bar{z},x,y;\ \bar{y},z,x \\ \bar{x},y,\bar{z};\ \bar{z},x,\bar{y};\ \bar{y},z,\bar{x};\ x,\bar{y},z;\ z,\bar{x},y;\ y,\bar{z},x \\ \bar{x},\bar{y},z;\ \bar{z},\bar{x},y;\ \bar{y},\bar{z},x;\ x,y,\bar{z};\ z,x,\bar{y};\ y,z,\bar{x} \end{cases}$	$\begin{cases} A_g + E_g + 3T_g \\ + A_u + E_u + 3T_u \end{cases}$

Table 7. Brillouin zone center vibrational modes for the carbon atoms in space group T_h^6 and T_h^5

I_h	T_h, T_h^5†	T_h^6‡
$2A_g$	$2A_g$	$2A_g + 2T_g$
$3F_{1g}$	$3T_g$	$3T_g + 3(A_g + E_g + 2T_g)$
$4F_{2g}$	$4T_g$	$4T_g + 4(A_g + E_g + 2T_g)$
$6G_g$	$6A_g + 6T_g$	$6(A_g + T_g) + 6(A_g + E_g + 3T_g)$
$8H_g$	$8E_g + 8T_g$	$8(E_g + T_g) + 8(A_g + E_g + 4T_g)$
$1A_u$	$1A_u$	$A_u + T_u$
$4F_{1u}$	$4T_u$	$4T_u + 4(A_u + E_u + 2T_u)$
$5F_{2u}$	$5T_u$	$5T_u + 5(A_u + E_u + 2T_u)$
$6G_u$	$6A_u + T_u$	$6(A_u + T_u) + 6(A_u + E_u + 3T_u)$
$7H_u$	$7E_u + 7T_u$	$7(E_u + T_u) + 7(A_u + E_u + 4T_u)$

†The modes associated with translations and rotations of the center of mass of the two C$_{60}$ molecules must also be counted for the space group, giving additional $T_g + T_u$ modes, one of which (T_u) is the acoustic mode.

‡The modes associated with translations and rotations of the center of mass of the four C$_{60}$ molecules must also be counted for the space group, giving additional $A_g + E_g + 3T_g + A_u + E_u + 3T_u$ modes, one of which (T_u) is the acoustic mode.

symmetry T_u, and there are nine other C$_{60}$ molecular displacive modes of symmetries $A_g + E_g + 2T_g$. The 12 molecular rotational modes have symmetries $T_g + A_u + E_u + 2T_u$. At general points in the Brillouin zone, the low frequency translational and rotational modes will hybridize and form librational modes.

To find the symmetry of the low frequency intermolecular modes for the T_h^5 space group, we take the direct product $(T_g + T_u) \otimes A_g$ since there is only one molecule per unit cell in this case (see footnote (‡) in Table 7). In this case the single acoustic mode T_u corresponds to translations and the T_g mode to rotations.

Further discussion of the symmetry of the doped C$_{60}$ compounds is given in section 4 in connection with the observed Raman spectra.

4. EXPERIMENTAL RESULTS AND DISCUSSION

4.1. C$_{60}$

Oxygen-free C$_{60}$ *(FCC)*. The first Raman spectrum of solid C$_{60}$ was reported by Bethune and Meijer [24, 25], who carried out studies at $T = 300$ K on film samples sublimed in vacuum onto Suprasil substrates. Their samples were studied in air, and no X-ray characterization of their films was reported. They identified 10 Raman lines, in agreement with the number $(2A_g + 8H_g)$ predicted by group theory for an isolated molecule and consistent with weak ball–ball interactions. In Fig. 6 we display our first reported polarized spectrum of a C$_{60}$ film on Si [26]. The samples were handled in air, but the data were taken with N$_2$ gas flowing over the sample surface and were found in good agreement with those of Bethune and Meijer [24, 25]. Both of these spectra can be compared to those reported more recently by our group (Table 8) on samples shown to be oxygen-free in the bulk by resonant alpha scattering [23]. This spectrum on oxygen-free C$_{60}$ is shown in Fig. 7a. The inset to Fig. 7 shows X-ray diffraction data taken with Cu(K$_\alpha$) radiation using a Philips 3100 Diffractometer for a 6000 Å C$_{60}$ film deposited on a Si(111) substrate, also unexposed to oxygen, which exhibits the fcc diffraction pattern and the same Raman spectrum as found for oxygen-free C$_{60}$ and shown in Fig. 7a. The X-ray peak positions were found to be consistent with an fcc structure with lattice constant $a = 14.2$ Å, and the peak widths indicate a structural coherence length of ~ 130 Å [23]. The later results of Zhou *et al.* [23] differ only slightly from those of Wang *et al.* [26], but are taken under more closely controlled conditions on material known to be free of oxygen in the bulk.

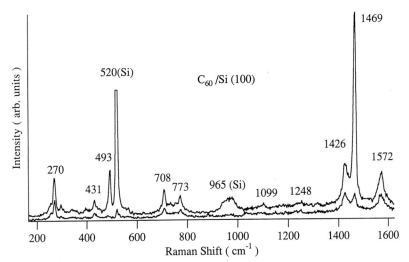

Fig. 6. Polarized Raman spectra for C$_{60}$.

Table 8. Experimental and calculated C_{60} Raman-active: (a) mode frequencies and infrared-active (b) mode frequencies

Table 8a

Zhou et al. [23]†	Wang et al. [26]†	Meijer and Bethune [25]†	Negri et al. [12]‡	Stanton and Newton [13]‡	Jishi et al. [45]§	Mode
270	270.0 [4.2, 0.52]	273	258	263 (69.3)	269	H_g
353						
431	430.5 [5.5, 0.40]	437	440	447 (90.0)	439	H_g
493	493.0 [2.5, 0.02]	496	513	610 (100)	492	A_g
708	708.0 [7.5, 0.40]	710	691	771 (96.5)	708	H_g
773	772.5 [9.0, 0.38]	774	801	924 (30.5)	788	H_g
—	1099 [7,-]	1099	1154	1261 (9.6)	1102	H_g
1248	1248 [7,-]	1250	1265	1407 (2.1)	1217	H_g
1318						
1426	1426.0 [7.5, 0.44]	1428	1465	1596 (0.8)	1401	H_g
1469	1468.5 [1.5, 0.10]	1470	1442	1667 (0.0)	1468	A_g
1573	1573.0 [9.5, 0.52]	1575	1644	1722 (1.2)	1572	H_g

Table 8b

This work ‖	Vassallo et al. [36]¶	Meijer and Bethune [25]‖	Krätschmer et al. [3]‖	Negri et al. [12]‡	Stanton and Newton [13]‡	Jishi et al. [45]§	Mode
527	528	527	528	544	577	505	F_{1u}
576	577	577	577	637	719	589	F_{1u}
1183	1183	1183	1183	1212	1353	1208	F_{1u}
1428	1429	1428	1429	1437	1628	1450	F_{1u}
1539	1537						

Numbers in brackets are natural line width (cm^{-1}) and depolarization ratio (I_{HV}/I_{HH}), respectively.
Number in parentheses is percentage of radial mode character.
†Experimental Raman scattering mode frequencies.
‡Calculations based upon the force constants for benzene.
§Calculation based on force constants adjusted to fit i.r.- and Raman-active mode frequencies.
‖Experimental i.r. mode frequencies (transmission FTIR).
¶Experimental emission FTIR mode frequencies.

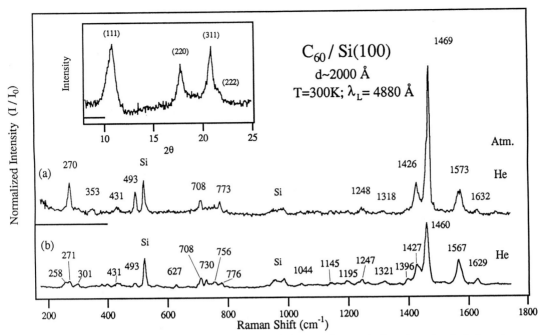

Fig. 7. Normalized Raman spectra of a C$_{60}$ film on Si(100) taken at the laser intensity Φ_L indicated using 4880 Å radiation: (a) oxygen-free [65] film in 1 atm of He; $\Phi_L = 5$ W cm^{-2}, (b) oxygen-free [65] film in 1 atm of He irradiated first at 75 W cm^{-2} for 30 min; $\Phi_L = 75$ W cm^{-2} (see Ref. 48). Inset shows XRD scan of oxygen-free C$_{60}$/Si(111). This sample exhibited the same Raman spectra as shown in Figs 7a and 9a.

For comparison, results of vibrational mode calculations for an isolated C$_{60}$ molecule are also included in Table 8. The mode-assignment in the table (e.g. H_g or A_g) uses the symmetry labels for the isolated molecule, and agree with the polarization analysis of the experimental spectra. The strongest Raman line (A_g) was observed at 1469 cm^{-1}, and, consistent with calculations, should be identified [12, 13, 15] with the 'pentagonal pinch' mode [44]. The agreement between experiment and theory is particularly good for the *ab initio* calculations of Negri *et al.* [12] and the model calculations of Jishi *et al.* [45], who adjusted nearest neighbor bond-bending and bond-stretching force constants to best fit the optical modes. The good agreement between the experimental (solid state) frequencies and the frequencies calculated for an isolated molecule also supports the conclusion that solid state interactions in C$_{60}$ are weak. Finally, it should be noticed that Stanton and Newton [13] have reported the radial character (%) of the C$_{60}$ molecular eigenmodes obtained in their calculations (Table 8a and b). As can be seen from the table, predominantly tangential (radial) carbon atom displacements are found for modes with frequencies above (below) 700 cm^{-1}.

In Fig. 4 we display an FTIR transmission spectrum for the region 400–1600 cm^{-1} for one of our films. The data were taken on a 5000 Å C$_{60}$ film deposited on KBr at $T = 300$ K (Table 8b). Four

F_{1u}-symmetry molecular mode frequencies are predicted for an isolated C$_{60}$ molecule, and five are observed in the spectrum. The four strongest lines are in good agreement with Meijer and Bethune [25]. The fifth mode at 1539 cm^{-1} is weak, and agrees with emission data [36]. We attribute the presence of this weak mode to solid state interactions. Thus, as found for the Raman-active modes, the principal features in the i.r. spectrum can also be associated with vibrational modes of isolated molecules. Finally, it is seen that good agreement is also obtained between theory and experiment for the i.r. mode frequencies (Table 8).

We now return to the question of the disparity in the number of observed Raman-active modes in solid C$_{60}$ and the number predicted by group theory. Because the solid state interactions are weak, most of the Raman-allowed modes for the T_h^6 space group are either insufficiently strong to be observed, and/or they give rise to small, unresolved splittings of the 10 main Raman-allowed intramolecular modes ($2A_g + 8H_g$). The symmetry analysis given in section 3 suggests the following mode scheme for interpreting the Raman results for solid C$_{60}$. The isolated molecule exhibits 10 Raman-active modes ($2A_g + 8H_g$). When placed in the lower T_h symmetry of the solid, each of the five-dimensional H_g modes are split into a two-dimensional E_g and a three-dimensional T_g mode. In addition, the three F_{1g}, four F_{2g} and six G_g modes become weakly Raman active because all even parity modes in T_h symmetry are Raman-allowed. The G_g modes are also split as indicated in the second column of Table 7. Finally, when the lower symmetry intermolecular interactions are considered, the odd-parity modes are all turned on and further splittings occur. The experiments indicate that the intensities of the odd-parity modes activated by intermolecular interactions are unobservably small, as are most of the splittings of the G_g- and H_g-derived modes.

It should be noted that low temperature ($T = 40$ K) Raman spectra reported by van Loosdrecht *et al.* [46] on single crystal C_{60}, taken below the orientational ordering transition temperature, however, show ~ 45 lines, including three broad lines at low frequency: 57, 82 and 108 cm^{-1}, indicating that solid state interactions can be observed at low temperature. Their lowest frequency lines can be compared with the lattice mode calculations of Tomanek and co-workers [16] shown in Fig. 3. From these dispersion curves we estimate that low frequency optical modes should be observed in the range 50–66 cm^{-1}, which might account for the broad line observed at 57 cm^{-1}. However, the $T = 300$ K frequency obtained for the pentagonal pinch mode by van Loosdrecht *et al.* in the same crystal is 1464 cm^{-1}, or 5 cm^{-1} less than observed by Meijer and Bethune [25] and our group [26, 27]. We make this comment because film samples subjected to high laser flux have been observed to exhibit a downshift in the pentagonal mode frequency by as much as 11 cm^{-1} [28, 47]. This photo-effect is discussed in detail in the next section.

Some concluding remarks about the depolarization ratios observed for the Ag symmetry modes in C_{60} are in order. We note from Table 8 that the *radial* Ag mode is very strongly polarized; i.e. $I_{HV}/I_{hh} = 0.02$, comparable to our instrumental polarization leakage ($\sim 2\%$). However, the observed *tangential* Ag mode polarization in that study is $I_{HV}/I_{HH} = 0.10$ as compared to a theoretical value of zero, suggesting that the site symmetry is only approximately T_h. As seen in the next section, it is possible to cause a photo-transformation in C_{60} films, even at the low laser flux level typically used to collect the Raman spectrum. Therefore, it may be difficult to prevent small amounts of laser-induced disorder from occurring, and thereby increasing the depolarization ratio of the tangential mode. It is not yet clear why the tangential mode might be more sensitive to the assumed photo-induced disorder than the radial Ag mode. In Fig. 8 we show the results of a lower laser flux study of the depolarization ratio of the 1469 cm^{-1} line. The data were taken on a C_{60} film deposited on Suprasil at the low laser flux of 2.5 W cm^{-2}. The upper and lower spectra refer, respectively, to scattering in the (H,H) and H,V) geometry, and the data (dots) were taken at room temperature on a spot not previously illuminated. The solid lines are a least squares fit of Lorentzian lineshapes to the data, and the dashed lines represent the least squares result for the individual components. In the (H,H) spectrum the 1469 cm^{-1} peak exhibits a shoulder at 1459 cm^{-1} which we identify with photo-transformed C_{60}, as discussed in detail in the next section. The (H,V)

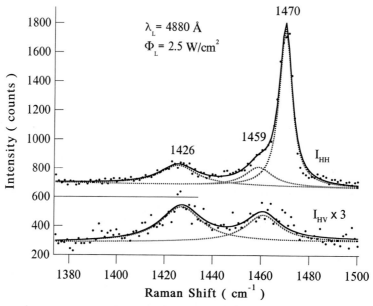

Fig. 8. Polarized Raman spectra ($T = 300$ K) of C_{60} film (1500 Å) on Suprasil taken with low laser flux 2.5 W cm^{-2} at 4880 Å. The solid lines are the result of a least squares fit of Lorentzian lineshapes to the data. At this laser flux the A_g symmetry pentagonal pinch mode is observed to be completely polarized. At higher laser power this symmetry is destroyed.

spectrum, which is shown magnified by a factor of three (3x), is well fit by the two lines at 1426 and ~1459 cm^{-1}. There is possibly some evidence for (H,V) scattering from the 1469 cm^{-1} line. If we take the r.m.s. noise of the scattering above the wing of the 1459 cm^{-1} peak for the upper bound for the intensity of the 1469 cm^{-1} line in this geometry, then we obtain a depolarization value 1469 cm^{-1} peak of ~2%, comparable to our instrumental detection limit. Therefore, to within experimental error, sufficiently low laser flux allows a perfectly polarized tangential A$_g$ mode to be observed in FCC C$_{60}$.

Photo-transformed C$_{60}$. Two fundamentally different Raman spectra ($T = 300$ K) have been reported for solid C$_{60}$. In this section we present evidence that these differing results are a consequence of the strength of the laser flux (W cm^{-2}) needed to collect the spectrum. That is, we have observed a photo-transformation [23] of C$_{60}$ from the fcc phase to a second, as yet unidentified, phase at moderate laser flux.

Duclos *et al.* [28, 47] reported a notably different and richer spectrum for pristine C$_{60}$ at $T = 300$ K than discussed in the previous section. Their spectrum exhibited more lines (21 as compared to 10 or 11 lines) and the pentagonal pinch mode frequency [44] associated with polarized scattering was found at 1458 cm^{-1}, downshifted by 11 cm^{-1} from 1469 cm^{-1}, the frequency obtained by Meijer and Bethune [25] and by our group [26, 27]. To try and understand why two different spectra might be obtained for C$_{60}$, we carried out a set of experiments on samples carefully excluded from exposure to oxygen and subjected to varying strength laser flux with wavelength 4880 Å [23]. Depending on the strength of the laser flux used to acquire the Raman spectrum, we were able to observe the two different spectra for C$_{60}$ reported in the literature.

In Fig. 7 we display these two Raman spectra (200–1700 cm^{-1}) for a C$_{60}$/Si(100) film ($d \sim 2000$ Å) [23]. The spectra were collected from the same film in ~1 atm of He gas (O$_2$ and H$_2$O exposure was avoided, see section 4.2) using low and high laser flux. The spectrum in Fig. 7a was collected using a low laser flux $\Phi_L \sim 5$ W cm^{-2} from a position on the film surface not illuminated previously. X-ray diffraction data on another film with the same Raman spectrum (shown in the inset to Fig. 7) can be indexed as fcc C$_{60}$. Therefore, we conclude the spectrum in Fig. 7a should be identified with fcc C$_{60}$. With the laser still incident at the same location on the film, the laser flux Φ_L was raised to ~75 W cm^{-2} for 30 min and then the spectrum in Fig. 7b was taken at that flux. As can be seen by comparing the Raman spectra in Fig. 7a and b, the oxygen-free, fcc C$_{60}$ film has undergone a

photo-transformation and a new Raman spectrum is observed. For the photo-transformed sample, several new Raman lines were detected, line-intensities changed, and a few lines shifted in frequency [48]. As shown in Fig. 9a (low Φ_L) and Fig. 9b (high Φ_L), the strongest line in the pristine C$_{60}$ film spectrum, identified with the pentagonal pinch mode [13, 15, 49], downshifts by ~9 cm^{-1} when the laser flux was increased to ~75 W cm^{-2}. A comparison of the 'high flux' Raman spectrum for solid C$_{60}$ shown in Fig. 7b with that reported by Duclos *et al.* [28, 47] taken at 50 W cm^{-2} indicates they are quite similar, although we found the pentagonal pinch mode at 1460 cm^{-1} slightly higher than the value at 1458 cm^{-1} they reported. Furthermore, we observed the same dependence on laser flux for several other oxygen-free C$_{60}$ films grown on both Suprasil and Si(100) substrates and studied both in vacuum and in He gas. Thus we conclude the C$_{60}$ spectra reported by Duclos *et al.* [28, 47] should be identified with the photo-transformed phase in our films.

Note that the spectra in Figs 7 and 9 [23] are normalized to the incident laser intensity I_0, and therefore reflect changes in the Raman cross-section σ. The laser photon energy used was 2.54 eV, which is above the threshold for strong interband absorption [33, 50]. Comparing the integrated intensity of the 1460 and 1469 cm^{-1} lines, we obtained a cross-section ratio $\sigma_{1469}/\sigma_{1460} \sim 3$, suggesting a possible difference in the electronic band structure for the

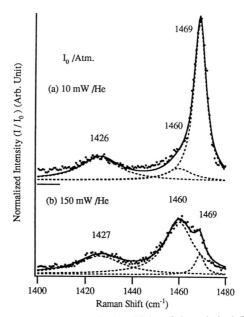

Fig. 9. Raman spectra in the vicinity of the polarized C$_{60}$ 'pentagonal pinch' mode near 1469 cm^{-1}: (a) low laser flux (5 W cm^{-2}); (b) high laser flux (75 W cm^{-2}). The dots represent the data and the solid and dashed lines represent results of Lorentzian lineshape analyses. Spectra in Fig. 9a and b correspond to those in Fig. 7a and b.

fcc and photo-transformed phases. It should be noted that experimental evidence for both an fcc [4, 21, 22] and hexagonal close packed (hcp) [51] phase in solid C_{60} has been presented. However, we do not have X-ray diffraction data on the photo-transformed phase of C_{60}, and have no evidence to assign the Raman spectra shown in Fig. 7b with a hcp phase of C_{60}.

The influence of pressure and temperature on the Raman scattering from single crystals of C_{60} were reported by Tolbert *et al.* [52] using a diamond anvil cell. They reported a pentagonal pinch mode frequency of 1458 cm^{-1} at room temperature, in agreement with the results of Duclos *et al.* [28, 47] and in agreement with the value we obtained at high laser flux on C_{60} films at $T = 300$ K. Tolbert *et al.* [52] reported the temperature-dependence of the pentagonal pinch mode frequency. With decreasing temperature, their data exhibit a sharp increase in mode frequency of ~ 4–5 cm^{-1} at $T = 240$ K, near the temperature $T = 249$ K reported in X-ray diffraction studies [22] of the orientational-ordering transition. Below ~ 100 K, the mode frequency remained at 1466 cm^{-1}, independent of temperature. Tolbert *et al.* [52] also reported the pressure studies of the pentagonal pinch mode and two high frequency H_g modes at ~ 1426 cm^{-1} and at ~ 1573 cm^{-1}. They found that the frequencies of these three Raman modes increase monotonically with increasing pressure up to 18 GPa. The mode-Grüneisen parameters, which describe how the frequency of a given Raman mode scales with the volume, are very small and compatible with there being very little change in

intramolecular bond length for a change in intermolecular spacing. This is consistent with the molecular character of C_{60} solid. van Loosdrecht *et al.* [46] also studied the temperature-dependence of the pentagonal pinch mode frequency in single crystal samples and found essentially the same results as Tolbert *et al.* [52].

4.2. Doped C_{60}

$C_{60}O_x$. We first consider the case of oxygen-doped C_{60}. Zhou *et al.* [23] have shown that photo-assisted oxygen doping of C_{60} films can be accomplished easily at room temperature with visible or u.v. irradiation ($\Phi_L \sim 75$ W cm^{-2} for 30 min) in the presence of O_2 at ~ 1 atm. The oxygen concentration in the films used for the Raman scattering experiments was inferred from Rutherford α-particle backscattering on C_{60} prepared on Au-coated graphite substrates and exposed to light and oxygen in the same way. It was found that these films exhibited an oxygen to carbon ratio $[O]/[C] = 4.3 \pm 0.4\%$, equivalent to a *bulk* stoichiometry of $C_{60}O_{2.6}$. Furthermore, a resonance scan of the incident α-particle incident energy verified that the oxygen was distributed throughout the bulk. A Raman spectrum was then taken on the irradiated spot using a low laser flux ($\Phi_L \sim 5$ W cm^{-2}) found previously [23] not to initiate photo-transformation of pristine C_{60} (see section 4.1). The spectrum for oxygen-doped C_{60} is shown in Fig. 10b, and for comparison in Fig. 10a we show the room temperature Raman spectra for fcc C_{60}. The oxygen doped spectrum was found to exhibit a narrow line at 1468 cm^{-1} [23],

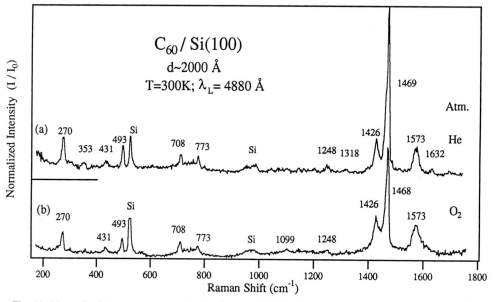

Fig. 10. Normalized Raman spectra of a C_{60} film on Si(100) taken at the Φ_L indicated using 4880 Å radiation: (a) oxygen-free [65] film in 1 atm of He; $\Phi_L = 5$ W cm^{-2}, (b) film in 1 atm of O_2 and irradiated first at 75 W cm^{-2} for 30 min; $\Phi_L = 5$ W cm^{-2}.

downshifted $\sim 1\,cm^{-1}$ from the pentagonal pinch mode of fcc C$_{60}$. Overall, the C$_{60}$O$_x$ Raman spectrum is found nearly identical to that shown in Fig. 10a for oxygen-free, fcc C$_{60}$ taken at low Φ_L [23]. However, in the region above $1500\,cm^{-1}$, which contains the peak position for the O$_2$ mode (gas phase) at $1555\,cm^{-1}$ [53], small differences are noted, including the asymmetric broadening of the $1573\,cm^{-1}$ peak and the disappearance of the $1632\,cm^{-1}$ peak under the high energy wing of the broadened $1573\,cm^{-1}$ line.

In sharp contrast to the behavior of oxygen-free, fcc C$_{60}$ films studied in He gas or in vacuum, very high laser flux (i.e. $\Phi_L > 150\,W\,cm^{-2}$) in the presence of O$_2$ did not affect the normalized spectrum, i.e. did not induce any photo-transformation to the C$_{60}$ phase exhibiting the $1458\,cm^{-1}$ polarized line.

Duclos *et al.* [47] in Raman studies ($\Phi_L = 50\,W\,cm^{-2}$ and $\lambda_L = 5145\,\text{Å}$) on C$_{60}$ films grown and studied in vacuum, identified a polarized $1458\,cm^{-1}$ line as the characteristic Raman line for the pentagonal pinch mode in oxygen-free fcc C$_{60}$. When exposed to O$_2$, they noticed that their samples exhibit an upshift of this line to $1469\,cm^{-1}$, which they attributed to an uptake of oxygen in the film. Our α-backscattering results show directly that their hypothesis of oxygen uptake is indeed correct. Furthermore, our studies indicate that the diffusion of

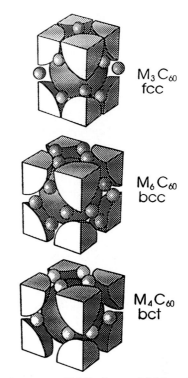

Fig. 11. Structures of M$_x$C$_{60}$ for $x = 0,3,6$ from Ref. 43.

oxygen into the bulk also requires photo-assistance. It should be recalled that several groups [54–56] have noticed that the reaction of oxygen with C$_{60}$ in solution [54, 55] or in solid form [56] requires

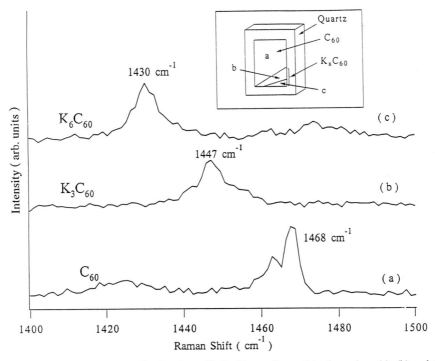

Fig. 12. Raman spectra ($T = 300\,K$) taken for a K$_x$C$_{60}$ film on Suprasil in the regions (a), (b) and (c) indicated schematically in the inset. A potassium concentration gradient between region (a) pure C$_{60}$, (b) K$_3$C$_{60}$ and (c) K$_6$C$_{60}$. The spectra shown are typical of a number of spectra taken in regions (a), (b) and (c).

236   P. C. EKLUND *et al.*

photo-assistance. In these studies, they cite evidence for significant chemical changes as a result of the photo-assisted reaction with O_2, such as a loss of a C_{10} unit to obtain C_{50} [55] or the possible formation of amorphous carbon [56]. In contrast to this behavior, our Raman results suggest that the skeletal structure of the C_{60} molecule is not strongly perturbed by the photo-assisted incorporation of oxygen into the lattice, at least under the doping conditions described in our experiments.

M_6C_{60} (M = K,Rb *and* Cs). The diffusion of the alkali metals M (M = K,Rb and Cs) in solid C_{60} is quite rapid at 200°C, which is ~150°C less than the sublimation temperature for C_{60} in vacuum. This allows the synthesis of M_xC_{60} to proceed by intercalation from M-vapor in a sealed ampoule. The saturated phase (x = 6) can therefore be synthesized quite easily, although caution must be exercised to prevent excess alkali metal from condensing on the external surface of M_6C_{60}. For x < 6, some means of terminating the reaction at the approximate

stoichiometry must be employed to obtain, for example, the x = 3 (fcc [22]) and x = 4 (bct [43]) stoichiometries. In Fig. 11 a schematic representation of the structures for the x = 3,4 and 6 M_xC_{60} compounds is shown [43]. The large and small spheres represent, respectively, the C_{60} shell and M^+ ions. The x = 3 phase is the most conductive phase, which is also superconducting, and *in situ* resistance measurements [57] can be carried out to terminate the reaction at x = 3. Likewise, *in situ* Raman scattering can also be used to determine x, since the high frequency A_g mode (pentagonal pinch mode) is sensitive to charge transfer [8, 26–28].

In Fig. 12 we show Raman spectra collected from a sample removed prematurely from a two-zone oven ($T_{hot} = 200°C$ and $T_{cold} = 150°C$) that exhibits the phase separation of C_{60}, K_3C_{60} and K_6C_{60}. The quartz substrate/C_{60} film and K metal were at the hot and cold ends, respectively, of a sealed Pyrex tube. The reaction was stopped when it was noticed that three distinct regions were evolving in the film with time.

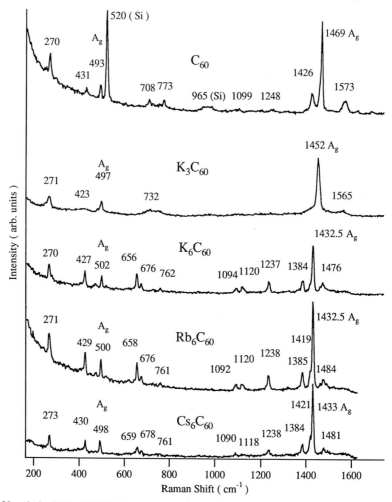

Fig. 13. Unpolarized (T = 300 K) Raman spectra for pristine C_{60}, K_3C_{60}, K_6C_{60}, Rb_6C_{60} and Cs_6C_{60}. The tangential and radial A_g symmetry modes are identified, as are the features associated with the Si substrates.

Apparently, intercalation had begun at a corner of the film closest to the K source. Raman spectra (Fig. 12) were collected in the vicinity of the charge-transfer-sensitive, tangential A_g mode from these three regions (indicated a,b,c). In the region furthest from the K source (a), a doublet is observed with the strongest line at 1468 cm⁻¹. The low frequency component of the doublet is centered at ~1460 cm⁻¹, the frequency associated with the photo-transformed phase A_g mode (see section 4.1). In the region closest to the K source (c), the A_g mode has downshifted by 38 cm⁻¹ to 1430 cm⁻¹. Finally, in (b) the central region, the A_g mode lies approximately midway in frequency between that for C_{60} and for K_6C_{60}, suggesting that the phase is K_3C_{60}. Thus there is approximately a 6 cm⁻¹ frequency downshift per electron transferred from the K(4s) band to the conduction band (t_{1u}) of C_{60}. If K_4C_{60} were to have formed in the sample, the A_g mode would be expected at ~1442 cm⁻¹. We have not collected any M_xC_{60} spectra exhibiting a peak near 1442 cm⁻¹.

Figure 13 shows low resolution (6 cm⁻¹) unpolarized (H,N = none) Raman spectra of homogeneous pristine C_{60}, K_3C_{60} and M_6C_{60} M = (K,Rb,Cs) [26, 27, 31]; films for M_6C_{60} were ~1000 Å in thickness. Spectral data from this figure and from higher resolution scans are collected in Table 9. The K_3C_{60} spectrum [31] will be discussed separately in the next section. Weak lines were observed in the C_{60} spectrum at 1099 and 1248 cm⁻¹ and are not due to a minority inclusion of C_{70}, since no Raman lines are observed in our spectra near the strong C_{70} line frequencies reported by Meijer and Bethune [25]. Furthermore, the counterparts of these C_{60} lines appear in the M_6C_{60} films at 1094 and 1237 cm⁻¹. The weak lines seen in Fig. 13 above 1573 cm⁻¹ vary in strength from that shown in Fig. 13, to even weaker intensities in other samples. Inspection of the figure leads to the conclusion that the doped spectra are all remarkably similar, exhibiting very little dependence on the mass, radius or electronic levels of the respective alkali metal constituents. Features at 520 and 965 cm⁻¹ in the C_{60} spectrum (Fig. 13) are identified with the silicon substrate [58] and vanish in the doped films, indicating higher optical absorption near the laser wavelength (4880 Å) in M_6C_{60} than in pristine C_{60}. Two (three) strongly polarized modes are observed for C_{60} (M_6C_{60}), and are identified below on the basis of symmetry considerations. In contrast to the case of C_{60}, two of the M_6C_{60} lines at ~271 and ~429 cm⁻¹ are resolved as doublets at higher resolution. Although, the C_{60} spectrum in Fig. 13 differs from that reported by Duclos et al. [28], it is interesting to note that the M_6C_{60} spectra in the figure are in good agreement with their results on alkali metal

Table 9. Experimentally observed Raman-active mode frequencies, depolarization ratios of M_xC_{60} (x = 0,3,6)

I_h mode	T_h mode	C_{60} (cm⁻¹)	$I_\perp/I_\parallel$	K_3C_{60} (cm⁻¹)	$I_\perp/I_\parallel$	K_6C_{60} (cm⁻¹)	$I_\perp/I_\parallel$	Rb_6C_{60} (cm⁻¹)	$I_\perp/I_\parallel$	Cs_6C_{60} (cm⁻¹)	$I_\perp/I_\parallel$
A_g	A_g	1468.5	0.10	1453	0.18	1432.5†	0.13	1432.5	0.10	1433.0	0.10
A_g	A_g	493.0	0.02	497	0.02	502.0	0.10	499.5	0.12	497.5	0.11
H_g	$E_g + T_g^‡$	1573.0	0.52	1547	0.52	1476.0	0.48	1483.5	0.80	1480.5	0.54
H_g	$E_g + T_g$	1426.0	0.44	1408	—	1383.5	0.50	1419.5 / 1385.0	0.14 / 0.60	1421.0 / 1384.0	0.28 / 0.52
H_g	$E_g + T_g$	1248.0	—	—	—	1237.0	0.88	1238.5	0.57	1238.0	0.44
H_g	$E_g + T_g$	1099.0	—	—	—	1120.0 / 1094.0	0.38 / 0.42	1120.5 / 1092.0	0.35 / 0.78	1118.0 / 1090.0	0.58 / 0.91
H_g	$E_g + T_g$	772.5	0.38	—	—	761.5	0.75	760.5	0.68	761.0	0.50
H_g	$E_g + T_g$	708.5	0.40	723	0.63	676.5 / 656.0	0.00 / 0.49	676.5 / 657.5	0.00 / 0.74	678.0 / 658.5	0.00 / 0.47
H_g	$E_g + T_g$	430.5	0.40	431	0.5	427.0 / 419.5	0.71 / 0.75	428.5 / 421.5	0.65 / 0.61	429.5 / 424.0	0.63 / 0.67
H_g	$E_g + T_g$	270.0	0.52	271	0.74	281.0 / 269.5	0.91 / 0.71	277.0 / 271.5	0.85 / 0.54	— / 272.5	— / 0.48

† Raman lines at 1430 cm⁻¹ and at 1447 cm⁻¹ for K_6C_{60} and K_3C_{60}, respectively have been reported by Haddon et al.

‡ For the modes which show a splitting, it would be expected that the mode with the smaller value of $I_\perp/I_\parallel$ would correspond to the E_g symmetry mode.

doped films. As discussed in section 4.1, we identify the C_{60} spectrum in their work [28, 47] with a photo-transformed phase of C_{60} [23], not with an impurity phase or a mixture of higher fullerenes. The M_6C_{60} spectra obtained by Duclos *et al.* [28] are in good agreement with those shown in Fig. 13.

It is worth mentioning where that the M_6C_{60} compounds have been observed to exhibit a significantly higher tolerance to high laser flux than pristine C_{60} [27]. For example, we have subjected these materials to laser flux greater than 150 W cm^{-2}, nearly 15 times greater than flux values which initiate an irreversible photo-transformation of C_{60} (see section 4.1) and not observed any change in the Raman spectrum.

From Fig. 13 and Table 9, it is clear that the downshift of the highest frequency tangential modes in M_6C_{60} is independent of the nature of the alkali metal dopant. This is interpreted as a clear indication that the charge transfer-induced elongation of the average C–C bond length dominates the effect, and the downshift is not significantly mediated by inter-ball or M–ball interactions. Haddon *et al.* [8] reported previously the x-dependent downshift of the A_g C_{60} mode at 1469 cm^{-1} [47] to 1445 cm^{-1} in highly conducting K_3C_{60} and to 1430 cm^{-1} in the highly doped but insulating compound K_6C_{60}. In agreement with their view, we also identify this softening with electron donation to the ball. In addition, we find that other high frequency tangential modes also soften upon doping to $x = 6$. The sold C_{60} lines at 1426 cm^{-1} and at 1469 cm^{-1} downshift by ~ 40 cm^{-1}, and the 1573 cm^{-1} line downshifts by ~ 100 cm^{-1} in solid M_6C_{60}. Previous theoretical studies by Chan *et al.* [59] of charge transfer effects on the Raman-active intralayer modes in graphite intercalation compounds (GICs) report a charge transfer shift $(\Delta\omega_{ct}/q_C) = -880$ cm^{-1}/e, here q_C is the average charge per C-atom donated to the graphene layers

in units of electrons per C-atom. Assuming six electrons transferred per C_{60}, and using an average mode downshift of 60 cm^{-1} for the highest three tangential modes, we arrive at a charge transfer shift of $(\Delta\omega_{ct}/q_C) = -600$ cm^{-1}/e, or $\sim 70\%$ of the effect in GICs [59].

In GICs, a universal relationship between C–C bondlength and q_C was proposed first by Pietronero and Strassler [60] on the basis of empirical calculations. Later Chan *et al.* [61] performed *ab initio* calculations of the effect, and found a weaker relationship between the C–C bond length (d_{CC}) and q_C. X-Ray and neutron diffraction studies in the alkali metal GICs do show a clear universal relationship between d_{CC} and q_C. A value $q_C = -0.1$ e in GICs is associated with an elongation $\Delta d_{CC} = 0.010$ Å [61]. In contrast to the high frequency tangential C_{60} modes, the low frequency radial C_{60} mode at 493 cm^{-1} (A_g), *increases* by ~ 7 cm^{-1} upon doping to M_6C_{60} indicating that a competing effect overwhelms the effect of C–C bond elongation. This conclusion is born out by recent theoretical calculations indicating that an electrostatic interaction arising from the charged C_{60} ball is partly responsible for the upshift [29].

Figure 14 shows the polarization study for a Rb_6C_{60} sample at room temperature. Other M_6C_{60} ($M = K$, Cs) films exhibit nearly the same spectra (all values for depolarization ratios (I_{HV}/I_{HH}) are listed in Table 9) indicating only a small dependence on alkali metal species. For the tangential A_g mode, the observed polarization ratio remains at $I_{HV}/I_{HH} \sim 0.10$ for both the pristine and M_6C_{60} samples, whereas the strong selection rule for the radial A_g mode in the M_6C_{60} compounds is weakened to the $I_{HV}/I_{HH} \sim 0.10$ level as a result of the alkali metal doping, so that I_{HV}/I_{HH} becomes similar in magnitude to that of the tangential A_g mode.

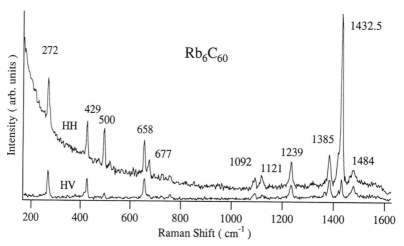

Fig. 14. Polarized Raman spectra ($T = 300$ K, resolution ~ 6 cm^{-1}) for Rb_6C_{60}.

This violation of the polarization selection rule would result from the change in the ball–alkali metal distances during the radial vibrations and suggests that the symmetry-lowering perturbation is most sensitive to this change in the ball–alkali metal distances.

The softening of the tangential A_g mode and the stiffening of the radial A_g mode upon alkali metal doping can be explained by charge transfer-related effects [29].

(1) The bond-stretching force constants decrease as a result of a lattice expansion [45].

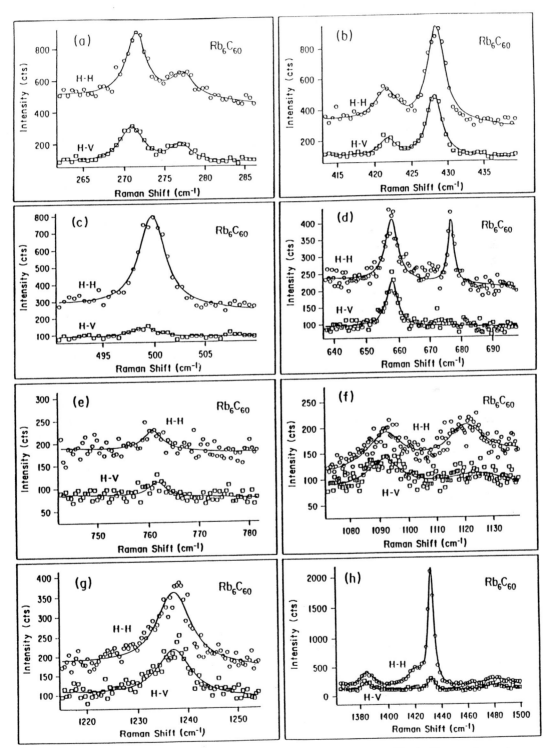

Fig. 15. Higher resolution ($\sim 2.5\ \text{cm}^{-1}$), polarized, Raman spectra for Rb_6C_{60}. The doublets seen in panels a, b, d and f are not resolved in Fig. 13.

(2) There is an even larger decrease in the bond angle bending force constants as a result of this lattice expansion [45].

(3) Modes with dominantly radial character are downshifted less in frequency than modes with predominantly tangential character. In fact, the radial A_g breathing mode is actually upshifted in frequency as explained in [29].

Thus, in determining the effect of the doping on the different modes, it is important to consider the radial or tangential character of the mode as far as the atomic displacements are concerned, especially the magnitude of the contribution to the normal mode of the angle bending force constants. The radial character of these modes has been supplied by Stanton and Newton [13] and their results for the Raman-active modes are shown in Table 8.

In Fig. 15 we show higher resolution (~ 2.5 cm^{-1}), polarized Raman data for Rb$_6$C$_{60}$ [27] in eight panels centered in frequency around positions of spectral features taken at 6 cm^{-1} resolution in Fig. 13. At this higher resolution several H_g-derived lines are resolved into doublets (panels a, b, d and f). The frequencies, FWHM and depolarization ratios for the lines evident in Fig. 15 are listed in Table 9. As can be seen from the table, similar frequencies and depolarization ratios were observed for K$_6$C$_{60}$ and Cs$_6$C$_{60}$, and for this reason the high resolution spectra for these samples are not shown here. The polarized A_g modes are shown in panel (c) (radial, $\omega = 499.5$ cm^{-1}) and (h) (tangential, $\omega = 1432.5$ cm^{-1}). The lines in the other panels are identified with C$_{60}$ H_g-derived modes for which we cannot detect a splitting in the cubic crystal field of M$_6$C$_{60}$. We argue that in these cases the T_g–E_g splitting caused by the cubic crystal field is too small to be observable. The observation of T_g and E_g modes for both ($//,//$) and ($//,\perp$) polarizations is also consistent with this interpretation. For the remaining

five H_g-derived modes, the splitting is small for the two lowest frequency H_g modes; for these H_g-derived modes, small but significant polarization effects are observed, presumably because of hybridization of the T_g and E_g partners. This, once again, may be attributed to the symmetry being lower than T_h in the materials under observation. The third H_g-derived mode at 708.5 cm^{-1} in pristine C$_{60}$ is strongly split, and shows a strong polarization selection rule for the E_g mode which is not downshifted as much as the T_g mode. We further note that the downshift for the E_g mode is less than that for the T_g mode, consistent with a large amount of bond stretching relative to bond bending for the E_g mode [13, 23]. The spectral features associated with H_g-derived modes at 1099 cm^{-1} in C$_{60}$ show a splitting without much change in center frequency. The small degree of polarization for these components is comparable to that observed for the two lowest frequency lines, presumably because of the hybridization of the T_g and the E_g partners associated with the symmetry-lowering perturbation. Finally the features derived from the 1426 cm^{-1} H_g mode in C$_{60}$ are again significantly downshifted, exhibit a large splitting, and a significant polarization selectively for the E_g mode which shows much less softening than the T_g mode.

K$_3$C$_{60}$. In Fig. 13 we also show the Raman spectrum ($T = 300$ K) of a typical K$_3$C$_{60}$ film (~ 2000 Å thick) in comparison with that of C$_{60}$ and M$_6$C$_{60}$ films [26, 27, 31]. The sample was obtained by annealing a C$_{60}$/K/C$_{60}$ three layer structure as described in section 2. Seven Raman-active modes were observed in various K$_3$C$_{60}$ films [31]. Three of these modes (at 271 cm^{-1}, 497 cm^{-1} and 1452 cm^{-1}) are in agreement with those reported by Duclos *et al.* [28]. Polarization studies were carried out to identify these seven modes with their counterparts in C$_{60}$ and K$_6$C$_{60}$ (Table 9). The polarization of the radial A_g symmetry modes

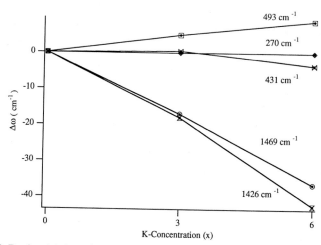

Fig. 16. Doping (x) dependence of representative Raman modes in K$_x$C$_{60}$ ($0 \leq x \leq 6$).

at 497 cm^{-1} is strong, consistent with our observations in the M_6C_{60} and C_{60} samples. The tangential A_g symmetry mode at 1453 cm^{-1}, however, exhibits a depolarization ratio $(I_{HV}/I_{HH}) \sim 0.18$, somewhat higher than the values of 0.10–0.13 observed in the M_6C_{60} compounds. In the M-doped C_{60} compounds, the increase observed in the depolarization ratio over that found in the pristine C_{60} is probably due to disorder in the M-sublattice.

Figure 16 shows the doping (x) induced change in the frequencies of representative radial and tangential A_g and H_g C_{60}-derived modes. The radial [13] modes (upper three curves) and tangential [13] modes (lower two curves) exhibit contrasting behavior. As discussed recently by Jishi and Dresselhaus [29], the radial A_g mode upshifts slightly in a nearly linear fashion with increasing x, whereas the tangential A_g mode exhibits a strong linear downshift. Note that the 70% radial [13] H_g-derived 271 cm^{-1} mode is nearly independent of x.

We now turn our attention to the larger linewidths in the Raman spectrum observed for the K_3C_{60} compounds. Varma et al. [17] predicted that the two highest frequency H_g modes of C_{60} at 1426 cm^{-1} and 1573 cm^{-1} should be most strongly coupled to t_{1u} electrons, giving rise to a linewidth of ~ 300 cm^{-1} and a 5% softening of the mode frequencies (~ 70–80 cm^{-1}) in K_3C_{60}. Another calculation by Johnson et al. [18] predicted that only the lowest frequency H_g mode is strongly coupled to the half-filled t_{1u} band electrons. Our observations reveal a downshift of ~ 18 cm^{-1} for the 1426 cm^{-1} mode and ~ 26 cm^{-1} for the 1573 cm^{-1} mode. The fact that all five H_g-derived modes observed in this work exhibit large linewidths (ranging from 12.5 cm^{-1} to 150 cm^{-1}) is consistent with strong coupling to the electrons in the conduction band (t_{1u}). However, temperature-dependent studies through the superconducting transition are needed to confirm this interpretation, since K-sublattice disorder can also broaden the linewidths. The H_g-derived modes have widths considerably larger than the widths for the A_g modes (Table 9). Mitch et al. [62] in in situ, interference-enhanced Raman scattering experiments on thin Rb_xC_{60} ($0 \leq x \leq 3$) films reported broadening of all the H_g-derived modes for $x \sim 3$, in agreement with our data on K_3C_{60} films. They also attributed this line broadening to a strong electron–phonon interaction.

Of special interest is the Breit–Wigner–Fano lineshape of the mode at 271 cm^{-1} in K_3C_{60}, and the Lorentzian lineshapes of the corresponding mode in C_{60} and K_6C_{60} (Fig. 17, higher resolution (3.5 cm^{-1})). Polarization studies of these modes are shown. The solid lines in the lower portion of the

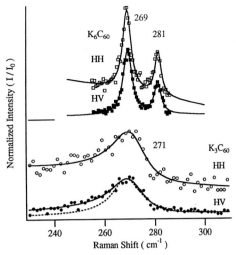

Fig. 17. Higher resolution (3.5 cm^{-1}), polarized, Raman spectra in the vicinity of 271 cm^{-1} for K_3C_{60} and K_6C_{60}. Lower two spectra (K_3C_{60}): solid lines represent a least squares fit to a Breit–Wigner–Fano lineshape on a linear background, and the dashed line in the bottom spectrum is a Lorentzian lineshape drawn to indicate the asymmetry in the peak. Upper two spectra (K_6C_{60}): solid lines represent a least squares fit to a Lorentzian lineshape on a linear background.

figure represent a least squares fit of K_3C_{60} experimental data to a Breit–Wigner–Fano resonance lineshape [63, 64]

$$I(\omega) = I_0 \left[1 + \frac{\omega - \omega_0}{q\Gamma} \right]^2 \Bigg/ \left[1 + (\frac{\omega - \omega_0}{\Gamma})^2 \right],$$

with fitting parameters $1/q = -0.22 \pm 0.02$, $\omega_0 = 271 \pm 1$ cm^{-1}, $\Gamma = 8.0 \pm 0.2$ cm^{-1}, where q is the coupling constant, ω_0 is the decoupled discrete mode frequency and Γ is a line width parameter. A negative value for q indicates that the Raman-active continuum that is coupled to the discrete mode lies lower in frequency than ω_0. The dashed line in the bottom spectrum of Fig. 17 represents a Lorentzian lineshape fit to the data, and is included to draw attention to the observed asymmetry of the 271 cm^{-1} peak in K_3C_{60}. Due to the high degree of radial character of the 271 cm^{-1} mode [13] in an isolated C_{60} molecule ($\sim 70\%$ radial character), this mode is also expected to couple strongly to the nearby potassium atoms, which do indeed have discrete low frequency modes. The Fano lineshape could either arise from the coupling to a nearby continuum of low energy electronic excitations, or to a continuum of K-related multi-phonon states associated with two or more of these modes. For comparison, we show in the upper portion of Fig. 17 two polarized spectra under higher resolution K_6C_{60} taken in the vicinity of 271 cm^{-1}, showing that the asymmetrical lineshape at 271 cm^{-1} in Fig. 13 (6 cm^{-1} resolution) is actually an unresolved doublet with Lorentzian components [26].

This doublet, for the case of Rb_6C_{60} is shown in Fig. 15.

Previous Raman studies on M_6C_{60} (M = K,Rb,Cs. Space group: T_h^5 [43]) report only 15 Raman-active modes [26]. As discussed above, even fewer lines (seven) are observed here for K_3C_{60}, thus suggesting that the H_g modes in K_3C_{60} split by a very small amount, smaller than for the case of K_6C_{60}, and thus the solid state interactions contribute predominately to the linewidth. This argument accounts for the absence of the expected polarization selectivity of the E_g and T_g components in T_h point group symmetry. The low scattering intensity observed is probably caused by two factors: (1) the absence of resonant Raman scattering which enhances the observed spectra features in C_{60} and K_6C_{60}; (2) the decrease in the optical penetration depth in metallic K_3C_{60} relative to that in insulating K_6C_{60} and C_{60}.

5. CONCLUSIONS

In this paper, we have reviewed our experimental studies of Raman active modes in pristine and doped C_{60} thin solid films. A group theoretical analysis of the molecular and solid state vibrational modes applicable to cubic symmetry has also been presented and compared with experimental results.

For pristine fcc C_{60}, we find that the material is particularly sensitive to visible light. Moderate laser flux has been observed to initiate a photo-transformation of the material to a lower symmetry system. Another interesting photo-effect we have observed is the photo-assisted diffusion of oxygen into the lattice. Surprisingly, only a few changes in the Raman spectrum above $\sim 1500\,cm^{-1}$ could be identified with oxygen-doping, which suggests that the oxygen may not have bonded to the C_{60} molecules, but rather occupies interstitial sites as O_2. Further investigations will be necessary to confirm this point.

Solid C_{60} at $T = 300\,K$ exhibits only a few additional optical modes over those anticipated from a group theoretical analysis of an isolated molecule. This is interpreted as strong evidence for weak intermolecular coupling in solid C_{60}. After saturation doping with alkali metals (M = K,Rb,Cs) to form M_6C_{60} compounds, a few additional lines are observed which appear to be C_{60} modes split by a weak interaction between the C_{60} and cation sublattice. Little dependence in the Raman spectra of M_6C_{60} compounds on the mass or ionic radii of the particular alkali metal dopant is observed, suggesting the primary role of the doping is to transfer electrons to C_{60}-derived energy bands, which for the case of the M_3C_{60} compounds produce superconductors with $18\,K < T_c < 33\,K$.

The modes that are observed for K_3C_{60} and K_6C_{60}, for example, can be identified with A_g- or H_g-derived C_{60} modes of an isolated molecule, supporting the view that the M-doped material is also a molecular solid. The doping (x) dependence observed for the radial and tangential modes of M_xC_{60} is very different. The high frequency modes exhibit a softening with increasing x ($\sim 6\,cm^{-1}$ per transferred electron for the $1469\,cm^{-1}$ A_g mode), analogous to K-intercalated graphite where K doping elongates the C–C intralayer bonds. Thus in C_{60} too, it appears that electron transfer into the aromatic rings dilates the ring structure and softens the tangential mode frequencies. The radial C_{60}-derived modes exhibit contrasting behavior, stiffening slightly with increasing x, suggesting that a competing mechanism is present which affects the radial modes. Finally, our $T = 300\,K$ studies of homogeneous K_3C_{60} films revealed a broadening of the H_g-derived C_{60} modes in agreement with theoretical suggestions of a strong *intraball* electron–vibration interaction for these materials which would enhance T_c.

Acknowledgements—The work at the University of Kentucky was funded, in part, by the Center for Applied Energy Research and the Electrical Powder Research Institute (EPRI No. RP7911-20), whereas the researchers at M.I.T. gratefully acknowledge support by the National Science Foundation (No. 88-1896-DMR). Special thanks are due to Ying-Wang, Matthew Holden, Song-lin Ren and Dr A. M. Rao at the University of Kentucky for invaluable assistance with various phases of the research. We thank Professors Jack P. Selegue and Mark S. Meier in the Chemistry Department (U.KY) for their assistance in purifying the fullerenes.

REFERENCES

1. Kroto H. W., Heath J. R., O'Brien S. C., Curl R. F. and Smalley R. E., *Nature* **318**, 162 (1985).
2. The name is derived from the famous architect and engineer Buckminster Fuller, creator of the geodesic domes which exhibit interconnected pentagonal and hexagonal patterns.
3. Krätschmer W., Lamb L. D., Fostiropoulos K. and Huffman D. R., *Nature* **347**, 354 (1990).
4. Fleming R. M. *et al.*, *Mater. Res. Soc. Symp. Proc.* **206**, 691 (1990).
5. David W. I. F. *et al.*, *Nature* **353**, (1991).
6. Hedberg K. *et al.*, *Science* **254**, 410 (1991).
7. Johnson R. D. *et al.*, *J. Am. Chem. Soc.* **112**, 8983 (1990).
8. Haddon R. C. *et al.*, *Nature* **350**, 320 (1991).
9. Rosseinsky M. J. *et al.*, *Phys. Rev. Lett.* **66**, 2830 (1991).
10. Hebard A. F. *et al.*, *Nature* **350**, 600 (1991).
11. Kelty S. P., Chen C. C. and Lieber C. M., *Nature* **352**, 223 (1991).
12. Negri F., Orlandi G. and Zerbetto F., *Chem. Phys. Lett.* **144**, 31 (1988).
13. Stanton R. E. and Newton M. D., *J. Phys. Chem.* **92**, 2141 (1988).
14. Wu Z. C., Jelski D. A. and George T. F., *Chem. Phys. Lett.* **137**, 291 (1987).

15. Weeks D. E. and Harter W. G., *J. Chem. Phys.* **90**, 4744 (1989).
16. Wang Y., Tomanek D. and Bertsch G. F., *Phys. Rev. B., Rapid Commun.* **44**, 6562 (1991).
17. Varma C. M., Zaanen J. and Raghavachari K., *Science* **254**, 989 (1991).
18. Johnson K. H., McHenry M. E. and Cloughtery D. P., *Physica C*, in press.
19. Dresselhaus M. S. and Dresselhaus G., *Advances in Physics* **30**, 139 (1981).
20. Dresselhaus G. *et al.*, *Phys. Rev. B*, **45**, 6923 (1992).
21. Fischer J. E. *et al.*, *Science* **252**, 1288 (1991).
22. Heiney P. A. *et al.*, *Phys. Rev. Lett.* **66**, 2911 (1991).
23. Zhou P. *et al.*, *Appl. Phys. Lett.*, **60**, 2871 (1992).
24. Bethune D. S. *et al.*, *Chem. Phys. Lett.* **174**, 219 (1990).
25. Meijer G. *et al.*, 1990 *MRS Symp. Proc.* **206**, 619 (1991).
26. Wang K.-A. *et al.*, *Phys. Rev. B, Rapid Commun.* **45**, 1955 (1991).
27. Zhou P. *et al.*, *Phys. Rev. B*, 46, 2595 (1992).
28. Duclos S. J. *et al.*, *Science* **254**, 1625 (1991).
29. Jishi R. A. and Dresselhaus M. S., *Phys. Rev. B*, **45**, 6914 (1992).
30. Dresselhaus G., Dresselhaus M. S., Saito R. and Eklund P. C., *Elementary Excitation in Solids*, 387 (Edited by J. L. Birman, C. SeBenne and R. F. Wallis). Elsevier, Amsterdam (1992). See also Ref. 20.
31. Zhou P. *et al.*, *Phys. Rev. B, Rapid Commun.* **45**, 10838 (1992).
32. Jishi R. A. and Dresselhaus M. S., *Phys. Rev. B, Rapid Commun.* **45**, 2597 (1992).
33. Wang Y., Holden J. M., Rao A. M., Ren S. L. and Eklund P. C., *Phys. Rev. Lett.* **45**, 14396 (1992).
34. Meier M. S. and Selegue J. P., *J. Org. Chem.* **57**, 1925 (1992).
35. Hebard A. F., Haddon R. C., Fleming R. M. and Kortan A. R., *Appl. Phys. Lett.* **59**, 2109 (1991).
36. Vassallo A. M., Pang L. S., Cole-Clark P. A. and Wilson M. A., *J. Am. Chem. Soc.* **113**, 7820 (1991).
37. C$_{60}$ was calibrated by comparing quartz microbalance and ellipsometry determined film thickness. Alkali metal calibration involved weighing a thick film on a 10 μg sensitivity balance located in the glove box and comparing with the microbalance.
38. Heiney P. A. *et al.*, *Phys. Rev. Lett.* **67**, 1468 (1991).
39. Zhou O. *et al.*, *Nature* **351**, 462 (1991).
40. Johnson R. D., Meijer G., Salem J. R. and Bethune D. S., *J. Am. Chem. Soc.* **113**, 3619 (1991).
41. Tycko R. *et al.*, *Science* **253**, 701 (1991).
42. Stephens P. W. *et al.*, *Nature* **351**, 623 (1991).
43. Fleming R. M. *et al.*, *Nature* **352**, 701 (1991).
44. The 'pentagonal pinch' mode involves 100% tangential displacement of the C-atoms on the ball so as to shrink the pentagons and expand the hexagons.
45. Jishi R. A., Mirie R. M. and Dresselhaus M. S., *Phys. Rev. B.* **45**, 13685 (1992).
46. van Loosdrecht P. H. M. *et al.*, *Phys. Rev. Lett.*, unpublished (1992).
47. Duclos S. J., Haddon R. C., Glarum S. H. and Lyons K. B., *Solid State Commun.* **80**, 481 (1992).
48. Experiments were also carried out in which the sample was illuminated with 2.54 eV photons at 75 W cm^2 for 30 min and the Raman spectrum was then collected at low flux, i.e. at $\Phi_L \sim 5$ W cm^{-2}. The spectrum exhibited the same features as in Fig. 7b.
49. Negri F., Orlandi G. and Zerbetto F., *J. Am. Chem. Soc.* **113**, 6037 (1991).
50. Ren S. L. *et al.*, *Appl. Phys. Lett.* **59**, 2678 (1991).
51. Tong W. M. *et al.*, *J. Phys. Chem.* **95**, 4709 (1991).
52. Tolbert S. H. *et al.*, *Chem. Phys. Lett.* (1991).
53. Rosatti F., *Phys. Rev.* **34**, 367 (1929).
54. Taylor R. *et al.*, *J. Chem. Soc. Commun.* 1423 (1990).
55. Kalsbeck W. A. and Thorp H. H., *J. electroanal. Chem.* **314**, 363 (1991).
56. Kroll G. H. *et al.*, *Chem. Phys. Lett.* **181**, 112 (1991).
57. Kochanski G. P., Hebard A. F., Haddon R. C. and Fiory A. T., *Science* **255**, 184 (1992).
58. Uchinokura K., Sekine T. and Matsuura E., *Solid State Commun.* **11**, 47 (1972).
59. Chan C. T., Ho K. M. and Kamitakahara W. A., *Phys. Rev. B.* **36**, 3499 (1987).
60. Pietronero L. and Strasszler S., *Phys. Rev. Lett.* **47**, 593 (1981).
61. Chan C. T., Kamitakahara W. A., Ho K. M. and Eklund P. C., *Phys. Rev. Lett.* **58**, 1528 (1987).
62. Mitch M. G., Chase S. J. and Lannin J. S., *Phys. Rev. Lett.*, submitted (1992).
63. Eklund P. C. *et al.*, *Phys. Rev. B* **16**, 3330 (1977).
64. Eklund P. C. and Subbaswamy K. R., *Phys. Rev. B* **20**, 5157 (1979).
65. By "oxygen-free" we mean no oxygen would be detected in the deep bulk, but surface oxides are possible.

ELECTRON ENERGY LOSS SPECTROSCOPY OF C_{60} FULLERITE FILMS

A. A. LUCAS

Institute for Studies in Interface Sciences, Facultés Universitaires Notre-Dame de la Paix, 61 rue de Bruxelles, B-5000, Namur, Belgium

(*Received* 16 *February* 1992)

Abstract—The fundamental principles of Electron Energy Loss Spectroscopy will first be summarized. Then selected spectra for C_{60} fullerite films will be presented in the energy range extending from the infrared to the vacuum ultraviolet. When possible, comparison with optical and photoelectron spectra will be made.

Keywords: Fullerenes, electron spectroscopy.

1. PRINCIPLES OF ELECTRON ENERGY LOSS SPECTROSCOPY

(A) *General*

The basic idea of EELS [1, 2], as in any particle scattering experiment, is to bombard a target with a monoenergetic and well-collimated electron beam and, by application of conservation laws, to deduce the excitation spectrum of the target from the scattered intensity as a function of energy and momentum transfers. Depending on whether the scattered electrons are collected in directions forward or backward with respect to the beam and solid target surface, one speaks of Transmission EELS (Fig. 1) or of Reflection EELS (Fig. 2). Because electrons are strongly scattered by atoms and molecules, T-EELS requires primary electron energies in the tens of keV range to enable the beam to traverse a self-supporting thin film (at least a few hundreds and at most a few thousands Å thick). On the other hand, R-EELS places no restriction on target thickness and works best for a very low primary energy of a few eV; but this energy cannot exceed a few hundred eV, otherwise there would remain no measurable intensities. As a consequence T-EELS is generally more appropriate for studies of bulk excitations, whereas R-EELS has been particularly successful in detecting near-surface excitations [1], especially those of adsorbed molecular layers such as the present fullerene films.

An electron interacting with the target may suffer no scattering at all or may experience elastic scattering events or/and inelastic collisions. A typical EELS spectrum will therefore look like the sketch of Fig. 3, where one plots the intensity of electrons of initial energy E_0 scattered in a given direction with final energy E, having lost a given amount of energy $\hbar\omega = E_0 - E$ to the target. There is typically a large peak around $\hbar\omega = 0$, the so-called elastic peak (or no loss line), and weaker, discrete or continuous structures, the so-called loss peaks, for $\hbar\omega > 0$. The position and intensity of the loss peaks relative to the elastic line give information on the spectrum of individual and collective excitations in the target. At nonzero temperature, there is also a gain spectrum on the $\hbar\omega < 0$ side. Apart from providing a sophisticated, indirect measurement of the target surface temperature, the gain spectrum is of no additional spectroscopic interest as it merely consists of the mirror image of the loss spectrum with respect to the elastic line scaled down by the Boltzmann factor $\exp(-\hbar\omega/k_B T)$ [3].

It is not always clearly appreciated that the occurrence, indeed the frequent dominance of the no loss line in the spectrum, is an entirely quantum mechanical phenomenon. More generally, the quantization of the number of inelastic scattering events is of quantum origin. There is indeed no way for a classical electron–target system to produce such statistical outcomes which are consequences of the wave properties of the electron (responsible for elastic diffraction) and the energy quantization of the target elementary excitations (responsible for the loss spectrum).

Regarding the elastic scattering, the target will produce sharp Laue transmission spots (Fig. 1a) or Bragg reflection spots (Fig. 2a) if it has high crystalline order and diffuse intensity distributions if it has a disordered atomic structure (Figs 1b and 2b). The interpretation of the EELS spectrum depends

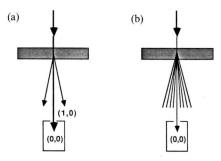

(a)   (b)

Fig. 1. Principles of EELS in transmission through an ordered film: (a) producing Laue diffraction beams and through a disordered film and (b) producing diffuse scattering. Here, the scattered electrons are collected around the forward direction of the incident beam.

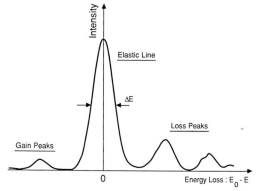

Fig. 3. The general shape of the EELS spectrum with its elastic or no loss line of FWHM ΔE and with the loss and gain features containing the information about the elementary excitations of the target with energy $\hbar\omega = E_0 - E$.

strongly on such coherent or diffuse elastic scatterings, as will be illustrated in the case of C_{60} fullerite thin films. In principle, with a sufficiently narrow spectrometer aperture, the EELS spectrum can be taken in an arbitrary scattering direction with respect to the primary beam. Then, *provided there is a well-defined diffraction pattern* (Figs 1a and 2a), the method allows to measure the loss intensities for fixed momentum transfer, i.e. the full dispersion relations of the excitations of the target. The necessary kinematic equations expressing energy and momentum conservations are given in the books [1, 2]. In that way, T-EELS has been, for a long time, a powerful method for the study of bulk excitations in solids [2]. On the other hand, in the early 1970s, R-EELS spectra of well-ordered crystalline surfaces used to be taken in directions where there is substantial intensity, namely around a Bragg diffraction spot, most often the specular one (note that in R-EELS, because the translation symmetry is broken normal to the surface, only the parallel component of the momentum is conserved). With the small acceptance aperture of the spectrometer (typically $\theta \approx 2 \times 10^{-2}$ rad), this restricts the possible inelastic momentum transfers to a small fraction $k \leq \theta k_0$ of the primary momentum k_0, itself of the order of reciprocal lattice vectors. Then only excitations of long wavelengths on the scale of interatomic distances could be detected by

this so-called *specular* R-EELS. With the improvement in the sensitivity of the spectroscopy however, off-specular spectra [1] could be taken on well-ordered surfaces by setting the spectrometer aperture in the diffuse background of LEED (low energy electron diffraction), i.e. between Bragg spots. This allows to measure the full dispersion relations of the surface excitations of the target. A beautiful example, relevant to the present fullerite theme, of nonspecular EELS as applied to the graphite surface is given in [4]. Off-specular, angular resolved EELS was the true breakthrough for surface studies by the method of electron inelastic scattering.

For disordered targets producing no clear diffraction patterns (Figs 1b and 2b), elastic diffuse scattering sends electrons quasiisotropically in all directions of a broad solid angle so that the EELS spectrum becomes essentially independent of the scattering direction. In this situation, the method loses its ability to select a given inelastic momentum transfer.

Two energy parameters are central to the EELS technique: the resolution ΔE and the energy range covered. In T-EELS, the instruments currently in use typically have $\Delta E > 0.1$ eV and thus are adequate for detecting electronic excitations from the near infrared to the vacuum ultraviolet and core level energies. By contrast the best R-HEELS instruments (HREELS) have a high resolution of $\Delta E \leq 0.003$ eV and can therefore be used to explore surface vibrational spectra down to very low energies. The upper energy loss range of R-EELS is limited, by energy conservation, to the primary electron energy E_0 and, as will be shown below, the VUV range can therefore be investigated by increasing E_0.

As previously mentioned, multiple inelastic scattering events can occur. This is a consequence of Coulomb scattering and the related fact that the coupling constant of the spectroscopy, involves the 'fine structure constant' $\alpha = e^2/\hbar v = (c/v)/137$, where

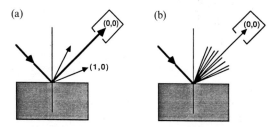

(a)   (b)

Fig. 2. Principles of EELS in reflection at an ordered surface: (a) producing Bragg diffraction beams and at a disordered surface and (b) producing diffuse scattering. Here, the scattered electrons are collected around the forward direction of the specular beam.

v is the electron velocity. How, then, does one decide whether a loss peak at $\hbar\omega$ represents the excitation of a genuine target degree of freedom or whether it is just the double excitation of $\hbar\omega_1 = \hbar\omega/2$ or the combination $\hbar\omega = \hbar\omega_1 + \hbar\omega_2$, etc...? In T-EELS this is not so severe since at 25 keV primary energy, $\alpha \approx 0.02$ and the loss intensities can be corrected for weak multiple losses or the latter can be eliminated by using thinner targets [2]. But in R-EELS at 10 eV, $\alpha \approx 1$ and multiple loss peaks are inherently present and may require deconvolutions. Multiple inelastic scattering events obey Poisson statistics [3] so that, in favorable conditions of resolution and sensitivity, such deconvolution into first order, second order, etc... can be carried out.

As mentioned above, the Coulomb interaction between the electrons and the electrons and ions of the target is responsible for both the elastic diffraction and the EELS spectrum. Scattering by the fixed ion cores produces diffraction or diffuse patterns, whereas Coulomb scattering by the mobile ions and the valence electrons leads to the inelastic loss spectrum from the i.r. to the v.u.v. The elastic and inelastic scattering events generally cannot be disentangled in the cross-section since, for example, a given momentum transfer can be effected either elastically by the fixed cores or inelastically by the elementary excitations. Yet, elastic and inelastic scatterings are often treated separately by dedicated, quantitative theories. For instance LEED theory is very advanced but includes the inelastic effects only through the optical potential approximation; on the other hand, the dielectric theory of EELS to be described below, is well developed but completely ignores the calculation of the diffracted intensities. Such separation is afforded by the fact that, in many situations, the elastic and inelastic events proceed independently on vastly different time and length scales. Then, their probabilities factor out in the cross-section so that the EELS single loss probability is given by the ratio of the first order to the zeroth order (elastic line) intensities in the spectrum [3].

(B) Dielectric theory of EELS

Let us now describe in some detail the dielectric theory of EELS [1–5], beginning with a short review of its modern historical development.

As far as T-EELS of thin films is concerned, its foundations date back to the works of Bohr, Fermi, Lindhard, etc... on stopping power. More recently the pioneering theoretical work of Ritchie appeared in the late 1950s [6] and the remarkable work of the German school on high resolution T-EELS of alkali-halide films appeared in the 1960s [7]. The interpretation of the T-EELS spectra was then mostly

based on classical physics, the electrons following a classical trajectory and the target excitations being described as classical oscillators embodied in a dielectric function (hence the name 'dielecric theory'). The existence of multiple losses in this approach was accounted for by the additional, ad hoc concept of electron mean free path, a concept justified in the homogeneous bulk of the target but difficult to visualize at the entrance and exit surfaces on account of the existence of localized surface excitations. To remove this assumption, the theory was, so to speak, semiquantized [8, 9] by maintaining classical electron trajectories but by taking into account the quantized nature of the elementary (collective) excitations which is the true origin of the discrete multiple energy losses.

As mentioned before, the coupling constant of R-EELS for inelastic scattering may be a hundred times stronger than in T-EELS and it then becomes crucial to have a quantitative theory of the multiple losses. This was the principal motivation to adapt the semiclassical, dielectric theory to the reflection geometry [3, 9]. Today, the theory of specular R-EELS is very advanced as it has been successfully extended to targets of layered materials [5, 10], anisotropic materials [11], molecular adsorbates, etc... The semiclassical approach was criticized [12] on the theoretical grounds that, at the eV energies of R-EELS, the electrons should always be treated quantum mechanically, as indicated by the diffraction patterns. But it turns out that, when elastic and inelastic scatterings proceed independently on different length scales, as is the case in specular R-EELS for which the semiclassical theory was devised, the inelastic cross-section can be calculated separately from the elastic scattering on the assumption of a classical electron trajectory. The semiclassical approximation was vindicated not only by its quantitative agreement with experiment but also by the nearly identical results of the full quantum theory [12].

The semiclassical dielectric theory starts with the classical step of computing the total work done on the electron trajectory by the electric polarization field induced in the target material by the Coulomb field of the electron

$$W = \int_{-\infty}^{+\infty} dt \vec{v}_e(t) \cdot e\vec{E}[\vec{r}_e(t)]. \qquad (1)$$

The electron trajectory $\vec{r}_e(t)$ is an assumed one, most often (but not necessarily) the straight trajectory of the transmitted $(0, 0)$ beam in T-EELS (Fig. 1a) or the broken trajectory of the $(0, 0)$ specular beam in R-EELS (Fig. 2a). The induced field $\vec{E}[r(t)]$ is computed from the nonretarded Maxwell equations in which the trajectory is injected as an external, time-dependent current source.

The physical properties of the target enter only via its long wavelength, frequency-dependent dielectric function $\epsilon(\vec{k}, \omega)$. By Fourier transforms, W is resolved into frequency and momentum components as

$$W = \int d\vec{k} \int_0^\infty d\omega \hbar\omega P_{cl}(\vec{k}, \omega), \qquad (2)$$

where $\vec{k}$ is a two-dimensional wavevector parallel to the target surface. This relation serves as the definition of the frequency-resolved classical loss function

$$P_{cl}(\omega) = \int d\vec{k} P_{cl}(\vec{k}, \omega), \qquad (3)$$

in which $P_{cl}(\vec{k}, \omega)$ appears as a classical probability of transferring energy $\hbar\omega$ and momentum $\hbar\vec{k}$ to the target. The integral over $\vec{k}$ in (3) extends over a range from 0 to k_{max} consistent with the spectrometer aperture which rejects electrons having been scattered outside a narrow angular cone around the chosen unperturbed direction. The total energy lost W must be a small fraction of the primary energy E_0, and the maximum momentum transfer k_{max} a small fraction of a typical reciprocal lattice vector g which itself is smaller then the electron momentum

$$\hbar\omega < W \ll E_0, \quad k_{max} \ll g \approx 2\pi/a < k_0. \qquad (4)$$

Otherwise inelastic recoil would invalidate the use of an unperturbed elastic trajectory in Maxwell's equations. These are the conditions for the validity of the dielectric theory of EELS. The conditions in (4) imply that, in real space, the 'formation zone', i.e. the length travelled by the electron during one period of the target elementary excitation to be produced is much larger than the elastic scattering length of the ion cores responsible for diffraction. In such conditions, it is intuitively clear that the elementary excitation 'feels' the passing electron not as a wave but as a point charge particle acting as a classical time-dependent external force.

Without going into the fine details of the theory (see Ref. 3), it will suffice here to quote the principal results. In T-EELS, when the film thickness L is large enough to allow for the neglect of surface effects, the energy loss probability turns out to be

$$P_{cl}^B(\vec{k}, \omega) = B(\vec{k}, \omega) \, \mathrm{Im}[-1/\epsilon(\vec{k}, \omega)]. \qquad (5)$$

The bulk kinematic prefactor B, which is proportional to

$$B(\vec{k}, \omega) \approx Le^2/\hbar(\omega^2 + k^2v^2), \qquad (6)$$

gives a spatial distribution of the inelastic intensity which is peaked in the forward direction ($k = 0$) of the electron beam of velocity v. The second factor in (5) contains the spectroscopic information about the bulk elementary excitations of the target. It has peaks at frequencies ω_b which render the dielectric constant vanishingly small

$$\epsilon(\vec{k}, \omega_b) = 0 \qquad (7)$$

$$\mathrm{Im}[-1/\epsilon(\vec{k}, \omega)] \approx \delta[\omega - \omega_b(\vec{k})]. \qquad (8)$$

By definition, these are the frequencies of the bulk collective motion, e.g. the longitudinal plasmons, as distinct from the individual one-electron interband excitations, or the bulk longitudinal phonons, etc. ...

In R-EELS on a thick target, the energy loss probability for the specular, nonpenetrating trajectory is found to be

$$P_{cl}^S(\vec{k}, \omega) = S(\vec{k}, \omega) \, \mathrm{Im}[-1/(\epsilon(\vec{k}, \omega) + 1)]. \qquad (9)$$

The surface kinematic prefactor S can be written as

$$S(\vec{k}, \omega) = (4/\pi^2)(e^2/\hbar v)(kv_\perp)^2/k^2[(\omega - \vec{k} \cdot \vec{v}_\parallel)^2$$
$$+ (kv_\perp)^2]^2, \qquad (10)$$

where $\check{v} = (\check{v}_\parallel, v_\perp)$ are the components of the electron velocity parallel and perpendicular to the surface. S is also seen to be strongly peaked close to the forward direction along the specular beam ($k \approx 0$), as sketched in Fig. 4. The angular aperture of the forward scattering lobe turns out to be of order $\hbar\omega/2E_0 \ll 1$ by virtue of condition (4). The spectro-

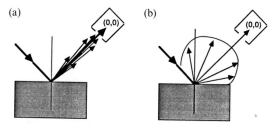

Fig. 4. Schematic illustration of the angular distribution of inelastically scattered electrons produced by the long range dipole (a) vs the start range impact (b) scattering mechanisms. The former gives a narrow forward scattering lobe whereas the latter produces a broad spatial distribution.

scopic loss function in (9) has peaks at frequencies of surface excitations defined by

$$\epsilon(\vec{k}, \omega_s) = -1 \qquad (12)$$

$$\mathrm{Im}[-1/(\epsilon(\vec{k}, \omega) + 1)] \approx \delta[\omega - \omega_s(\vec{k})]. \qquad (13)$$

Lifetime effects or band dispersion effects confer an imaginary part to $\epsilon(\vec{k}, \omega)$ and broaden the δ-functions of (8) and (13) into Lorentzians or continua. One notes that in the classical theory, the spectrum merely consists of single peaks at the frequencies of the relevant excitation modes, there being no elastic line and no multiple excitation peaks. The mechanism of energy transfer is via the classical resonance between the elementary material oscillators and the electron acting like a white source of external electromagnetic perturbation.

The first classical step of the dielectric theory must be complemented by a second quantum mechanical description of the multiple excitations emitted or absorbed by the electrons [5]. This is accomplished by quantizing the material eigenmodes and formulating the problem in terms of quantum oscillators driven by the same classical, time-dependent external perturbation. The result which is exact is the following expression for the full energy loss probability for a target material at temperature T in terms of the classical loss spectrum

$$P(\omega, T) = (1/2\pi)P_0(T) \int_{-\infty}^{+\infty} \exp[P_{cl}](t, T)e^{i\omega t}] \, dt,$$
$$(14)$$

where

$$P_{cl}(t, T) = (1/2) \int_0^{+\infty} d\omega' \, P_{cl}(\omega')\{e^{-i\omega't}$$

$$\times [\coth(\hbar\omega'/2k_B T) + 1]$$

$$+ e^{+i\omega't}[\coth(\hbar\omega'/2k_B T) - 1]\}. \qquad (15)$$

The prefactor in (14) is given by

$$P_0(T) = \exp\left\{-\int_0^{+\infty} d\omega' P_{cl}(\omega')\coth(\hbar\omega'/2k_B T)\right\}.$$
$$(16)$$

It represents the relative intensity of the elastic line and ensures the normalization to one of the total loss probability at all temperatures. When one expands (14) in powers of P_{cl} and performs the indicated time-integration, one sees that the zeroth order term

is $P_0(T)$ while the first order term is just $P_0(T)$ times the classical spectrum $P_{cl}(\omega)$. The latter therefore represents the ratio, accessible to measurement, of the first order spectrum to the no loss line intensity. Higher power terms yield the multiple loss peaks whose relative intensities approximately follow Poisson statistics. The first terms in the curly bracket of (15) generates the loss spectrum while the second term generates the gain spectrum.

The theoretical EELS spectrum of eqn (14) can further be convoluted with the instrumental resolution function [5]. This is especially useful at low resolution and mandatory in the i.r. energy range where the widths of the vibrational peaks are often resolution limited.

The formulation of the EELS spectra in terms of a long wavelength, macroscopic dielectric function of the target is legitimate when such a function can be meaningfully defined, e.g. when the target is homogeneous or when its inhomogeneities are over a scale large compared to interatomic distances. It is not appropriate for example when dealing with a monolayer or submonolayer of molecules adsorbed on a surface. For such a case, it is better to formulate EELS in terms of the scattering properties of the individual adsorbed molecules [1, 13]. In the same kinematic conditions as in eqn (4), there is again a narrow forward scattering lobe in the specular R-EELS spectrum of the molecular internal degrees of freedom. In the i.r., the spectrum is dominated by the electric-dipole active modes which present a component of the transition dipole moment perpendicular to the surface. The reason is that the common substrates such as Si or metals, on account of their large i.r. dielectric constants, tend to enhance such perpendicular vibrations while suppressing the dipole activity of the vibrations parallel to the surface by creating equal and opposite image dipoles. This so-called dipole selection rule has played a key role in mode assignment and adsorption site determination for polyatomic molecules. Its role is however weakened in the visible and u.v. regions of the R-EELS spectrum (e.g. for the plasmon modes in C$_{60}$) for which the screening by the substrate is much less effective.

(C) Dipole scattering vs impact scattering

It is customary in the EELS trade to clearly distinguish between the dipole and impact scattering mechanisms.

A peak in the EELS spectrum is said to arise from dipole scattering in the circumstances described in the previous section, namely when the inelastic intensity is due to the long-range dipole interaction between the electrons and the polarizable target. This results either from an intrinsic property of the target, e.g. the strong optical activity of a molecular motion such as the stretching mode of the CO molecule, or from the instrumental selection of a small momentum transfer. As we shall see for C_{60}, the induced polarization may comprise many higher multipolar components but then the long wavelength scattering conditions specified by eqn (4) in effect favors the longest range, dipole component. As was discussed before, this selection rule operates only for atomically well-ordered targets and yields spatial distribution of the scattered intensity which is strongly peaked in the forward direction, as shown schematically in Fig. 4a for R-EELS.

An EELS peak results from impact scattering when it is caused by the short range interaction of the electron with the individual atoms and molecules of the target. As a result of the complementarity between direct and reciprocal spaces, short range scattering necessarily implies large momentum transfer. This means that the inelastic scattering intensity will be broadly distributed over angles, as sketched in Fig. 4b. Then, for well-ordered targets having a clear diffraction pattern, the electron is sent off the specular direction by the inelastic event and the peak can be detected in off-specular directions where the dipole scattering lobe has vanished. In this way, it has been possible to map out completely the dispersion relations of surface phonon bands [1, 4] and to detect optically inactive molecular vibrations of adsorbed species [1].

The short range interaction is what is left when taking out the dipole part of this material polarization. Clearly the range of the multipolar components decreases gradually with multipolar order so that, for targets such as C_{60} with large multipolar polarizabilities, there may not be such a sharp distinction between dipole and impact scatterings.

The impact scattering regime is unfortunately not amenable to a semiclassical treatment similar to the one presented above. In T-EELS, one must incorporate the electron recoil due to scattering, for example by using the first Born approximation. In R-EELS, the scattering act may be a direct or exchange Coulomb scattering or may even involve transient occupation of continuum resonance levels. The calculation of the inelastic spectrum then requires a full quantum mechanical formulation to take account of the strong admixture of the target wavefunctions with the primary electron wavefunction. Such calculations have been attempted in some particular cases [14], but, in view of the peculiarity of each system, it is doubtful whether a theory of general applicability can be developed in a similar manner to the semiclassical

treatment of dipole scattering. For further details on these questions, we refer to the review paper by Gadzuk [15].

(D) *Relations to other spectroscopies*

It is clear from our previous description of EELS that the information obtainable from the *specular* EELS set-up, namely a loss function involving the long wavelength dielectric function $\epsilon(\vec{k}, \omega)$ of the target (see eqns (5) and (9)), is basically the same as that obtainable from optical probes such as Absorption, Reflectance and Attenuated Total Reflectance Spectroscopies [10] whose output is some other function of ϵ. Indeed, it has been current practice for the interpretation of, say, specular R-EELS spectra to start from a dielectric function constructed by Kramers–Kronig analysis of optical absorption or reflectivity data [5]. And vice versa, EELS spectra can serve as data for a KK determination of the optical constants of solids [16]. Although the same basic dipole selection rule operates, it does so more strictly in optics than in EELS in the sense that the momentum transfers of optics are ten to a hundred times smaller than those characteristic of the dipole scattering lobe of EELS. Excitation modes of all wavelengths down to the mesoscopic range ($\lambda \approx 100$ Å) are those to which specular EELS is most sensitive whereas in optics the only excited wavelengths are macroscopic, being of order ω/c. This entails that specular EELS has a somewhat better sensitivity in cases involving microscopic targets such as adsorbed submonolayers [15]. As another consequence, the probing depth (of order λ) of EELS is much shallower than in optics, which may be a further EELS advantage for the study of thin films and modern heterostructures [5, 10]. The big advantages of optical methods are, of course, their vastly superior resolution and the possbility of applying them outside the stringent UHV conditions.

If performed on disordered targets, off-specular EELS follows no selection rule at all while, as explained before, if there is a sharp diffraction pattern, it excludes only the dipole active modes seen in the specular direction. Thus EELS spectroscopy really gives access to all possible excitations, individual or collective, dipole active or not, including those accessible to the other classic optical method, Raman spectroscopy (RS). The EELS mechanism for exciting Raman active modes is just the same as in RS: for example, in the i.r., it involves the modulation imparted by the nuclear vibrational motions to the target electronic polarizabilities and hence to the electron multipolar scattering potential. Here again, the excited modes need not have optically-long wavelengths.

With its ability to explore all momentum transfers, angular resolved, off-specular EELS is more akin to Neutron Inelastic Spectroscopy (NIS) than to optical methods for the study of vibrational modes. The principal advantage of NIS is, of course, the simplicity and exactness of the scattering potential by the nuclei as compared to the complexities of Coulomb scattering in EELS. But this turns into a big advantage for EELS for measuring the electronic excitations, totally inaccessible to NIS.

Some analogies and differences worth bearing in mind also exist between EELS and the ultraviolet (UPS), X-ray (XPS) photoelectron spectroscopies (PS), or inverse UPS (IUPS). Detailed results obtained on the pure and doped fullerites by photoelectron spectroscopies are discussed by Weaver elsewhere in this issue. The energy distribution curves (EDC) of PS ideally reflect the density of initial or final states occupied by the photoelectron. Since the target serves either as the source (UPS, XPS) or as a sink (IUPS) of phototelectrons, these spectroscopies provide information on either the N-1 or the $N + 1$ electrons ionized systems. By contrast EELS, like photoabsorption or Raman, is a two states spectroscopy involving transitions between initial and final states of the neutral N electrons target. To some extent, the EDCs include electron energy loss processes due to inelastic scatterings suffered by the photoemitted electrons but presumably these can be separated from density of state features.

The closest PS analogue to an EELS spectrum is probably realized in the satellite spectrum of core lines in XPS. Here, a fast electron is photoejected from a deep level of an atom and appears at the spectrometer as a sharp line at an energy equal to the photo energy minus the binding energy of the core level. Satellite features develop at lower energies because of onsite relaxation and scattering effects and offsite energy loss processes. If the latter are dominant, the spectrum should resemble an EELS spectrum for electrons of primary energy equal to the core line energy. This will be confirmed below when we discuss the close resemblance between EELS and XPS C1s spectra of C_{60}.

2. EELS OF FULLERITE FILMS

(A) *Infrared*

The first HREELS spectra were taken on thin C_{60} fullerite films evaporated on a clean Si(100) surface [17]. The films, 50–60 Å thick, were deposited at room temperature and did not show any sharp diffraction pattern in LEED. In such conditions, there is no selection rule at all and all 46 distinct modes of C_{60}

can make some contribution to the loss spectrum in the range between 30 meV and 200 meV, which are the expected lower and higher bounds of the intramolecular vibrational frequencies.

Figure 5 shows the i.r. energy losses in the range from 25 meV to 0.4 eV with the sample at room temperature. The resolution was 10.5 meV. From 30 meV to 200 meV there are 11 distinct features, the sharpness of which is resolution limited. The most intense HREELS band is centred at 66 meV (532 cm^{-1}) and falls near the two strongest lines seen in absorption spectroscopy (AS) [18] at 527 and 577 cm^{-1} which are too close to be resolved here. One expects that all four i.r., dipole active modes of C_{60} should contribute to the loss spectrum but the two high energy AS modes at 1183 (148 meV) and 1428 cm^{-1} (178 meV) have weaker oscillator strengths (approximately one sixth of the combined strength of the 66 meV band) and are not seen as distinct peaks in Fig. 5. Due to the film disorder, most of the rest of the spectrum is in fact dominated by dipole inactive modes. Even the 66 meV band is partly due, on its low energy wing, to a strong Raman line, the radial-breathing A_g mode at 496 cm^{-1} (62 meV). A more detailed discussion is given in [17]. For lack of a better energy resolution and of a sufficiently accurate theory of the vibrational frequencies, we do not know yet with any certainty which modes or group of modes of C_{60} are responsible for the HREELS bands. Perhaps a qualitative indication as to the general nature of the carbon frame motions giving rise to the bands can be gained by examining the phonon dispersion relations on a basal surface of crystalline graphite observed by specular and off-specular HREELS [4]. Here, there is a large density of vibrational states around 58 meV,

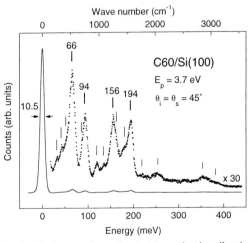

Fig. 5. The loss spectrum due to intramolecular vibrations of C_{60} for a disordered fullerite film of ≈ 60 Å thickness on Si(100) [17]. The spectrum is taken in the 45° specular direction and with primary electrons of 3.7 eV.

90–100 meV, 155–175 meV and 180–200 meV. These four frequency domains closely correspond to our four C_{60} strongest bands (Fig. 5), as if the vibrational modes of a graphite layer which respond to EELS survived when, as it were, the layer is warped into the C_{60} spherical geometry. A comparison between neutron inelastic spectra (NIS) from C_{60}, glassy carbon and graphite powder measured recently [19] corroborates this simple view. The conclusion is that, also for its i.r. elementary excitations, C_{60} deserves the unusual name of 'graphiterole' [20].

The weak bands in the i.r. HREELS spectrum beyond 200 meV in Fig. 5 are due to combinations of the four strong bands below 200 meV. The clear band at 360 meV also receives a contribution from the C–H bond stretching motion of a contamination of the fullerite film by hydrocarbons which come from the solvent used in the powder preparation or/and from the residuals of the vacuum chamber.

If the molecules formed a well-ordered film on the surface, the off-specular spectrum should exclude dipole scattering and one could then at least sort out the active modes and their EELS strengths. While this has not been possible so far on the Si(100) surface, it turns out that C_{60} fullerene does order epitaxially on other substrates such as GaAs, GaSe, MoS_2, Mica, etc. . . HREELS measurements on well-ordered C_{60} films on GaSe [21] succeeded in detecting all four i.r. active modes. An i.r. spectrum of ordered C_{60}/GaSe film is shown in Fig. 6 [21] with resolution of 7.5 meV. A detailed discussion can be found in the paper by Gensterblum et al. in this issue [21]. First note the large peak around 30 meV, which is assigned to one of the strongly active GaSe optical phonons, more

precisely to the corresponding GaSe/C_{60} interface phonon. Although this interface is buried under several hundred Å of a uniform C_{60} layer, the interface phonon comes out sharp and clear in the EELS spectrum of specularly reflected electrons, notwithstanding the fact that the latter may not have penetrated more than 5 Å into the C_{60} film. This, incidentally, illustrates the capability of specular HREELS to investigate, as it were from a distance, the vibrational properties of buried interfaces. From eqn (3), the probing depth of the spectroscopy turns out to be at least $1/k_{max} \approx 1/(k\theta) \approx 100$ Å, where k is the incident wavevector and where θ is the spectrometer numerical aperture. The intensity of this 30 meV GaSe peak can in fact be used to calibrate quantitatively the thickness of the epitaxial C_{60} layer [21].

The structural order of the film causes several large differences with the C_{60}/Si spectrum of Fig. 5. The strong band at 66 meV is more prominent and almost pure dipole scattering. It can be deconvoluted into the two dipole mode components seen in AS. The other two dipole active modes at 148 and 178 meV are now dominant over the impact scattering modes. The strong band at 94 meV is almost pure impact scattering, apart from a weak dipole contribution arising from the combination of the 30 and 66 meV peaks. Figure 6 shows the accuracy with which the dipole scattering part of the spectrum can be simulated by the dielectric theory formulae of the previous section (the details of the simulation are given in [21]). This allows its substraction from the specular spectrum, leaving the contribution of the pure impact excited modes such as the bands marked by a bar in Fig. 6, for which theoretical simulations have not been attempted yet.

The energy resolution of HREELS is still relatively modest for the infrared so that it cannot compete with AS and Raman which are able to detect all 14 active modes in C_{60} [18]. Nevertheless, the HREELS technique remains of primary interest for characterizing the nonoptical modes and for its relevance to the general question of the electron–phonon interaction, e.g. in the superconducting fullerides.

(B) Near i.r.-vis-u.v.

With its high resolution, the R-EELS instrument is at its best for measuring the fullerene excitations in the visible (vis) and ultraviolet (u.v.) energy ranges. Figure 7 shows a spectrum taken in the specular direction for a film on Si(100) and with primary electrons of 10 eV (resolution: 40 meV). There is a broad region extending from 0.8 to 1.4 eV which shows only a flat background of extremely low noise intensity. This is the gap in the excitation spectrum of

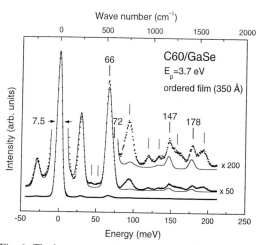

Fig. 6. The loss spectrum due to intramolecular vibrations of C_{60} for an ordered fullerite film of ≈ 350 Å thickness on GaSe(0001) [21]. The full line is the theoretical simulation of the dipole scattering part of the spectrum. The spectrum is taken in the 45° specular direction and with primary electrons of 3.7 eV.

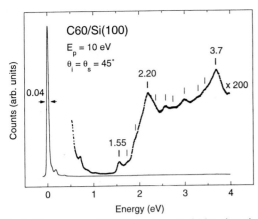

Fig. 7. The specular R-EELS spectrum in the i.r.-vis region of the C_{60} fullerite band gap taken with primary electrons of 10 eV. The i.r. at 0.72 eV is due to the double excitation of the C–H stretching mode (from hydrocarbon contamination) at 0.36 eV seen in the unmagnified spectrum. The 1.55 eV peak corresponds to the energy gap and is common to both C_{60} and C_{70} fullerites.

the C_{60} fullerite. It is followed by a peak at 1.55 eV which appears to be the first electronic excitation in C_{60}, at least the excitation of lowest energy observed so far with any spectroscopy. The HOMO–LUMO level separation of the C_{60} molecule is theoretically predicted at about 1.8 eV [22, 23] and the C_{60} fullerite band gap is calculated at 1.5 eV [24]. Our observation is not inconsistent with these values. The spectrum shown in Fig. 7 was for a fullerite film made from a mixed powder containing at least 85% C_{60} and at most 15% C_{70}. Subsequent unpublished measurements [25] on purified C_{60} and C_{70} fullerite films on Si(100) showed that the 1.55 eV feature is present and of comparable strength in both films and thus appears to be the true electronic excitation gap of these fullerites. In fact the shape of the onset spectrum from 1.4 to 2 eV is quite similar for C_{60} and C_{70} [25] and comprises the several peaks and shoulders shown in Fig. 7. The fundamental gap is dipole forbidden in the molecule and in the solid and hence cannot be detected in AS. The first optically allowed transition in the molecule from its ^1A_g ground state to the lowest $^1T_{1u}$ excited state occurs at 3.04 eV [26]. This transition which has a weak oscillator strength is probably responsible for the 3.0 eV weak peak in Fig. 7. Note that the excitation spectra near the gap, particularly the 1.55 eV peak, have been revealed only by R-EELS because of the excellent resolution and high sensitivity attainable by the spectroscopy in the low energy, reflection set-up and because of the absence of selection rules for disordered films.

The HREELS vis-u.v. spectrum comprises several further peaks or shoulders up to 7 eV. This is shown in Fig. 8 where several spectra are displayed for increasing primary electron energies. The three out-

standing peaks at 2.2, 3.7 and 4.8 eV are observed to remain nearly independent of the primary energy. Our previous theoretical considerations indicate that these peaks must be generated by impact scattering from π–π^* single electron excitations. Their energy positions are in quite close agreement with those observed in T-EELS [27–30] and in XPS [31, 32], notwithstanding the difference in relative intensities of the peaks as observed in the different spectroscopies. The rather large 2.2 eV peak in Fig. 8 is seen only as a weak feature in T-EELS but is also prominent as a C1s satellite feature at 1.9 eV in XPS [32], where it is interpreted as a monopole-allowed π–π^* shake-up. It is not seen in AS. The two large peaks at 3.7 and 4.8 eV are, however, seen in AS with comparable lineshapes and large strengths. They must therefore correspond to further dipole-allowed, ^1A_g–$^1T_{1u}$, single electron π–π^* transitions [26, 33, 34].

The constancy of the intensity of the first three large peaks in Fig. 8 for varying primary electron energy is in sharp contrast to the behaviour of the next group of intense peaks around 6 eV. This group is in fact made from at least two peaks: a first band of nearly constant intensity around 5.5–5.8 eV and a second, initially smaller one at 6.3 eV (197 nm) whose intensity grows continuously when the primary electron energy increases above 10 eV. For primary energies above 30 eV the 6.3 eV peak intensity stabilizes and dominates the 5.5–5.8 band which persists as its shoulder. At that point, the 6.3 eV shouldered band strikingly resembles in both position and lineshape the corresponding feature observed in

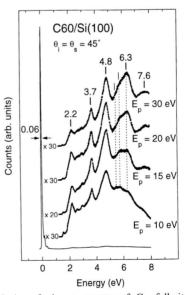

Fig. 8. Series of vis-u.v. spectra of C_{60} fullerite in the specular direction for increasing primary electron energy E_p. Note the near constancy of the first three main peaks at 2.2, 3.7 and 4.8 eV, as opposed to the variability of the 6.3 eV peak which is identified as the π-plasmon quantum [17].

high-resolution T-EELS [29, 30]. This lineshape is also very similar to the AS shouldered band at 214 nm (5.8 eV) [26, 33, 34], notwithstanding the 0.5 eV blue shift. We argue that, together with the 4.8 peak, the 6.3 eV band form the familiar 'camel back' structure so characteristic in the originally observed AS spectrum of C_{60} [33]. This doublet will remain the ubiquitous structure in all further HREELS spectra taken at higher primary electron energies (see below, Fig. 9). Note that the doublet is also seen in EELS of the gas-phase C_{60} at approximately the same position as the AS doublet [35].

Our interpretation [17] of the doublet structure seen in HREELS is the following: the 4.8 eV peak is the one-electron transition discussed above; the constant intensity shoulder at 5.5–5.8 eV corresponds to further ^1A_g–$^1T_{1u}$ one-electron transition of weak oscillator strength, observed in AS at 5.46 eV [26]; the 6.3 eV peak is the so-called π-plasmon, the quantum of collective motion of the π-electrons subsystem in the molecule. There are two justifications for this latter assignment. First, on account of the analogy between planar graphite and the fullerenes in their π-valence bonding structures, one expects that the prominent π-plasmon observed by EELS and AS in graphitic carbon [36] at about 6–7 eV should also appear as a common feature in the u.v. spectrum of all fullerenes. Second, the collective nature of the elementary excitation is positively identified by its variable intensity on increasing the primary electron energy. Indeed the giant dipole fluctuations of the π-electron density over the whole molecular sphere scatter the EELS electrons according to the mechan-

ism of long-range dipole scattering as opposed to the short range, impact scattering mechanism of one-electron transitions. This results in the formation of a forward scattering lobe (Fig. 4a) whose opening angle $\hbar\omega_\pi/2E_p$ shrinks into the spectrometer aperture when the primary electron energy E_p increases. The observed blue shift of the HREELS π-plasmon at 6.3 eV in C_{60} fullerite with respect to the corresponding 5.8 eV feature of the AS camel back [33] and the 6 eV feature in the gas-phase EELS [35] can be understood from the different response functions of the spectroscopies. Basically, AS of bulk solid C_{60} measures $\mathrm{Im}\epsilon(\omega)$, T-EELS measures $\mathrm{Im}\alpha(\omega)$ in the gas phase and R-EELS measures $\mathrm{Im}[1 + \epsilon(\omega)]^{-1}$ at the solid surface where $\alpha(\omega)$ is the C_{60} dipole polarizability to which the solid dielectric function $\epsilon(\omega)$ is related via the Clausius–Mosotti relation. By simulating $\alpha(\omega)$ with just two Lorentzians for the π and σ dipole plasmons on the molecule, it can easily be shown that these response functions along with the solid state effect explain the observed trend in the relative peak positions of the three spectroscopies. One should further note that our R-EELS peak at 6.3 eV incorporates contributions from higher-energy π-plasmons of angular momenta higher than $l = 1$. This should also contribute to the blue shift and broadening of the plasmon feature as compared to the narrower peak seen in AS at 5.8 eV.

(C) v.u.v.

A spectrum taken with a primary electron energy of 100 eV is shown in Fig. 9 [17]. One recognizes the camel back feature which, on this expanded energy loss scale, looks more like rabbit ears. Following this feature, the first peak at 7.6 eV probably reflects the threshold ionization energy of the C_{60} molecule as observed both in the gas phase and for thin fullerite films by Lichtenberger et al. [37] with UPS. The next peaks occur at positions also closely correlated to UPS lines as observed by using a He source [37] or synchrontron radiation [38]. In general, the HREELS bands in this v.u.v. region reflect π–σ^* and σ–σ^* one-electron transitions of the N-electron system and need not exactly coincide with the UPS lines which measure the occupied density of state of the N-1 system.

Finally Fig. 9 reveals the existence of a family of high energy elementary excitations in the C_{60} fullerites film giving rise to the broad energy loss continuum around 28 eV. A similar feature appears as a strong shake-up satellite to the XPS C1s core line [31, 32] and has also been seen in T-EELS [27–30]. Given its similarity with a corresponding EELS peak in the v.u.v. spectra of graphitic carbon [36], it is understood as due to the excitation of multipolar 'σ-plasmons',

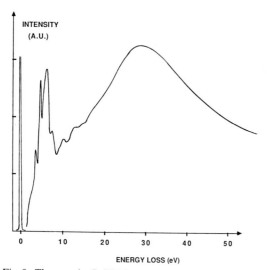

Fig. 9. The specular R-EELS spectrum of C_{60} fullerite in the u.v.-v.u.v. region for a primary electron energy of 100 eV, resolution of 0.06 eV. The large hump around 28 eV is assigned to the excitation of multipole σ-plasmons of the entire C_{60} valence shell [17].

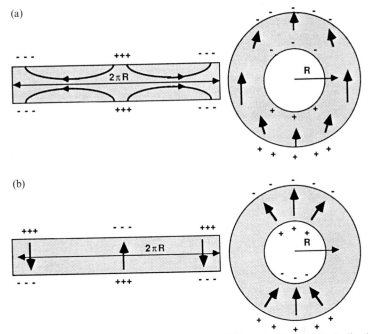

Fig. 10. The polarization eigenvectors of the tangential (a) and radial (b) dipole ($l = 1$) plasmons of a spherical dielectric shell of radius R simulating the C$_{60}$ fullerene valence shell [42]. The arrows indicate the directions of charge displacements. These modes correspond, respectively, to the symmetric and antisymmetric modes of the wavelength $2\pi R$ in a thin film (left illustrations). Both modes are optically active on the sphere but are inactive in the film.

the quanta of collective motions of the entire 240 valence electrons of the molecular sphere. It cannot show up in the HeI-UPS but it could possibly appear with a HeII-source as causing a broad autoionization continuum. The collective nature of the electron motions was experimentally confirmed [17] by the sensitivity of their excitation strength to the primary electron energy, a behavior completely analogous to the π-plasmon case seen in Fig. 8. That the collective excitations are characteristic of the molecular shell itself rather than a solid state effect was further confirmed by the detection of a giant optical absorption around 21 eV in the gas phase of both C$_{60}$ and C$_{70}$, via photoionization spectroscopy using synchrotron radiation [38]. The absorption is caused by the $l = 1$, dipole member of the multipolar plasmon family. RPA microscopic calculations [39, 40] have confirmed the existence of such giant resonances. The peak positions have been rationalized in terms of an empirical three-dimensional dielectric shell model [41] similar to the two dimensional 'graphiterole' model of Barton *et al.* [20]. The 3-D model predicts that, due to the mere fact that C$_{60}$ and other fullerenes are hollow, each multipolar plasmon must be split into one mostly tangential and one mostly radial modes. The polarization eigenvectors of the $l = 1$ mdoes in C$_{60}$ are shown in Fig. 10. The big peak of the photoionization spectrum is caused by the tangential mode (Fig. 10a) and it is predicted [41] that the radial

motion (Fig. 10b) could also be seen in spite of its smaller oscillator strength.

The excitability of the whole set of multipolar plasmons by fast electrons is clearly demonstrated by XPS and EELS data but the quantitative simulation of an EELS spectrum such as in Fig. 9 will require further theoretical work.

Acknowledgments—I am particularly indebted to J. Weaver for many stimulating discussions on the content of this paper. This work was funded by the Belgian National Program of Interuniversity Research Projects initiated by the State Prime Minister Office (Science Policy Programming) and by the Belgian National Science Foundation. The work was done in collaboration with G. Gensterblum, J. J. Pireaux, P. A. Thiry, R. Caudano, J. P. Vigneron, Ph. Lambin and W. Krätschmer. Correspondence and discussions with J. E. Fischer, A. Rosen and K. Michel are gratefully acknowledged.

REFERENCES

1. Ibach H. and Mills D. L., *Electron Energy Loss Spectroscopy*. Academic Press, New York (1982).
2. Rather H., *Excitations of Plasmons and Interband Transitions by Electrons*. Springer Verlag, Berlin (1980).
3. Lucas A. A. and Sunjic M., *Phys. Rev. Lett.* **26**, 229 (1971); *Phys. Rev.* **3**, 719 (1971); *Prog. in Surface Sci.* **2**, 75 (1972); *Surface Sci.* **32**, 439 (1972).
4. Oshima C. *et al.*, *Solid State Commun.* **65**, 1601 (1988).
5. Lambin Ph. *et al.*, *Phys. Rev.* **B32**, 8203 (1985); Lambin Ph. *et al.*, *Phys. Rev. Lett.* **56**, 1842 (1986); Lucas A. A. *et al.*, *Int. J. Quant. Chem.: Quantum Chemistry Symposium* **19**, 687 (1986).
6. Ritchie R. H., *Phys. Rev.* **106**, 874 (1957).

7. See e.g. Boersch H. *et al.*, *Phys. Rev. Lett.* **17**, 379 (1966) and ref. therein. See also: Lucas A. A. and Kartheuser E., *Phys. Rev.* **B1**, 3588 (1970).

8. Lucas A. A. *et al.*, *Solid State Commun.* **8**, 1075 (1970); *Phys. Rev.* **B2**, 2488 (1970); Lucas A. A. and Sunjic M., *Solid State Commun.* **8**, 1889 (1970).

9. Sunjic M. and Lucas A. A., *Phys. Rev.* **B3**, 719 (1971).

10. Lucas A. A. *et al.*, *Physica Scripta* **T13**, 150 (1986); Lambin Ph. *et al.*, *Physica Scripta* **35**, 343 (1987).

11. Lucas A. A. and Vigneron J.-P., *Solid State Commun.* **49**, 327 (1984).

12. Evans E. and Mills D. L., *Phys. Rev.* **B5**, 4126 (1972).

13. Delanaye F. *et al.*, *Surface Sci.* **71**, 629 (1978); Willis R. F. *et al.*, in *Chemical Physics of Solid Surfaces and Heterogeneous Catalysis* (Edited by King D. A. and Woodruff D. P.), Vol. 2. Elsevier, Amsterdam (1983).

14. Davenport J. *et al.*, *Phys. Rev.* **B17**, 3115 (1978); Li C. H. *et al.*, *Phys. Rev.* **B21**, 3057 (1980); Aers G. C. *et al.*, *J. Phys.* **C14**, 3995 (1981).

15. Gadzuk J. W., in *Vibrational Spectroscopy of Molecules on Surfaces* (Edited by Yates Y. T., Jr. and Madey T. E.). Plenum Press, New York (1987).

16. Hageman H. G. *et al.*, DESY Report No SR-74/7 (1974).

17. Gensterblum G. *et al.*, *Phys. Rev. Lett.* **37**, 2171 (1991); Lucas A. A. *et al.*, *Phys. Rev.*, in press (1992).

18. Bethune D. S. *et al.*, *Chem. Phys. Lett.* **179**, 183 (1991).

19. Kamitakahara W. A. *et al.*, to be published.

20. Barton G. *et al.*, *J. Chem. Phys.* **95**, 1512 (1991).

21. Gensterblum G., Li-Ming Yu, Pireaux J. J., Thiry P. A., Caudano R., Lambin Ph., Lucas A. A., Krätschmer W. and Fischer J. E., *J. Phys. Chem. Solids* **53**, xx (1992).

22. Mintmire J. W. *et al.*, *Phys. Rev.* **B43** *Rapid Comm.* 14,281 (1991); Dunlap B. I. *et al.*, *J. Phys. Chem.* **95**, 8737 (1991).

23. Wästberg B. and Rosén A., *Physica Scripta* **44**, 276 (1991); Rosén A. and Wästberg B., *Proc. ECOSS 12 Conf.*, 9–12 September (1991), *Surface Sci.*, in press.

24. Saito S. and Oshiyama A., *Phys. Rev. Lett.* **66**, 2637 (1991).

25. Gensterblum G. *et al.*, unpublished results obtained from thin films prepared from purified C_{60} and C_{70} powders kindly provided by J. Fischer and R. Smalley, respectively.

26. For a detailed analysis of the electronic excitations of C_{60} in *n*-hexane solution seen in optical absorption spectroscopy, see Leach S. *et al.*, *Chem. Phys.*, in press.

27. Sohmen E. *et al.*, *Europhys. Lett.* **17**, 51 (1992).

28. Hansen P. L. *et al.*, *Chem. Phys. Lett.* **181**, 367 (1991).

29. Saito Y. *et al.*, *Japan J. appl. Phys.* **30**, L1068 (1991).

30. Kuzuo R. *et al.*, *Japan J. appl. Phys.*, in press.

31. Weaver J. H. *et al.*, *Phys. Rev. Lett.* **66**, 1741 (1991).

32. Benning P. J. *et al.*, *Phys. Rev.* **B44**, 1962 (1991).

33. Krätschmer W. *et al.*, *Chem. Phys. Lett.* **170**, 167 (1990).

34. Hare J. P. *et al.*, *Chem. Phys. Lett.* **177**, 394 (1991).

35. Keller J. W. *et al.*, *Chem. Phys. Lett.* (submitted).

36. Diebold U. *et al.*, *Surf. Sci.* **197**, 430 (1988); Zeppenfeld, Z. *Phys.* **211**, 391 (1968); Klucker R., Interner Bericht, DESY F41-72/1 (1972).

37. Lichtenberger D. L., Nebesny K. W., Ray Ch. D., Huffman D. R. and Lamb L. D., *Chem. Phys. Lett.* **176**, 203 (1991); Lichtenberger D. L., Jatcko M. E., Nebesny K. W., Ray Ch. D., Huffman D. R. and Lamb L. D., to be published.

38. Vertel I. V. *et al.*, *Phys. Rev. Lett.* **68**, 784 (1992).

39. Bertsch G. F. *et al.*, *Phys. Rev. Lett.* **67**, 2690 (1991).

40. Bulgac A. *et al.*, *Phys. Rev. Lett.* submitted.

41. Lambin Ph. *et al.*, *Phys. Rev.*, in press.

HIGH RESOLUTION ELECTRON ENERGY LOSS SPECTROSCOPY OF EPITAXIAL FILMS OF C$_{60}$ GROWN ON GaSe

G. Gensterblum,[†] Li-Ming Yu,[†] J. J. Pireaux,[†] P. A. Thiry,[†] R. Caudano,[†] Ph. Lambin,[†] A. A. Lucas,[†] W. Krätschmer[‡] and John E. Fischer[§]

[†]Institute for Studies in Interface Sciences, Facultés Universitaires Notre-Dame de la Paix, 61 rue de Bruxelles, B-5000 Namur, Belgium

[‡]Max Planck Institut für Kernphysik, P.O. Box 103980, D-6900 Heidelberg 1, Germany

[§]Laboratory for Research on the Structure of Matter, University of Pennsylvania, Philadelphia, PA 19104-6272, U.S.A.

(*Received* 19 *February* 1992)

Abstract—The growth of thin C$_{60}$ films on GaSe(0001) has been studied by infrared High Resolution Electron Energy Loss Spectroscopy (HREELS) in specular reflection geometry. In contrast to previously studied disordered films on Si(100), it was found that deposition on a GaSe substrate heated to 120°C leads to well-ordered, epitaxial C$_{60}$ films. The structural order, manifested by a sharp LEED pattern, was confirmed by the clear detection in the specular beam of all four T_{1u} dipole-active vibrations of the C$_{60}$ molecule. Several other infrared inactive modes, which were prominent in the C$_{60}$/Si HREELS spectrum, are still seen in C$_{60}$/GaSe but with reduced intensities. Application of the dielectric theory of reflection EELS allows a quantitative determination of the contribution of the dipole-active modes to the total spectrum. Although the *relative* oscillator strengths of the four infrared modes are in good agreement with optical absorption measurements, their *absolute values*, as determined by theoretical simulation, are found to be three times larger than the optical results.

Keywords: Expitaxial growth, C$_{60}$, infrared active modes, dielectric theory, dipole scattering, impact scattering, Electron Energy Loss Spectroscopy.

INTRODUCTION

Since the discovery of the fullerenes [1, 2] the structural, vibrational, electronic and other physical properties [3–9] of the C$_{60}$ prototype molecule have been investigated, but so far mostly on polycrystalline powders or thin films. For a thorough understanding of the material, it is desirable to experiment with single crystals. Recently, Meng *et al.* [10] have reported the successful preparation of macroscopic single crystals, thus opening the way for new investigations.

Since they are insulating, thick, undoped fullerene crystals are not suitable for low-energy electron spectroscopies such as High Resolution Electron Energy Loss Spectroscopy (HREELS) or Inverse Photoelectron Spectroscopy (IPES), which require either conducting surfaces or sufficiently thin films on conducting substrates to alleviate charging effects [11]. Well-ordered or epitaxial thin films may also be needed for possible future applications.

The epitaxial growth of C$_{60}$ has already been observed by Sakurai *et al.* [12] on a MoS$_2$ substrate using RHEED, and by Schmicker *et al.* [13] and Krakow *et al.* [14] on mica using helium atom scattering and electron diffraction, respectively. In the present paper, we report on the successful epitaxial growth of C$_{60}$ fullerene films on a heated GaSe(0001) substrate and on measurements of the HREELS spectrum in the infrared range. Our choice of GaSe was dictated by its being easily cleaved in UHV and by the existence of a sharp and well-characterized surface optical phonon to which we could refer as a quantitative monitor of film thickness. Furthermore, since GaSe is a layered material with weak van der Waals interlayer bonding, the cleaved surface should present a low surface energy, thus favouring the ordering of the deposited C$_{60}$ molecules. We compare the C$_{60}$ spectrum with earlier measurements on disordered C$_{60}$ films grown on Si(100) [8]. In keeping with our expectations as discussed in [8], we find that the structural order of the C$_{60}$/GaSe system indeed results in a strong enhancement of the dipole-scattering part of the spectrum over impact scattering, which was the dominant mechanism in the C$_{60}$/Si system. Using the dielectric theory of EELS [15] in conjunction with recent optical infrared absorption measurements [16], we are able to evaluate quantitatively the dipole contribution. The dipole mode oscillator strengths measured by HREELS are found to be larger than the IRAS values by a factor of three.

Separate HREELS experiments starting from disordered films deposited on a cold GaSe substrate further reveal that ordering can also be achieved by thermal annealing. This is in sharp contrast to the C_{60}/Si case, where substrate heating or film annealing had no ordering effect as seen by HREELS or by the absence of a LEED pattern. This behavior is interpreted in terms of a weaker binding of the first layer of adsorbed fullerene molecules on the GaSe surface, allowing sufficient mobility for rearrangement at moderate temperatures.

EXPERIMENTAL

The experiments were performed in a two-chamber UHV system. The base pressures in the analysis and preparation chambers were 1×10^{-8} and 1×10^{-7} Pa, respectively. The analysis chamber was equipped with a high resolution electron spectrometer (ISA-Riber) and a LEED diffractometer. The HREELS spectrometer consisted of a fixed monochromator and a rotatable analyser. The incident and analysis angle could be set at any desired value between 0 and 90°. The half angle aperture of the analyser was about 1.5°.

Carbon soot containing C_{60} and C_{70} was produced by the contact-arc method of Krätschmer *et al.* [2]. The fullerenes were extracted from the soot by solution in toluene. C_{60} was subsequently purified by using a liquid chromatography process on alumina diluted with mixtures of hexane. The purity of the resulting C_{60} was better than 99%. The purified C_{60} powder was outgassed in UHV for 1 h at 250°C. The C_{60} was then sublimed at a pressure of 5×10^{-7} Pa at 300°C on to a clean substrate. As substrate we used either a Si(100) surface annealed at 1000 K prior to deposition or a GaSe(0001) surface cleaved in UHV. The substrate surfaces were checked by LEED and HREELS prior to deposition. For the sublimation of C_{60} we used a Knudsen cell with a boron nitride crucible. The typical evaporation rate was about 5 Å min^{-1}. The thickness of the deposited film was estimated by a quartz crystal oscillator and by the intensity of the Fuchs–Kliewer phonon in the case of the GaSe substrate (see below). Evaporations were performed for different substrate temperatures ranging from room temperature up to 150°C.

RESULTS AND DISCUSSION

For comparison with the new results on the C_{60}/GaSe system, we first show in Fig. 1 the energy loss spectrum of a 60 Å thin film of C_{60} on a Si(100) surface with the substrate held at room temperature during sublimation [8]. As the Si surface is infrared

inactive in the frequency range considered, all the loss peaks are due to C_{60}. There are four prominent peaks at 66 (532), 94 (758), 156 (1258) and 194 meV (1565 cm^{-1}) and six weaker bands or shoulders [8]. In comparison to the optical absorption spectrum of C_{60} [5] we see that the 66 meV (532 cm^{-1}) band corresponds to two of the four T_{1u} dipole active modes at 526 and 576 cm^{-1}, which are too close to be resolved by HREELS without further processing of the data (see below). These modes are presumably excited by a dipole-scattering mechanism. The other three strong bands in Fig. 1 correlate better with dipole-inactive Raman [5] or neutron [7] lines and represent vibrational modes excited by impact scattering. The question arises as to why the two higher frequency infrared active T_{1u} modes at 1183 and 1430 cm^{-1} are not clearly seen. We believe that this is due to the diffuse nature of the electron scattering which occurs on disordered targets. It should also be kept in mind that the dipole activity of these two modes is much smaller than that of the two infrared phonons around 66 meV (532 cm^{-1}).

There are several pieces of evidence which indicate that our C_{60}/Si films are disordered or clumped in islands. First, they showed no LEED pattern. In particular, the measured off-specular *elastic* intensity was not very different from the specular one. Second, the off-specular *inelastic* spectrum was also very similar to the specular one. If the film was ordered, the off-specular spectrum would exclude the dipole-scattering intensities, and at least the 66 meV (532 cm^{-1}) peak should strongly decrease, which it did not.

We tried to induce ordering of our C_{60}/Si films by heating the substrate during deposition or by annealing after deposition. Heating had no significant effect on the spectra, apart from a slight decrease

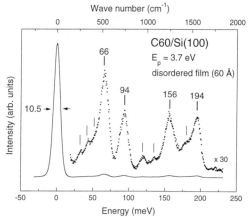

Fig. 1. HREELS spectrum of a disordered C_{60} film/Si(100). The substrate was held at room temperature during deposition. Peak positions are indicated by bars.

of the intensity of the impact-excited modes. The strong intensity of these impact modes relative to the dipole-active pair of phonons at 66 meV (532 cm^{-1}) suggests an efficient process for their electron excitation which is not yet understood. Such coupling of molecular vibrations to low-energy electrons may be relevant in understanding the nature of the electron–phonon coupling believed to be at the origin of superconductivity in the doped fullerides.

To investigate further the inelastic scattering processes, we need first to evaluate quantitatively the dipole scattering strength by working on well-ordered films. We decided to try growing films on a freshly-cleaved GaSe substrate. GaSe is a semiconducting layer compound with D_{3h} hexagonal symmetry and a lattice parameter of 3.75 Å. As in the case of graphite, MoS$_2$ or mica, the interlayer interaction is of the van der Waals type. The GaSe(0001) surface is a rather passive one compared to the Si(100) surface with its unsaturated dangling bonds. Since C_{60} is itself a van der Waals solid, it is likely that the weak interaction between the C_{60} molecules and the GaSe substrate makes the growth of a well-ordered film possible on this substrate as well as on MoS$_2$ and mica. To favour epitaxial growth, the GaSe substrate was heated to about 120°C during sublimation and deposition. The heating of the surface ensures a high mobility for the C_{60} molecules during nucleation and growth of a well-ordered first monolayer. In the case of the highly reactive Si(100) surface, the interaction of the C_{60} molecules with the surface Si atoms appears to be so strong as to inhibit surface mobility and preclude the formation of a well-ordered first monolayer even when the temperature is raised to 150°C. At higher temperatures, the uppermost layers of C_{60} start to sublime and the formation of a film becomes impossible.

For low coverages (1–3 monolayers) we obtained LEED patterns of the GaSe substrate and of the C_{60} overlayer which indicate that the C_{60} surface corresponds to the (111) face of the fcc fullerene in registry with the hexagonal structure of the substrate surface. The LEED study will be published elsewhere together with a STM study.

Figure 2 shows the energy loss spectrum obtained on a 350 Å thick ordered film. Unlike Si, GaSe is partly ionic and has infrared active lattice vibrations. These give rise to a surface optical phonon (Fuchs–Kliewer mode) around 29.5 meV (238 cm^{-1}) which is the first outstanding loss feature in Fig. 2 (note the corresponding gain peak at −29.5 meV). The width of this peak, like that of all the others in the spectrum, is primarily determined by the 7.5 meV (60 cm^{-1}) instrumental resolution for the elastic peak. It is interesting to note that this interface phonon

is observed remotely at a distance of 350 Å with a non-penetrating low energy electron beam. This interface capability of HREELS is a consequence of the long range Coulomb interaction which is the dominant excitation mechanism for dipole-active vibrations. The rest of the spectrum differs from the C_{60}/Si case of Fig. 1 in several aspects which all point to an ordered structure for the present film. The strong peak at 66 meV (532 cm^{-1}) due to the two lowest T_{1u} dipole modes of C_{60} is more prominent than in Fig. 1. The three impact-excited bands at 94 (758), 156 (1258) and 194 meV (1565 cm^{-1}) are drastically reduced in intensity. The other two T_{1u} dipole modes are now clearly detected at 147 (1186) and 178 meV (1436 cm^{-1}) and even become the dominant ones in the upper part of the spectrum.

In addition to these observations, the angular distribution of the dipole lobes becomes very sharp. Indeed we have measured the intensity of the elastic peak and of the two peaks at 66 (532) and 94 meV (758 cm^{-1}) vs the angle of analysis away from the specular direction (45°). These angular profiles are shown in Fig. 3. The FWHM of the 66 meV (532 cm^{-1}) feature (5.3°) is only 0.6° broader than the FWHM of the elastic line (4.7°). This agrees with the expected theoretical width of the dipole lobe $\hbar\omega/2E_p \approx 10^{-2} \approx 0.6°$ and definitely confirms our assignment of the band to the two lowest i.r. lines at 526 and 576 cm^{-1}. On the contrary, the 94 meV (758 cm^{-1}) peak has a much broader, nearly pure impact-like angular distribution: over and above an isotropic distribution amounting to 70% of the specular intensity, there is only a weak and sharp dipole-like contribution. The latter anisotropic intensity can be ascribed, at least partly, to a

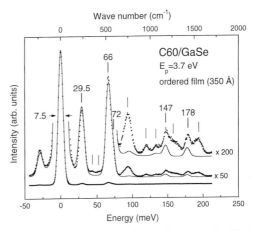

Fig. 2. HREELS spectrum of an ordered C_{60} film/GaSe. The substrate was heated at 120°C during deposition. Peak positions are indicated by bars. The solid curve is the result of the computation of the dipole part of the spectrum (see text).

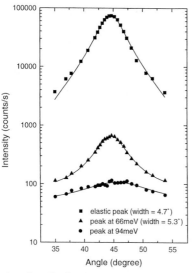

Fig. 3. Angular distribution curve around the specular direction (45°) for the HREELS elastic peak intensity and for the 66 (532) and 94 meV (758 cm^{-1}) peak intensities measured on an ordered film of C$_{60}$/GaSe.

combination band of the Fuchs–Kliewer interface mode at 29.5 meV (238 cm^{-1}) and of the C$_{60}$ 66 meV (532 cm^{-1}) mode, both exhibiting a sharp dipole lobe distribution. This will be confirmed later on by our theoretical analysis. Finally, on comparing the present C$_{60}$/GaSe spectrum with that of C$_{60}$/Si system, a decrease is observed in the background intensity and an increase in the elastic line absolute intensity by a factor of 10, indicative of a strong reduction of the diffuse scattering. From this body of observations, we can conclude unambiguously that our C$_{60}$ films on GaSe are highly ordered.

In order to establish the importance of the substrate temperature in determining the film structure, a post-deposition annealing experiment was conducted. It consisted of recording the HREELS spectrum before and after annealing to about 120°C for a film which had been sublimed on a GaSe substrate held at room temperature during deposition. The result is shown in Fig. 4. Before annealing, the spectrum looks very much like that of C$_{60}$/Si. However, the stronger prominence of the 66 meV band relative to the impact-scattering peaks appears to indicate that the film deposited at room temperature is already better ordered on GaSe than it was on Si. Upon annealing, one observes a decrease in intensity of the impact-excited modes and the emergence of the weaker dipole-active ones at 147 (1186) and 178 meV (1436 cm^{-1}). In addition, the absolute intensity of the elastic peak increases and the inelastic background decreases by about a factor of 4, while the spectral shape of the loss features tends toward that of the film deposited on a hot substrate (Fig. 2).

These effects are a clear consequence of the ordering of the C$_{60}$ film upon annealing.

The availability of a HREELS spectrum for a well-ordered system warrants the application of the dielectric theory of specular reflection EELS [15, 17] and enables the dipole scattering component of the inelastic spectrum to be determined quantitatively. The theory requires constructing models for the long wavelength, bulk dielectric functions for the substrate and for the film materials. For the GaSe anisotropic substrate, the correct function to use [15] is the geometrical average of the principle components of the uniaxial dielectric tensor:

$$\epsilon_{\text{eff}}^{s} = \sqrt{\epsilon_{\perp}^{s}(\omega)\epsilon_{\parallel}^{s}(\omega)}. \tag{1}$$

The components are represented by single-oscillator Lorentzian formulae

$$\epsilon_{\perp}^{s}(\omega) = \epsilon_{\perp}^{s}(\infty) + \frac{\rho_{\perp}\omega_{\text{TO},\perp}^{2}}{\omega_{\text{TO},\perp}^{2} - \omega^{2} - i\omega\gamma_{\text{TO},\perp}}$$

$$\epsilon_{\parallel}^{s}(\omega) = \epsilon_{\parallel}^{s}(\infty) + \frac{\rho_{\parallel}\omega_{\text{TO},\parallel}^{2}}{\omega_{\text{TO},\parallel}^{2} - \omega^{2} - i\omega\gamma_{\text{TO},\parallel}} \tag{2}$$

with parameters taken from the i.r. optical constants of GaSe [18, 19]. The oscillator strength $\rho_{\perp}$ had to be adjusted to reproduce the experimental HREELS spectrum of GaSe(0001) prior to film deposition [20]. The parameters are listed in Table 1. The small values of the damping parameters γ in eqns (2) and (3) below which were used in the calculations are not listed since the widths of the loss features are dominated by the 7.5 meV instrumental resolution.

Inserting into the Clausius–Mosotti relation a four-oscillator expression for the dipole polarizability of

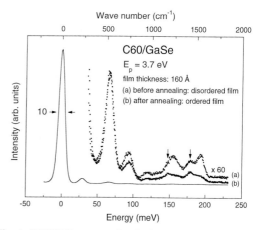

Fig. 4. HREELS spectra of a C$_{60}$/GaSe immediately after deposition on the substrate at room temperature (a) and after annealing at 120°C (b). The two weak dipole-active T_{1u} vibration modes at 147 (1186) and 178 meV (1436 cm^{-1}) which emerge after annealing are indicated by arrows.

Table 1. Infrared constants of GaSe

$\epsilon_\perp(\infty)$	$\epsilon_\parallel(\infty)$	$\rho_\perp$	$\rho_\parallel$	$\omega_{TO,\perp}$ (cm^{-1})	$\omega_{TO,\parallel}$ (cm^{-1})
7.44	5.76	2.17	0.55	215	236

Table 2. Infrared constants of C$_{60}$ as determined by IRAS and HREELS

Frequency ω_i (cm^{-1})	Absorption coefficient μ_i (cm^{-1})	Width Γ_i (cm^{-1})	IRAS oscillator strength ρ_i	HREELS oscillator strength	Relative oscillator strength
526	6600	2.7	0.021	0.07	100
576	3200	1.9	0.006	0.07	29
1183	1600	3.1	0.0012	0.004	6
1430	1500	4.0	0.0010	0.003	5

C$_{60}$ leads to the following expression for the dielectric function of C$_{60}$ fullerene in the i.r. range:

$$\epsilon^f(\omega) = \epsilon^f(\infty) + \sum_{i=1}^{4} \frac{\rho_i \omega_i^2}{\omega_i^2 - \omega^2 - i\omega\gamma_i}, \qquad (3)$$

where the ω_i are the resonance frequencies of the four T_{1u} dipole active modes of C$_{60}$. Approximate values for the oscillator strengths ρ_i were evaluated from recent measurements of the absorption coefficients μ_i of fullerene powders and the corresponding line widths Γ_i of the four, T_{1u} vibrations [16]. Assuming a Lorentzian shape for the absorption lines, we obtained the corresponding line intensities $A_i = \pi\Gamma_i\mu_i/2$, from which we could derive the expected oscillator strengths in eqn (3) via the formula

$$A_i = \frac{\pi\rho_i\omega_i^2}{2c\sqrt{\epsilon^f(\infty)}}, \qquad (4)$$

where the electronic dielectric contant $\epsilon^f(\infty)$ was taken as 4.3, slightly less than the static value $\epsilon^f(\infty)$ of 4.4 observed by Hebard [21]. The results of this fit are given in Table 2. We obtained a second set of

oscillator strengths from the HREEL spectrum in Fig. 2 by using a standard computing technique [17]. The thickness d of the C$_{60}$ film and the strengths ρ_i were found from the best fit to the spectrum. In the fit, the *relative* oscillator strengths were kept constant and equal to those taken from the infrared data. The quality of the simulated spectrum, including the interface peak, is evident from the comparison with the observed one in Fig. 2. The optimized thickness of 350 Å was found to be two times greater than the quartz balance estimate. The absolute oscillator strengths were 3.3 times larger than the optical results, as shown in Table 2. This discrepancy may arise from the difficulty in extracting absolute absorption strengths from the optical data published so far [16].

The validity of our assumption that the relative oscillator strengths are the same in IRAS and in HREELS was confirmed by a detailed analysis of the 66 meV (532 cm^{-1}) line shape. The spectrum from 20 to 110 meV was fitted by means of combinations of Lorentzian and Gaussian functions as shown in Fig. 5. For this fit, we have neglected the contribution

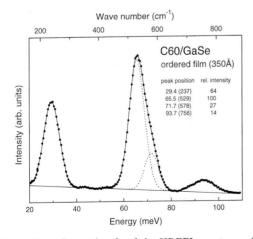

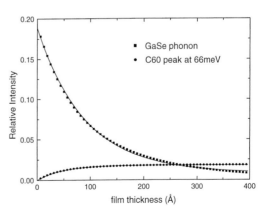

Fig. 5. Gauss–Lorentzian fit of the HREEL spectrum of an ordered C$_{60}$ film/GaSe in the range from 20 to 110 meV. The contribution of the double excitation of the GaSe Fuchs–Kliewer phonon has been neglected.

Fig. 6. Computation of the intensities (normalized to the elastic peak intensity) of the GaSe Fuchs–Kliewer phonon and of the dipole C$_{60}$ peak at 66 meV (532 cm^{-1}) vs C$_{60}$ film thickness for a primary energy of 3.7 eV.

from the double loss of the Fuchs–Kliewer phonon. One clearly sees that the 66 meV band arises from two contributions at 65.5 (529) and 71.7 meV (578 cm^{-1}) with relative strengths of 100 and 27, which agree with the optical values in Table 2.

Finally, we note that the intensity of the GaSe Fuchs–Kliewer phonon and of the 66 meV (532 cm^{-1}) C_{60} band may be used to obtain a measurement of the film thickness. From the dielectric theory, the predicted intensities of these two peaks as fractions of the elastic peak intensity are shown in Fig. 6 as a function of the film thickness. One sees that the calculated points follow a single exponential law almost exactly. The decay of the Fuchs–Kliewer peak allows an accurate determination of film thicknesses up to about 350 Å. The 66 meV (532 cm^{-1}) band closely follows a complementary exponential increase.

CONCLUSION

In this paper, we have demonstrated the possibility of growing epitaxial C_{60} films on GaSe. In comparison with previous results obtained on C_{60}/Si(100), it appears that substrates with low surface energy are ideal candidates for C_{60} epitaxy. The effect of substrate heating during the deposition has also been discussed. The epitaxial system C_{60}/GaSe has been investigated by HREELS. Quantitative interpretation of the spectra has been given in the framework of dielectric theory, and i.r. optical parameters have been determined for the four T_{1u} dipole active modes. These results have been compared to IRAS data. The epitaxy of C_{60} reported in this paper broadens the field of investigation and applications of fullerene thin films.

Acknowledgements—This work was funded by the Belgian national program of Inter-university Research Projects initiated by the State Prime Minister Office (Science Policy Programming). Ph. Lambin acknowledges the National Fund for Scientific Research of Belgium.

REFERENCES

1. Kroto H. W., Heath J. R., O'Brien S. C., Curl R. F. and Smalley R. E., *Nature (London)* **318**, 162 (1985).
2. Krätschmer W., Lamb L. D., Fostiropoulos K. and Huffman D. R., *Nature (London)* **347**, 354 (1990).
3. Heiney P. A., Fischer J. E., McGhie A. R., Romanow W. J., Denenstein A. M., McCauley J. P., Jr and Smith A. B., *Phys. Rev. Lett.* **66**, 2911 (1991).
4. Johnson R., Meijer G. and Bethune D., *J. Am. Chem. Soc.* **112**, 8985 (1990).
5. Bethune D. S., Meijer G., Tang W. C., Rosen H. J., Golden W. G., Seki H., Brown C. A. and Devries M. S., *Chem. Phys. Lett.* **179**, 181 (1991).
6. Weaver J. H., Martins J. L., Komeda T., Chen Y., Ohno T. R., Kroll G. H., Troullier N., Haufler R. E. and Smalley R. E., *Phys. Rev. Lett.* **66**, 1741 (1991).
7. Cappelletti R. L., Copley J. R. D., Kamitakahara W. A., Li Fand, Lannin J. S. and Ramage D., *Phys. Rev. Lett.* **66**, 3261 (1991).
8. Gensterblum G., Pireaux J. J., Thiry P. A., Caudano R., Vigneron J. P., Lambin Ph., Lucas A. A. and Krätschmer W., *Phys. Rev. Lett.* **67**, 2171 (1991).
9. Jost M. B., Troullier N., Poirier D. M., Martins J. L., Weaver J. H., Chibante L. P. F. and Smalley R. E., *Phys. Rev. B* **44**, 1966 (1991).
10. Meng R. L., Ramirez D., Jiang X., Chow P. C., Diaz C., Matsuishi K., Moss S. C., Hor P. H. and Chu C. W., *Appl. Phys. Lett.* **59**, 3402 (1991).
11. The use of the flood gun technique to provide transient conductivity, which was successful for MgO insulating surface, may also work for thick fullerenes, see for example Liehr M., Thiry P. A., Pireaux J. J. and Caudano R., *Phys. Rev. B* **33**, 5682 (1986).
12. Sakurai M., Tada H., Saiki K. and Koma A., *Japanese J. appl. Phys.* **30**, L565R (1991).
13. Schmicker D., Schmidt S., Skofronick J. G., Toennies J. P. and Vollmer R., *Phys. Rev. B* **44**, 10995 (1991).
14. Krakow W., Rivera N. M., Roy R. A., Ruoff R. S. and Cuomo J. J., IBM research report, RC 17345 (#76621) 10/30/91.
15. Lucas A. A. and Sunjic M., *Prog. Surf. Sci.* **2**, 75 (1972); Ibach H. and Mills D. L., *Electron Energy Loss Spectroscopy and Surface Vibrations*. Academic Press, New York (1982).
16. Chase B., Herron N. and Holler E., *J. Phys. Chem.*, submitted.
17. Lambin Ph., Vigneron J. P. and Lucas A. A., Computer Physics Communications **60**, 351 (1990).
18. Finkman E. and Rizzo A., *Solid State Commun.* **15**, 1841 (1974).
19. Landoldt-Bornstein III, Vol. 178, p. 24.
20. Degiovanni A., Yu Li-Ming, Dewandel J. L., Thiry P. A., Caudano R., to be submitted.
21. Hebard A. F., *Appl. Phys. Lett.* (in press).

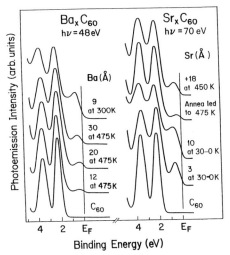

Fig. 13. Valence band spectra for Ba_xC_{60} and Sr_xC_{60} grown under a variety of conditions. Variations in the intensity of the band at E_F reflect surface stoichiometries. This band reflects hybridization of the LUMO and the s-state of the alkaline earth. Metallic character is evident in all cases. The top left spectrum shows the result of deposition of 9 Å of Ba on solid C_{60} at 300 K where a Ba-rich surface layer is produced.

inverse photoemission gives a center-to-center separation of ~ 3.7 eV for the HOMO and LUMO bands. This is ~ 2 eV greater than the lowest energy for HOMO–LUMO on-site excitations and the difference is a measure of the electron–electron interaction parameter U in solid C_{60}. From Fig. 12, the separation between feature A and the lowest conduction band feature for Na_2C_{60} is ~ 1.6 eV. This is close to U in magnitude and is therefore consistent with a band split by correlation effects.

4. ALKALINE EARTH FULLERIDES

Most studies of fullerides have focused on the alkali metals, Chen et al. [7] have demonstrated recently that alkaline earth deposition onto C_{60} also produces fullerides. Such mixing was anticipated because the alkaline earths have relatively small cohesive energies, small ionization energies, and they are found with formal 2 + valency in compounds with electronegative elements. The most significant change in the valence band spectra for Mg_xC_{60} is the appearance of emission from a band entirely below E_F, a band attributed to states derived from the LUMO of C_{60} and the s states of Mg. Analysis shows a feature centered 1.1–1.2 eV above HOMO that has a FWHM similar to that of HOMO, i.e. 1.0–1.1 eV. The peak-to-peak energy separation between it and HOMO is close to that for semiconducting Li_2C_{60} and Na_2C_{60} where a new band of states developed below E_F [58]. It is considerably smaller than the corresponding separation for K_3C_{60} (1.9 eV) or K_6C_{60}

(1.6 eV) [1–4] and Chen et al. [7] concluded that Mg doping produced only non-metallic fullerides.

Figure 13 shows valence band spectra for Ba_xC_{60}. The deposition of 12 Å at 475 K produced a band centered 1.9–2.0 eV above HOMO that was cut by the Fermi level, and there was broadening of the π-derived bands. Spectra acquired with higher resolution showed an asymmetric lineshape related to the convolution of the Ba-induced structure with the Fermi function. The band E_F grew with Ba deposition at 475 K, and the HOMO peaks continued to broaden. The LUMO-derived peak was centered 1.8 eV above HOMO by 30 Å deposition. The Ba $4d$ core level intensity for this coverage suggests a stoichiometry of $x = 1 \pm 0.3$. Further deposition at 475 K resulted in much stronger Fermi level emission but the intensity was time dependent, demonstrating that the Ba atoms were kinetically trapped near the surface. The highest Ba concentration before metal cluster formation (as judged by the Ba $4d$ core level lineshape) was $x = 6 \pm 2$ and was reached after 9 Å Ba deposition at 300 K. The valence band spectrum for 9 Å Ba coverage is shown in the top left of Fig. 13.

The right panel of Fig. 13 shows photoemission spectra for Sr deposited on a solid C_{60} film. Sr deposition of 3 Å at 300 K produced a feature centered ~ 1.6 eV above HOMO. The Sr-induced peak was stronger for 10 Å but it was also broader. The Sr $3d$ core level intensities suggest a composition of $x = 5 \pm 1.5$ at this stage. Annealing to 475 K allowed Sr diffusion into the bulk. The separation between HOMO and the Sr-induced feature also increased to 1.8 eV and the Sr $3d$ core level intensity gave a composition of $x = 0.6 \pm 0.2$.

For the alkaline earth fullerides, Sr_xC_{60} and Ba_xC_{60} were always metallic and Mg_xC_{60} was never metallic. Hence, a simple picture in which electrons are transferred to occupy the six-fold degenerate LUMO levels is not applicable. For alkaline earth fullerides, pure ionic bond formation requires the transfer of two s-electrons. Analysis based on the ionization energies of alkaline earths, the electron affinity of C_{60}, and the electrostatic (Madelung) energy of the crystal demonstrates that two-electron transfer is not favorable. Therefore, bonding must involve hybridization between C_{60} states and alkaline earth states. Note that the two electrons in each alkaline earth atom will be equally involved in the bonding.

Mixing of the LUMO-derived and the s-derived states of the alkaline earths is favored because they are close in energy and because the LUMO is unoccupied. In the fullerides, it is not known whether hybridization will result in separation of bonding and antibonding bands, but we speculate that they will overlap because the respective wave functions are not very localized. LUMO + 1 is likely to mix with alkaline earth s states, and the bands probably overlap with the LUMO-derived bands. Therefore, it is even less probable that a band gap can exist. For the same reason, a fortuitous band splitting driven by crystal symmetry is not likely to occur over the entire Brillouin zone.

Superconductivity is closely related to distribution of electronic states and those at E_F are derived almost entirely from the LUMO levels. Several studies have focused on electron–electron coupling via C_{60} phonons although they have not specifically considered the dopant [60]. Every C_{60} fulleride that exhibits metallic character also shows superconductivity at remarkably high temperature. Sr_xC_{60} and Ba_xC_{60}, the only metallic fullerides found outside the alkali family, provide an example of a hybrid conduction band. Mg_xC_{60} stands in sharp contrast because E_F is pinned at the valence band maximum. The existence of a gap is confirmed by inverse photoemission and current–voltage measurements using scanning tunneling microscopy [61]. We propose that the electronic structure of Mg_xC_{60} also reflects strong electron–electron repulsion, noting that the center-to-center separation between the occupied and unoccupied LUMO–s hybrid bands is $\sim 2\,eV$ [61]. This is close to the U value for the insulating fullerides.

The atomic and ionic sizes of Mg are much smaller than those for Sr and Ba, showing a similar trend when one compares Li and Na to K, Rb and Cs. The small Li and Na ions might not occupy the large octahedral interstitial sites of the fcc C_{60} lattice. The structure of Mg_xC_{60} could also be different than Sr_xC_{60} and Ba_xC_{60}. Such structural changes may not be large enough to split the conduction band in the independent-electron band description, but the system could be pushed to the insulating side of the metal-insulator transition by electron correlation.

Acknowledgements—This work was supported by the Office of Naval Research and the National Science Foundation. The synchroton radiation photoemission studies were conducted at Aladdin, a user facility operated by the University of Wisconsin and funded by the National Science Foundation. We are pleased to acknowledge stimulating discussions and collaborations with J. L. Martins, N. Troullier, Y. Z. Li, J. C. Patrin, T. R. Ohno, G. H. Kroll, D. M. Poirier, P. J. Benning, Y. Chen, M. B. Jost, F. Stepniak, R. E. Smalley and L. P. F. Chibante. Correspondence and discussions with A. F. Hebard, A. Rosén, S. Saito, J. Bernholc, A. A. Lucas, G. Sawatzky, J. R. Chelikowsky and J. E. Fischer contributed to this work.

REFERENCES

1. Benning P. J., Martins J. L., Weaver J. H., Chibante L. P. F. and Smalley R. E., *Science* **252**, 1417 (1991).
2. Wertheim G. K., Rowe J. E., Buchanan D. N. E., Chaban E. E., Hebard A. F., Kortan A. R., Makhija A. V. and Haddon R. C. *Science* **252**, 1419 (1991).
3. Poirier D. M., Ohno T. R., Kroll G. H., Chen Y., Benning P. J., Weaver J. H., Chibante L. P. F. and Smalley R. E., *Science* **252**, 6464 (1991).
4. Chen C. T. *et al.*, *Nature* **352**, 603 (1991) showed very high resolution valence band spectra with a distinct Fermi level cutoff for intermediate stoichiometries.
5. Troullier N. and Martins J. L., *Phys. Rev. B* **46**, 1754 (1992); Martins J. L. and Troullier N., *Phys. Rev. B* **46**, 1766 (1992); Troullier N, PhD thesis, University of Minnesota (1991).
6. Saito S. and Oshiyama A., *Phys. Rev. B* **44**, 11536 (1991); Xu Y.-N., Huang M.-Z., and Ching W. Y., *Phys. Rev. B* **44**, 13171 (1991); Erwin S. C. and Pederson M. R., *Phys. Rev. Lett.* **67**, 1610 (1991); Satpathy S., Antropov V. P., Andersen O. K., Jepsen O., Gunnarson O. and Liechtenstein A. I., *Phys. Rev. B* **46**, 1773 (1992).
7. Chen Y., Stepniak F., Weaver J. H., Chibante L. P. F. and Smalley R. E., *Phys. Rev. B* **45**, 8845 (1992).
8. Ohno T. R., Kroll G. H., Weaver J. H., Chibante L. P. F. and Smalley R. E., *Nature* **355**, 401 (1992).
9. Ohno, T. R., Chen Y., Harvey S. E., Kroll G. H., Weaver J. H., Haufler R. E. and Smalley R. E., *Phys. Rev. B* **44**, 13747 (1991).
10. STM results are discussed by Li Y. Z., Chander M. Patrin J. C., Weaver J. H., Chibante L. P. F. and Smalley R. E., *Science* **253**, 429 (1991) and Li Y. Z., Chander M., Patrin J. C., Weaver J. H., Chibante L. P. F. and Smalley R. E., *Science* **252**, 547 (1991). STM studies of monolayer films have been reported by Wilson R. J. *et al.*, *Nature* **348**, 621 (1990) and Wragg J. L., Chamberlain J. E., White H. W., Krätschmer W. and Huffman D. R., *Nature* **348**, 623 (1990).
11. See, for example, Cardona M. and Ley L., *Photoemission in Solids I and II, Topics in Applied Physics* **26**. Springer-Verlag, Berlin (1979); Feuerbacher B., Fitton B. and Willis R. F., *Photoemission and the Electronic Properties of Surfaces*. Wiley, New York (1978); Plummer E. W. and Eberhardt W. E., *Adv. Chem. Phys.*, Vol. 49 (Edited by I. Prigogine and S. A. Rice), p. 533. Wiley, New York (1982); Margaritondo G. and Weaver J. H. in *Methods in Experimental Physics: Surfaces* (Edited by M. G. Lagally and R. L. Park), p. 127. Academic Press, New York (1985).
12. See, for example, Himpsel F. J., *Surf. Sci. Reports* **12**, 4 (1990); Smith N. V., *Rep. Prog. Phys.* **51**, 1227 (1988); Dose V., *Surf. Sci. Reports* **5**, 337 (1985).
13. Weaver J. H., Martins J. L., Komeda T., Chen Y., Ohno T. R., Kroll G. H., Troullier N., Haufler R. E. and Smalley R. E., *Phys. Rev. Lett.* **66**, 1741 (1991).
14. Benning P. J., Poirier D. M., Troullier N., Martins J. L., Weaver J. H., Haufler R. E., Chibante L. P. F. and Smalley R. E., *Phys. Rev. B* **44**, 1962 (1991).
15. Jost M. B., Troullier N., Poirier D. M., Martins J. L., Weaver J. H., Chibante L. P. F. and Smalley R. E., *Phys. Rev. B* **44**, 1966 (1991).
16. Jost M. B., Benning P. J., Poirier D. M., Weaver J. H., Chibante L. P. F. and Smalley R. E., *Chem. Phys. Lett.* **181**, 423 (1991).
17. Benning P. J., Poirier D. M., Ohno T. R., Chen Y., Jost M. B., Stepniak F., Kroll G. H., Weaver J. H., Fure J. and Smalley R. E., *Phys. Rev. B* **45**, 6899 (1992).
18. Martins J. L., Troullier N. and Weaver J. H., *Chem. Phys. Lett.* **180**, 457 (1991).
19. Mott N. F., *Metal–Insulator Transitions*. Taylor & Francis, London (1990).
20. Krätschmer W., Lamb L. D., Fostiropoulos K. and Huffman D. R., *Nature* **347**, 354 (1990); Skumanich A., *Chem. Phys. Lett.* **182**, 486 (1991); Cheville R. A. and Halas N., *Phys. Rev. B* **45**, 4548 (1992); Hebard A. F.,

20. Haddon R. C., Fleming R. M. and Kortan A. R., *Appl. Phys. Lett.* **59**, 2109 (1991); Kelly M. K., Etchegoin P., Fuchs D., Krätschmer W. and Fostiropoulos K., *Phys. Rev. B* (in press).

21. Ajie H. *et al.*, *J. Phys. Chem.* **94**, 8630 (1990); Hare J. P., Kroto H. W. and Taylor R., *Chem. Phys. Lett.* **177**, 394 (1991); Ren S. L., Wang Y., Rao A. M., McRae E., Holden J. M., Hager T., Wang K.-A., Lee W.-T., Ni H. F., Selegue J. and Eklund P. C., *Appl. Phys. Lett.* **59**, 2678 (1991); Leach S., Vervloet M., Despres A., Bréherot E., Hare J. P., Dennis T. J., Kroto H. W., Taylor R. and Walter D. R. M., *Chem. Phys.* (in press).

22. Braga M., Larsson S., Rosén A. and Volosov A., *Astron. Astrophys.* **245**, 232 (1991); Wästberg B. and Rosén A., *Physica Scripta* **44**, 276 (1991); Rosén A., *Surf. Sci.* (in press).

23. Zhang Q.-M., Yi J.-Y. and Bernholc J., *Phys Rev. Lett.* **66**, 2633 (1991).

24. Saito S. and Oshiyama A., *Phys. Rev. Lett.* **66**, 2637 (1991).

25. Mintmire J. W., Dunlap B. I., Brenner D. W., Mowrey R. C. and White C. T., *Phys. Rev. B* **43**, 14,281 (1991).

26. Ching W. Y., Huang M.-Z., Xu Y.-N., Harter W. G. and Chan F. T., *Phys. Rev. Lett.* **67**, 2045 (1991).

27. Lichtenberger D. L., Nebesny K. W., Ray C. D., Huffman D. R. and Lamb L. D., *Chem. Phys. Lett.* **176**, 203 (1991); Lichtenberger D. L., Jatcko M. E., Nebesny K. W., Ray C. D., Huffman D. R. and Lamb L. D., *Mat. Res. Soc. Symp. Proc.* **206**, 673 (1991).

28. Gao Y., Grioni M., Smandek B., Weaver J. H. and Tyrie T., *J. Phys. E* **21**, 489 (1988).

29. Neumann D. A. *et al.*, *Phys. Rev. Lett.* **67**, 3808 (1991).

30. Takahashi T., *et al.*, *Phys. Rev. Lett.* **68**, 1232 (1992) reported photoemission and inverse photoemission results for C_{60} and K_3C_{60}. The position of the two leading occupied state features are in agreement with Fig. 9 but the empty state features are shifted ~0.5 eV toward E_F. Upon doping to K_3C_{60}, their empty states broaden but do not shift significantly, in contrast to Fig. 9, and they showed little loss in intensity. From our results, we would conclude that the empty state spectra were more representative of K_1C_{60} than K_3C_{60}.

31. Lof R. W., van Veenendaal M. A., Koopmans B., Jonkman H. T. and Sawatzky G. A., *Phys. Rev. Lett.* **68**, 3924 (1992), discussed excitons and correlation energies for C_{60}, using photoemission and inverse photoemission results like those of Figs 2 and 3. They also measured the C 1s Auger lineshape, obtaining clear evidence for correlation and the magnitude of *U*.

32. Fink J., Sohmen E., Masaki A., Merkel M., Romberg H. and Knupfer M., *Proceedings of the International Winterschool on Electronic Properties of High Temperature Superconductors*, reported high resolution low temperature photoemission studies of K_xC_{60}.

33. Satpathy S., *Chem. Phys. Lett.* **130**, 545 (1986); Hale P. D., *J. Am. Chem. Soc.* **108**, 6087 (1986); Haddon R. C., Brus L. E. and Raghavachari K., *Chem. Phys. Lett.* **125**, 459 (1986).

34. McFeely F. R., Kowalczyk S. P., Ley L., Cavell R. G., Pollack R. A. and Shirley D. A., *Phys. Rev. B 9*, 5268 (1974).

35. Heath J. R., O'Brien S. C., Zhang Q., Liu Y., Curl R. F., Kroto H. W. and Smalley R. E., *J. Am. Chem. Soc.* **107**, 7779 (1985).

36. Vaughan G. B. M., Heiney P. A., Fischer J. E., Luzzi D. E., Rickets-Foot D. A., McGhie A. R., Hui Y.-W., Smith A. L., Cox D. E., Romanow W. J., Allen B. H., Coustel N., McCauley J. P. and Smith A. B., *Science* **254**, 1350 (1991).

37. Heiney P. A., Fischer J. E., McGhie A. R., Romanow W. J., Denenstein A. M., McCauley J. P. Jr., Smith A. B. III and Cox D. E., *Phys. Rev. Lett.* **66**, 2911 (1991).

38. Haufler R. E., Wang Lai-Sheng, Chibante L. P. F., Jin Changming, Conceicao J. J., Cahi Yan and Smalley R. E., *Chem. Phys. Lett.* **179**, 449 (1991).

39. Saito S. and Oshiyama A., *Phys. Rev. B* **44**, 11532 (1991).

40. Scuseria G. E., *Chem. Phys. Lett.* **180**, 451 (1991).

41. Sohmen E., Fink J. and Krätschmer W., *Europhys. Lett.* **17**, 51 (1992).

42. Gensterblum G., Pireaux J. J., Thiry P. A., Caudano R., Vigneron J. P., Lambin Ph., Lucas A. A. and Krätschmer W., *Phys. Rev. Lett.* **37**, 2171 (1991); and Lucas A. A., *et al.*, *Phys. Rev. B* **45**, 13694 (1992).

43. Saito Y., Shinohara H. and Ohshita A., *Jpn. J. Appl. Phys.* **30**, 1145 (1991); Kuzuo R., Terauchi M., Tanaka M., Saito Y. and Shinohara H., *Jpn. J. Appl. Phys.* (in press).

44. Sohmen E., Fink J., Baughman R. H. and Krätschmer W., *Z. Physik. B* **86**, 87 (1992).

45. Nordfors D., Nilsson A., Mårtensson N., Svensson S., Gelius U. and Lunell S., *J. Chem. Phys.* **88**, 2630 (1988).

46. Terminello L. J., Shuh D. K., Himpsel F. J., Lapiano-Smith D. A., Stöhr J., Bethune D. S. and Meijer G., *Chem. Phys. Lett.* **182**, 491 (1991).

47. Haddon R. C., *et al.*, *Nature* **350**, 320 (1991).

48. Hebard A. F., *et al.*, *Nature* **350**, 600 (1991).

49. Ohno T. R., Kroll G. H., Weaver J. H., Chibante L. P. F. and R. E. Smalley, *Phys. Rev. B* (in press).

50. Chen Y., Poirier D. M., Jost M. B., Gu C., Ohno T. R., Martins J. L., Weaver J. H., Chibante L. P. F. and Smalley R. E., *Phys. Rev. B* (in press).

51. Zhu Q., Zhou O., Coustel N., Vaughan G. B. M., McCauley J. P., Romanow W. J., Fischer J. E. and Smith A. B., *Science* **254**, 547 (1991).

52. Fleming R. M., *et al.*, *Nature* **352**, 701 (1991).

53. Ohno T. R., Chen Y., Harvey S. E., Kroll G. H., Benning P. J., Weaver J. H., Chibante L. P. F. and Smalley R. E., *Phys. Rev. B* (submitted).

54. Zhou O., *et al.*, *Nature* **351**, 462 (1991).

55. Kochanski G. P., Hebard A. F., Haddon R. C. and Fiory A. T., *Science* **255**, 184 (1992).

56. Abeles B., *Granular Metal Films*, Applied Solid State Science **6**, 1. Academic Press, New York (1976) and references therein; Morris J. E., Mello A. and Adkins C. J., *Mat. Res. Soc. Symp. Proc.* **195**, 181 (1990).

57. Kirkpatrick S., *Rev. Mod. Phys.* **45**, 574 (1973).

58. Gu C., Stepniak F., Poirier D. M., Jost M. B., Benning P. J., Chen Y., Ohno T. R., Martins J. L., Weaver J. H., Fure J. and Smalley R. E., *Phys. Rev. B* **45**, 6348 (1992).

59. Rotter L. D., *et al.*, *Nature* (submitted).

60. See, for example, Martins J. L., Troullier N. and Schabel M., Theory of Superconductivity in K_xC_{60}, unpublished, 9 May, 1991; Schlüter M., Lannoo M., Needels M., Baraff G. A. and Tomanek D., *Phys Rev. Lett.* **68**, 526 (1992).

61. Jost M. B., PhD dissertation, University of Minnesota, and to be published.

BUCKMINSTERFULLERENE C$_{60}$ AND ORGANIC FERROMAGNETISM

FRED WUDL† and JOE D. THOMPSON‡

†Institute for Polymers and Organic Solids and Departments of Chemistry and Physics, University of California, Santa Barbara, CA 93106, U.S.A. and ‡Los Alamos National Laboratory, Los Alamos, NM 87545, U.S.A.

(*Received* 20 *May* 1992)

Abstract—Interest in the electronic properties of C$_{60}$ led to the discovery of metals, superconductors and a solid with a transition to a magnetic state. The approaches to organic ferromagnets are reviewed briefly, with emphasis on the McConnell model. The review is followed by the description of the magnetic properties of C$_{60}$ TDAE at ambient pressure and under hydrostatic pressure, showing that the 16 K transition is extremely pressure sensitive and that the effect of pressure is reversible. An explanation of the properties of C$_{60}$ TDAE in terms of a modification of the McConnell model is described for the first time.

Keywords: Organic ferromagnetism, buckminsterfullerene C$_{60}$, organic charge transfer solid, soft ferromagnet, itinerant ferromagnet.

1. INTRODUCTION

A year ago, the truncated icosahedral molecule buckminsterfullerene C$_{60e}$ [1] caused considerable excitement in the scientific community, particularly after the discovery of superconductivity upon doping with alkali metals [2, 3]. The success of *n*-doping to the metallic state was presaged by two events: (1) structural studies which showed that the interstitial holes in the fullerene's fcc lattice could readily accommodate alkali metal ions [4]; and (2) cyclic voltammetry studies [6, 7], which showed that fullerenes C$_{60}$ and C$_{70}$ could easily be reduced (*n*-doped). The latter observation was in accord with the theoretically determined low lying lowest unoccupied molecular orbitals (LUMO) orbitals [2]. The same theoretical calculations also revealed that both the LUMO and the highest occupied molecular orbitals (HOMO) were triply and five-fold degenerate, respectively. Such a large number of degenerate levels was expected to lead to unusual condensed matter electronic and magnetic effects. In this article we show that the optimism regarding magnetism was not a fantasy.

The relatively esoteric branch of condensed matter science dedicated to organic ferromagnetism has attracted attention as a result of a Russian discovery of a 0.1% ferromagnetic component in a bulk-polymerized nitroxyl-substituted diacetylene [8], a roughly equally low amount of bulk ferromagnetism reported by IBM researchers for a 1,3,5-triamino

benzene-iodine solid [9], and the discovery of ferromagnetism in a C$_{60}$ compound [10]. Of these materials, the first two appear to suffer from the fact that the structure of the ferromagnetic component is unknown.

To place the achievement of a transition to a soft ferromagnetic state with C$_{60}$ in perspective, before we treat the magnetic properties of organic-cation-doped buckminsterfullerene C$_{60}$, the field of organic ferromagnetism will be reviewed.

ORGANIC FERROMAGNETS

Organic materials are molecular solids where the intermolecular distances are normally much longer than the covalent, interatomic distances observed in traditional metallic and ferromagnetic solids. The molecules themselves are, in general, diamagnetic, closed shell species, yet all unusual electronic and magnetic properties could only arise from open shell molecules. Fortunately there are two families of organic molecules whose members are stable, open shell entities. One consists of molecules containing neutral, localized unpaired electron-containing functional groups (free radicals) and the other consists of charge species (radical ions). In the former category are the nitroxyls (as, e.g. nitronyl nitroxide **1**), triphenyl verdazyls (e.g. **2**), sterically hindered hydrazyls (e.g. DPPH) and sterically hindered phenoxyls (e.g. galvinoxyl) and

a nitronyl nitroxide	a hindered hydrazyl	a hindered phenoxyl	a verdazyl
1	(DPPH)	(galvinoxyl)	**2**

in the latter are the radical ions which result from single electron transfer from the π molecular orbitals of an electron rich (D) molecule (eqn (1)) or transfer of an electron to the π molecular orbitals of an electron deficient (A) molecule (eqn (2)).

$$(1)$$

$$(2)$$

Throughout this paper, a heavy dot next to a particular atom (as in the upper left oxygen of **1**, above) represents an unpaired electron. Due to their flat parallelopiped geometry, the $D^{+\cdot}$ and $A^{-\cdot}$ species form crystal lattices which consist of two types of stacks: alternating and segregated. The former can be represented as $D \cdots A \cdots D \cdots A$, etc. and the latter as $A \cdots A \cdots A \cdots A$, etc. or $D \cdots D \cdots D \cdots D$, etc. Alternating stacks are the most numerous and invariably lead to organic insulators or semiconductors, whereas a majority of segregated stack solids are metallic conductors. Since in the $D \cdots D$ or $A \cdots A$ stacks the intermolecular, intrastack, distance is usually shorter than the sum of the van der Waals radii of the component molecules, the materials exhibit properties of low-dimensional solids and are subject to all the theoretically predicted solid phase transitions [11]. Single stack materials ($D \cdots D \cdots D \cdots$, etc.) with closed shell counterions can crystallize in a way that permits interstack overlap, leading to two-dimensional electronic interactions. The latter motif is usually found in the organic superconductors, although the higher T_c superconductors' structure (*kappa* phase) consists of dimers of Ds which interact in a plane [12].

Once these paramagnetic species are in the solid state, as a rule, they are Curie–Weiss paramagnets. If there is a tendency for the unpaired spins to order, they do so antiferromagnetically at low temperature. There is only one case (1) where ferromagnetic ordering at low temperature (< 1 K) was established [13]. But there are a number of examples where very

short-range ferromagnetic exchange was observed [14, 15].

In view of the above, a number of approaches to hypothetically achieve organic ferromagnetism have been proposed, some are simply fortuitous isolation of a particular polymorph of a neutral free radical [13, 14] or polymerization of open shell [8, 14] and non-Kekulé monomers [14, 16]. More systematic approaches are based on: (a) charge transfer complexes [14, 17, 18]; (b) very high spin polycarbene and polyradical solids [14, 19, 20] and (c) co-crystallization of high and low spin molecules in a ferrimagnetic array [14, 21]. The first approach was proposed by McConnell [18] and has been the main driving force behind designs by Le Page and Breslow [17] of charge-transfer ferromagnets. The second approach was developed by Itoh and Iwamura [14, 19] and the third approach was proposed by Buchachenko for organic [14, 21] and the same approach is successful for inorganic molecular ferromagnetism [14, 22, 23].

The principal design requirement is that the molecule should have a degenerate frontier orbital either as a charged species [17] or a neutral one [24]. The more speculative aspect of all but one approach [14, 21] to organic ferromagnetism is the assumption that the triplet or higher multiplicity species will interact in the solid state to produce ferromagnetically ordered electron spin domains. A number of variants of the McConnell model have been published [9, 17, 24] and are shown schematically in Fig. 1. In the figure, MC, B, T and FOM represent McConnell, Breslow, Torrance and ferromagnetic organic metal, respectively. Miller and Epstein have found that for the MC (or MC-derived) models to work, the requirement of a non-half-filled, degenerate POMO (partially occupied molecular orbital) must be

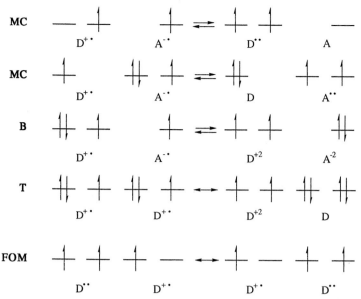

Fig. 1. Schematic representation of the electronic character of a heteromolecular stack consisting of donors (D) and acceptors (A), where the neutral donor or acceptor is a triplet diradical. MC is the McConnell model. B is the Breslow approach, where the redox potentials of D$^{+\cdot}$ and A$^{-\cdot}$ are matched. T is the Torrance modification consisting of a homomolecular stack. FOM is the McConnell modification consisting of ground state triplet D and its singly ionized doublet form D$^{+\cdot}$.

met [25]. Among the processes described in Fig. 1, FOM is the only one that is degenerate; the next closest is the T modification, hence these two modifications of the MC model have double-headed arrows connecting the two halves of the electron transfer process. In the MC model and the B modification, 'equilibrium' arrows are drawn to denote that the electron transfer process is not truly degenerate.

In the FOM modification, (a) the ensembles responsible for long-range ferromagnetic exchange are homomolecular; i.e. the two species (triplet and doublet) are the same molecular framework but in different oxidation states and (b) the triplets that interact with doublets are part of the same set of conduction electrons, and because of spin conservation rules and Hund's rule, are expected to give a net ferromagnetic moment to the ensemble [26]; however, see the discussion of Yamaguchi's MO calculation results below. This type of stacking is well precedented in organic metals of stoichiometry D$_2$X [e.g. (TMTSF)$_2$X], where X- is a closed shell anion. In short, the FOM modification of the MC model predicts the formation of a ferromagnetic organic metal, an itinerant ferromagnet [27], and it has the

advantage that there is no need to match the redox potentials of D and A. Thus the FOM modification is another formulation of a suggestion that nonlocalized electrons are necessary for ferromagnetic exchange [28]. No one has yet isolated a stable neutral triplet organic molecule but the model is equally applicable to a multiplet (e.g. triplet) dianion in combination with a doublet monoanion (Fig. 2).

Results of *ab initio* and semiempirical MO (GMO) calculations [29] of intermolecular effective exchange integrals between HTTM, TDMT and other proposed precursors to organic ferromagnets have shown that these molecules will most likely give rise to ferrimagnetism and not ferromagnetism. The GMO (generalized molecular orbital) method was developed by the authors [29] who claim that the problems associated with orbital and spin degeneracy instabilities encountered when restricted Hartree–Fock approaches are employed, are not met with this method. Since a ferrimagnet is essentially a solid with a predominant ferromagnetic moment (Fig. 3) and since the effect of metallic band formation is not clear from the calculations of intermolecular effective exchange interactions,

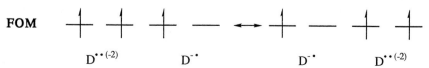

Fig. 2. The FOM modification of the McConnell model as applied to charged molecules where the dianion has a degenerate (triplet or higher) state.

Fig. 3. Schematic representation of spin magnetic moments
on a hypothetical ferrimagnetic stack.

research should continue on the synthesis and full
characterization of molecules possessing D_{3h} or
higher symmetry.

THE MAGNETIC PROPERTIES OF ORGANIC
ION-DOPED FULLERENE C_{60}

With the advent of buckminsterfullerene, the organic solid state chemist was awarded an acceptor
[5–7, 30] which, due to the high degeneracy and
partial filling of its (t_{1u}) LUMO [2], forms metals
and superconductors [31]. When the doped cluster
molecules were not allowed to interact strongly in the
solid state, as was the case with a large organic
counterion (PH$_4$P$^+$) [32], then a semiconductor was
obtained. The latter had unprecedented magnetic
resonance properties: a room temperature g value of
1.999 and a very steeply decreasing linewidth as
a function of decreasing temperature [32]. Its d.c.
magnetic susceptibility as a function of temperature
was that of an unexceptional, Curie–Weiss paramag-
net. These results were interpreted in terms of partial
occupation of the highly degenerate LUMO.

When a reaction analogous to a combination of
eqns (1) and (2) was carried out between the strong
donor TDAE and the acceptor C_{60}, the complex C_{60}
TDAE was formed:

magnetization without remanence [10]. This result,
which was interpreted as a transition to a soft,
itinerant ferromagnet [27], was recently reproduced
[34].

The data which led to the conclusion for itinerant
ferromagnetism in C_{60} TDAE are as follows [10]: (1)
although M increased sharply below $T_c \sim 16$ K, as
expected for a ferromagnet, the temperature dependence of M did not follow that of conventional mean
field theory; and (2) within experimental error, there
was no hysteresis between cooling and warming. The
structure in $M(T)$ below 10 K was field dependent;
similar measurements in applied fields of increasing
strength showed that with increasing field an
observed minimum in susceptibility at 8 K evolved
into only a change in slope for an applied field
of 1000 Oe. That the 16 K transition was one
to a ferromagnetic state also was suggested by susceptibility ($\chi \equiv M/H_a$) measurements as a function of
temperature with $H_a = 10$ kOe. When χT was plotted
as a function of temperature, above ~ 30 K, χT
increased approximately linearly with T, reflecting a
weakly varying χ at these temperatures. However,
below ~ 30 K, χT increased sharply as expected from
the onset of ferromagnetic correlations precursive
to a ferromagnetic transition at $T_c = 16$ K. The maximum in χT at 10 K and the decrease at lower
temperatures was considered a consequence of
moment saturation below T_c. The absence of high
temperature Curie–Weiss susceptibility with a positive θ greater than T_c, though unusual for local
moment ferromagnets, may be found in itinerant
ferromagnets [27]. Additional evidence supporting a

$$\text{TDAE} \longrightarrow \quad + \; e^{(-)} \quad (3)$$

$$C_{60} \quad + \quad e^{(-)} \quad \longrightarrow \quad C_{60}^{\bullet -} \quad (4)$$

$$\text{TDAE} \quad + \quad C_{60} \quad \longrightarrow \quad \text{TDAE}^{\bullet +} C_{60}^{\bullet -}$$

The new complex was found to be very atmos-
phere-sensitive, much like the alkali metal complexes,
a property which made initial elemental analyses
difficult and resulted in the incorrect reporting of the
stoichiometry as $(C_{60})_{1.16}$TDAE [10]. X-ray crystal
packing determination confirmed the correct stoichiometry as (C_{60})TDAE [33]. A.C. susceptibility
measurements indicated that the material underwent
a transition to a state in which the spins ordered
ferromagnetically and d.c. susceptibility measurements indicated that at 16 K the material showed

ferromagnetic interpretation for C_{60} TDAE came
from measurements of the field dependence of T_c,
defined by the intersection of linear extrapolations
of $M(T)$ from above and below the rapid rise in
$M(T)$. T_c increased to 17.2 K for an applied field
(H_a) of 1 kOe and then increased approximately
linearly with H_a to 24.3 K at 50 kOe. A field induced
enhancement of the Curie temperature is typical for
ferromagnets.

An alternative explanation for the data of Ref. 10
is that they arise from superparamagnetism. The field

variation of M is similar to that observed in super-paramagnetic systems in which H_a is perpendicular to the easy magnetic axis [35]. However, in those experiments the fine C_{60} TDAE powder was constrained in the sample holder only by gravity, and it was unlikely that the powder satisfied this condition since it would tend to orient itself with the easy axis parallel to Ha. Further, for superparamagnetism, M should scale as H/T below T_c [36]. Experimental $M(H)$ isotherms did not scale with H/T nor with H/T scaled by the saturated moment, which were assumed to be proportional to ΔM. Consequently, the conclusion that C_{60} TDAE was a very soft ferromagnet with a remanence of zero or nearly so was inescapable. However, the lack of remanence, which in the absence of crystal structural data, was attributed to a fully three-dimensional system devoid of domain pinning sites, must now be corrected since the structure showed the solid to be anisotropic [33]. Unlike crystalline C_{60} or superconductors K_3C_{60} and Rb_3C_{60} which have cubic crystal structures [37], C_{60} TDAE has a c-centered monoclinic unit cell with two formula units per cell and with dimensions a, 15.849 Å; b, 12.987 Å; c, 9.965 Å and $\beta = 93.31°$. Rietveld analysis of the X-ray spectrum gave a best fit to the data if the long axis of the TDAE molecule was assumed to be along the monoclinic c-axis. Surprisingly, at room temperature the intermolecular contacts are only 9.96 Å along the c-axis and 10.24 Å within the a–b plane. These separations are less than, or comparable to, those in metallic alkali-doped A_3C_{60} superconductors which have intermolecular distances of 10.07–10.15 Å [37]. The relatively short intermolecular separation in C_{60} TDAE suggested that there could be sufficient wavefunction overlap to form a partially filled band and consequently metallic-like conductivity. Further, the structural anisotropy would imply a low-dimensional band structure, which, combined with the observation of a small saturated moment, would favor an itinerant-ferromagnet interpretation.

To get a better handle on the nature of the magnetism in C_{60} TDAE, the effect of applied hydrostatic pressure on C_{60} TDAE was measured [38].

Results of the pressure measurements at 457 Hz were dramatic, at 0.1 kbar, the large a.c. response found at ambient pressure was suppressed substantially and T_c appeared to be shifted to lower temperature by 1–2 K. At an applied pressure of 1.6 kbar the a.c. response showed no evidence for a phase transition above 2 K. Additional measurements at 0.5 kbar, revealed a very weak anomaly in the response at ~ 11 K which could be associated with a magnetic transition. The pressure effects were fully reversible. Inspection of the data suggested that it was the magnetic moment itself that was most strongly suppressed by pressure and that the rapidly decreasing transition temperature did not decrease as rapidly with pressure as the moment.

There are at least four mechanisms to generate a small saturated moment: (1) superparamagnetism; (2) ferrimagnetism; (3) field-induced weak ferromagnetism and (4) itinerant ferromagnetism. The first possibility was ruled out above and in Ref. 10. Mechanisms 2 and 3 assume localized moments that are coupled dominantly by antiferromagnetic exchange. In fact, measurements of the magnetic susceptibility above 50 K indicate Curie–Weiss behavior with a negative paramagnetic temperature $\theta_p = 22.5$ K and an effective moment μ_{eff} 1.72 $\mu B/C_{60}$, a value close to that expected for a simple spin-1/2 system. Favoring a ferrimagnetic interpretation is the presence of a maximum in the magnetization near 10 K that could arise from a difference in magnitude of two different sublattice magnetizations. Generally, though, ferrimagnetism requires two inequivalent magnetic sites. Best refinement of the crystal structure of C_{60} TDAE suggests that all C_{60} sites are crystallographically equivalent, and furthermore the strong suppression of magnetism with pressure would require that the local moments be unstable with respect to modestly decreasing volume, an unlikely possibility. On the other hand, the lack of inversion symmetry in the crystal structure of C_{60} TDAE could allow field-induced weak ferromagnetism through anisotropic exchange, as discussed by Dzyaloshinski [39] and Moriya [40]. The strong pressure dependence then could be explained by a large, anisotropic compressibility [41] of C_{60} TDAE. However, the Dzyaloshinski–Moriya interaction leads to a relationship between the saturated magnetization M_s and the Lande g-factor: $M_s/g\mu_B S \sim \Delta g/g$. Taking $S = 1/2$, as inferred from the high-temperature susceptibility, and $g = 2.0008$, measured by EPR [10], gives $\Delta g/g = 4 \times 10^{-4}$, which is much smaller than $M_s/g\mu_B S \cong 0.1$. Further, there is no evidence from isothermal $M(H)$ curves for a field-induced transition, unless the field dependence of the magnetization near 10 K were interpreted this way. This leaves the possibility of itinerant ferromagnetism, as suggested originally [10] for C_{60} TDAE. In this case Coulomb correlations among electrons in a relatively narrow, partially-filled band produces a ferromagnetic ground state among the conduction electrons. Theoretical arguments [42] suggest that this might be possible in C_{60} TDAE. The strong pressure dependence of magnetism in this picture would arise from a strong volume dependence of the anisotropic band structure and of the effective electron–electron interaction. Indeed, Wohlfarth [27] has pointed out that, among weak itinerant ferromagnets based on Fe, $dT_c/dP \cong -\alpha/T_c$, where $\alpha = (2000 + 100)$ K^2/kbar. More generally, if the moment itself changes with pressure, we expect [43] $dT_c/dP = -2 K\chi_0 T_c(P = 0)$, where χ_0 is the high field susceptibility at $T = 0$ and k is a constant. Therefore, a very large negative pressure coefficient of T_c might be expected if C$_{60}$ TDAE were an itinerant ferromagnet. Further, the observation of a Curie–Weiss susceptibility above T_c is not necessarily at odds with the existence of itinerant ferromagnetism, as shown [44] recently from a Fermi-liquid analysis of itinerant-electron magnetism in low-carrier density systems.

All the observed properties of C$_{60}$ TDAE cannot be fit to any single mechanism discussed above. This may be due , in part, to the strong anisotropy of these properties. Such effects cannot be determined from studies on polycrystalline samples, yet are expected on the basis of the quasi-one-dimensional crystal structure of C$_{60}$ TDAE. In the absence of single crystals of C$_{60}$ TDAE, perhaps the most important experiment is to determine accurately the temperature dependence and magnitude of the conductivity of C$_{60}$ TDAE, a measurement that would help to decide between localized and itinerant descriptions. Certainly, single crystals will be essential for a complete understanding of magnetism in C$_{60}$ TDAE.

The exact nature of the ferromagnetism remains an open question. The observation of ambient temperature compressed pellet conductivity ($\sim 10^{-2}$ Scm^{-1}) and preliminary ESR results which show strong narrowing of the ESR line with decreasing temperature, imply that the material might be metallic; if so it might be an itinerant ferromagnet. Compounds such as ZrZn$_2$ and Ni$_3$Ga [45] show, qualitatively, features which are very similar to the ones described above for C$_{60}$ TDAE.

CONCLUSION AND FUTURE DIRECTIONS

Since it was concluded that C$_{60}$ TDAE is an itinerant ferromagnet, it is not unreasonable to assume that it constitutes the first corroboration of the FOM modifications of the MC model (see Figs 2 and 4). The spins associated with the TDAE$^{+\uparrow}$ radical cations could be responsible for the background paramagnetism above T_c observed in C$_{60}$ TDAE [38]. Even though the new ferromagnet has a Curie temperature of ~ 16 K (~ 27 times that of previous strictly organic molecular ferromagnets), it has no remanence.

This lack of retention of magnetism outside an applied field, makes the material interesting only from an academic point of view. The crystal structure

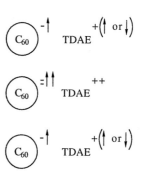

Fig. 4. Schematic representation of part of the C$_{60}$ TDAE structure. An excitation can transfer a second electron from TDAE$^{+\uparrow}$ to C$_{60}^{-\uparrow}$, resulting in diamagnetic TDAE^{++} and triplet C$_{60}^{-\uparrow\uparrow}$. The left side of the figure corresponds to the schematic representation of the FOM modification of the McConnell model shown in Fig. 2.

determination, magnetic measurements under pressure and further transport measurements, as well as Raman spectroscopy lend support to the conclusion that C$_{60}$ TDAE is an itinerant ferromagnet. Since TDAE is not the only organic strong reducing agent known and C$_{60}$ could be converted to fulleroids [46], enough pinning defects could be introduced for domain formation with concomitant creation of remanence. The future looks bright for novel organic materials based on fullerenes and fulleroids. The triply degenerate LUMO of C$_{60}$ when partially filled to the -2 and -1 oxidation states, makes this system an ideal example of the FOM variant of the McConnell model as depicted in Fig. 2.

Acknowledgements—The above was made possible by the great fortune of having colleagues such as Pierre-Marc Allemand, "Chan" Li, (electrochemistry, materials synthesis), Andrew Koch, Kishan Khemani, Andreas Hirsch (Fullerene preparation and purification, nucleophilic additions), Gordana Srdanov (fullerene preparation, fulleroid preparation, X-ray structure determination), K. Holczer (UCLA and C.N.R.S., Orsay for magnetic susceptibility measurements), and G. Sparn (Los Alamos National Laboratory for magnetic susceptibility measurements and interpretation of magnetism) and P. W. Stephens (SUNY Stony Brook for X-ray structural studies).

National Science Foundation is wholeheartedly acknowledged for support through Grants DMR-88-20933, DMR-91-11097 and CHE-89-08323.

REFERENCES

1. Krätschmer W., Lamb L. D., Fostiropoulos K. and Huffman D. R., *Nature* **347**, 354 (1990); Haufler R. E., Conceicao J., Chibante L. P. F., Chai Y., Byrne N. E., Flanagan S., Haley M. M., O'Brien S. C., Pan C., Xiao Z., Billups W. E., Ciufolini M. A., Hauge R. H., Margrave J. L., Wilson L. J., Curl R. F. and Smalley R. E. *J. Phys. Chem.* **94**, 8634 (1990); Koch A., Khemani K. C. and Wudl F., *J. Org. Chem.* **56**, 4543 (1991).
2. Haddon R. C., *Acc. Chem. Res.* **25**, 127 (1992).

3. Cox D. M., Behal S. M., Disko M., Gorun S., Greaney M., Hsu C. S., Kollin E., Miliar J., Robbins J., Robbins W., Sherwood R. and Tindall P., *J. Am. Chem. Soc.* **113**, 2940 (1991); Greaney M. A. and Gorun S. M., *J. Phys. Chem.* **95**, 7142 (1991).

4. Fischer J. E., Heiney P. A. and Smith A. B., III, *Acc. Chem. Res.* **25**, 112 (1992).

5. Haufler R. E., Conceicao J., Chibante L. P. F., Chai Y., Byrne N. E., Flanagan S., Haley M. M., O'Brien S. C., Pan C., Xiao Z., Billups W. E., Ciufolini M. A., Hauge R. H., Margrave J. L., Wilson L. J., Curl R. F. and Smalley R. E., *J. Phys. Chem.* **94**, 8634 (1990).

6. Allemand P.-M., Koch A., Wudl F., Rubin Y., Diederich F., Alvarez M. M., Anz S. J. and Whetten R. L., *J. Amer. Chem. Soc.* **113**, 1050 (1991).

7. Dubois D., Kadish K. M., Flanagan S. and Wilson L. J., *J. Am. Chem. Soc.* **113**, 7773 (1991).

8. Korshak Y. V., Ovchinnikov A. A., Shapiro A. M., Spector V. N., *JETP Lett.* **43**, 309 (1986); Korshak Y. V., Medvedeva T. V., Ovchinnikov A. A. and Spector V. N., *Nature* **326**, 370 (1987).

9. Torrance J. B., Oostra S. and Nazzal A., *Synth. Met.* **19**, 709 (1987).

10. Allemand P.-M., Khemani K. C., Koch A., Wudl F., Holczer K., Donovan S., Grüner G. and Thompson J. D., *Science* **253**, 301 (1991).

11. *Proceedings of the International Congress of Pacific Basin Societies*, Honolulu, Hawaii, 17–22 December 1989; *Mol. Cryst. Liq. Cryst.* (1990), 181. *Proceedings of the ICSM 1988 Synth. Met.* 27–29, (1988).

12. Williams J. M., Ferraro J. R., Thorn R. J., Carlson K. D., Geiser U., Wang H. H., Kini A. M. and Whangbo M. H., *Organic Superconductors*. Prentice Hall, Englewood Cliffs, New Jersey (1992).

13. Takahashi M., Turek P., Nakazawa Y., Tamura M., Nozawa K., Shiomi D., Ishikawa M. and Kinoshita M., *Phys. Rev. Lett.* **67**, 746 (1991).

14. *Proceedings of the Symposium on Ferromagnetic and High Spin Molecular Based Materials* (Edited by J. Miller and D. A. Dougherty), pp. 1–562 **176** (1989).

15. Allemand P.-M., Srdanov G. and Wudl F., *J. Amer. Chem. Soc.* **112**, 9392 (1990).

16. Dougherty D., *Mol. Cryst. Liq. Cryst.* **176**, 25 (1989).

17. Le Page T. and Breslow R., *J. Am. Chem. Soc.* **109**, 6412 (1987).

18. McConnell H. M., *Proc. Robert A. Welch Found. Conf. Chem. Res.* **11**, 144 (1967); McConnell H. M., *J. Chem. Phys.* **39**, 1910 (1963).

19. Iwamura H., *Pure and Appl. Chem.* **58**, 187 (1986).

20. Utamapanya S. and Rajca A. *J. Am. Chem. Soc.* **113**, 9242 (1991).

21. Buchachenko A. L., *Russ. Chem. Rev.* **59**, 307 (1990).

22. Pei Y., Kahn O., Nakatani K., Codjovi E., Mathoniere C. and Sletten J., *J. Am. Chem. Soc.* **113**, 6558 (1991); Nakatani K., Berjerat P., Codjovi E., Mathoniere C., Pei Y. and Kahn O., *Inorg. Chem.* **30**, 3978 (1991).

23. Caneschi A., Gatteschi D., Sessoli R., Rey P., *Acc. Chem. Res.* **22**, 392 (1989); Ferraro F., Gatteschi D., Sessoli R. and Corti M., *J. Am. Chem. Soc.* **113**, 8410 (1991).

24. Dormann E., Nowak M. J., Williams K. A., Angus R. O., Jr. and Wudl F., *J. Am. Chem. Soc.* **109**, 2594 (1987).

25. Miller J. S., Epstein A. J. and Reiff W. A., *Chem. Rev.* **88**, 201 (1988).

26. Calculations using the high-temperature expansion of the degenerate Hubbard model predict a ferromagnetic stage for three electrons distributed between two atoms: Lyon-Caen C. and Cyrot M., *J. Phys. C.* **8**, 2091 (1975).

27. Wohlfarth E. P., in *Itinerant-Electron Magnetism* (Edited by R. D. Lowde and E. P. Wohlfarth), p. 305. North-Holland, Amsterdam (1977); White M. and Geballe T. H., *Long Range Order in Solids*, p. 143. Academic Press, Boston (1979).

28. Kinoshita M. and Callen E. O. S. B. 1., 12, 5 *ONRFE Sci. Bull.* **12**, 12 (1987).

29. Yamaguchi K., Toyoda Y., Nakano M. and Fueno T., *Synth. Met.* **19**, 87 (1987).

30. Jehoulet C., Bard A. J. and Wudl F., *J. Am. Chem. Soc.* **113**, 5456 (1991); Cox D. M., Behal S., Disko M., Gorun S., Greaney M., Hsu C. S., Kollin E., Miliar J., Robbins J., Robbins W., Sherwood R. and Tindall P., *J. Am. Chem. Soc.* **113**, 2940 (1991).

31. Hebard A. F., Rosseinsky M., Haddon R. C., Murphy D. W., Schneemeyer L. F., Glarum S. H., Palstra T. T. M., Ramirez A. P. and Kortan A. R., *Nature* **350**, (1992) in press; Holczer K., Klein O., Huang S.-M., Kaner R. B., Fu K. J., Whetten R. L. and Diederich F., *Science* **252**, 1154 (1991).

32. Allemand P. M., Srdanov G., Koch A., Khemani K., Wudl F., Rubin Y., Diederich F., Alvarez M. M., Anz S. J. and Whetten R. L., *J. Amer. Chem. Soc.* **113**, 2781 (1991).

33. Stephens P. W., Cox D., Lauher J. W., Mihaly L., Wiley J. B., Allemand P.-M., Hirsch K., Holczer Q., Li Q., Thompson J. D. and Wudl F., *Nature* **355**, 331 (1992).

34. Tanaka K., Zakhidov A. A., Yoshizawa K., Okahara K., Yamabe T., Yakushi K., Kikuchi K., Suzuki S., Ikemoto I. and Achiba Y., *Phys. Lett. A*, (1992), in press.

35. For example, Morrish A. H., in *The Physical Principles of Magnetism*. pp. 332. Wiley, New York (1965).

36. Berkowitz A. E. and Kneller E., in *Magnetism and Metallurgy*, p. 393. Academic, New York (1969).

37. See Heiney P. A., *J. Phys. Chem. Solids* **53**, 1333 (1992).

38. Sparn G., Thompson J. D., Allemand P. M., Li Q., Wudl F., Holczer K. and Stephens P. W., *Sol. State. Commun.* **82**, 779 (1992).

39. Dzyaloshinski I., *J. Phys. Chem. Solids* **4**, 241 (1958).

40. Moriya T., *Phys. Rev.* **120**, 91 (1960).

41. Molecular crystals are considerably more compressible than traditional inorganic solids.

42. Chakravarty S., Gelfand M. P. and Kivelson S., *Science* **254**, 970 (1991). Varma C. M., Zaanen Z. and Raghavachari K., *Science* **254**, 989 (1991).

43. White M. and Gabelle T. H., *Long Range Order in Solids*, p. 143. Academic Press, Boston (1979).

44. Misawa S., Tanaka T. and Tsuru K., *Europhys. Lett.* **14**, 377 (1991).

45. Wohlfarth E. P., *Comments on Solid State Physics* **6**, 123 (1975).

46. Wudl F., *Acc. Chem. Res.* **25**, 157 (1992).

ELECTRONIC STRUCTURES OF C_{60} FULLERIDES AND RELATED MATERIALS

Atsushi Oshiyama, Susumu Saito,† Noriaki Hamada† and Yoshiyuki Miyamoto

Microelectronics Research Laboratories, NEC Corporation and †Fundamental Research Laboratories, NEC Corporation, 34 Miyukigaoka, Tsukuba 305, Japan

(Received 26 *February* 1992*)*

Abstract—We report microscopic total-energy electronic-structure calculations which provide the cohesion mechanism and the energy-band structures of new forms of solid carbon, C_{60} and related materials. We find that C_{60} clusters are condensed via van der Waals-like forces and the resulting face centered cubic C_{60} is a semiconductor with an energy gap of $\sim 1.5\,eV$. Both the valence-band top and the conduction-band bottom are at the zone boundary (X point), but rotation of the C_{60} clusters modifies the energy-band dispersion. We also find that alkali-atom doping transforms solid C_{60} into a strongly bonded ionic metal in which both Madelung and band energies contribute to its cohesion. The overall structure of the calculated density of states is in excellent agreement with the photoemission, inverse photoemission and X-ray emission data for both pristine and alkali-doped fullerides. A unique linear relation between the superconducting transition temperature and the Fermi-level density-of-states over a wide range of temperature is found for potassium-doped and rubidium-doped C_{60} under pressure. We predict that the high electronegativity of C_{60} hinders the hole injection with halogen-atom doping. Finally, systematic calculations for carbon micro-tubules have been performed using a realistic tight binding model. We find that micro-tubules exhibit a striking variation in electronic conduction, from a metal to narrow-gap and wide-gap semiconductors, depending on the diameter and the degree of helical arrangement of the tubule.

Keywords: C_{60}, fullerides, carbon micro-tubules, energy bands, superconductivity.

1. INTRODUCTION

In 1970, Osawa [1] mentioned the possible existence of an aromatic C_{60} molecule with a truncated icosahedral shape, i.e. similar to a soccer ball, with carbon atoms located at the 60 vertices of the ball. Spherical carbon clusters were indeed observed in amorphous carbon films by electron microscope [2]. Later, it was found that mass spectroscopy of a carbon cluster beam yielded two prominent peaks corresponding to 60 and 70 carbon atoms, strongly suggestive of the existence of cage-structure carbon clusters (Buckminsterfullerene or fullerene) [3]. The conjectures about this unique shape in these pioneering works [4] have been confirmed by the recent success in the production of macroscopic quantities of cage-structure clusters from carbon soot [5], and by subsequent photoabsorption [5, 6] and nuclear magnetic resonance (NMR) [7] measurements. In particular, X-ray diffraction studies [5, 8] have revealed that C_{60} clusters are condensed in a close-packed crystalline form (fullerides). Moreover, the discovery of superconductivity in alkali-doped fullerides, K_3C_{60} (transition temperature $T_c = 19\,K$ [9]), Rb_3C_{60} ($T_c = 29\,K$ [10]) and Cs_2RbC_{60} ($T_c = 33\,K$ [11]), has triggered a great expansion in research on the condensed matter properties of this new family of cage-structure solids [12].

In this paper we present microscopic total-energy electronic-structure calculations for C_{60}, alkali-doped C_{60} and halogen-doped C_{60}. The results clarify unique features in the cohesive and electronic properties of these new materials. In particular, we find that solid C_{60} weakly condensed via a van der Waals-like force is transformed upon alkali-atom doping into a strongly bonded ionic metal. We also find a unified linear relation between T_c and the density of states (DOS) near the Fermi level (E_F) in alkali-doped fullerides. We predict that halogen-atom doping is unlikely to produce holes in the valence bands of pristine C_{60} due to its high electronegativity. We also report tight-binding calculations for graphitic micro-tubules which have been recently observed [13] on the carbon electrode used for arc discharge. We predict that this fascinating new material will exhibit striking variations in electronic transport, from metallic to semiconducting, depending on the diameter and the degree of helical arrangement of the micro-tubules.

In section 2, we give a brief description of our total-energy electronic-structure calculations within the local density approximation. The cohesive properties and the energy bands of C_{60} and K_xC_{60} are presented in sections 3 and 4, respectively. The results are compared with other local density calculations and with experimental data. In section 5, the unusual

287

linear relation between T_c and the DOS near E_F in K_3C_{60} and Rb_3C_{60} under several pressures is reported, and its implication for the mechanism of super-conductivity is discussed. Calculations for bromine-doped C_{60} are presented in section 6. The results indicate an exceptionally high electronegativity for fullerides. Section 7 is devoted to graphitic micro-tubules and section 8 summarizes the results.

2. TOTAL-ENERGY ELECTRONIC-STRUCTURE CALCULATION

Three important ingredients in the present calcu-lations are norm-conserving pseudopotentials, local density approximation (LDA) in density functional theory and the Gaussian-orbital basis set [14]. First, interactions of valence electrons with nuclei and with core electrons are simulated by pseudopotentials which are constructed from first-principles all-elec-tron atomic calculations [15]. To assure high transfer-ability of the pseudopotentials, we choose small core radii (0.39 Å for C s and p, 1.4 Å for K s, 1.6 Å for Rb s, 0.60 Å for Br s and 0.66 Å for Br p valence states) from which the pseudo-valence orbitals are identical to the real valence orbitals. It is important to use these "hard" pseudopotentials to guarantee the reliability of the calculated structural properties.

Second, after placing the pseudopotential at each atomic site in a solid, interactions among the valence electrons are treated within LDA. We use the Ceper-ley–Alder form [16] for the exchange-correlation energy. It is well known that structural properties, such as the lattice constants and bulk moduli of covalent and metallic solids, are well reproduced by LDA, whereas the cohesive energy is overestimated to some extent and the electronic excitation energy, e.g. energy gap, is considerably underestimated [17].

Third, we use a Gaussian-orbital basis set to express the valence-electron wavefunctions and thus the electron density. The Gaussian-orbital basis set has an advantage due to its small size compared with a well-converged plane-wave basis set, when we treat rather compact electron orbitals such as the valence orbitals of the first-row atoms in the periodic table. On the other hand, it has a disadvantage in the sense that systematic improvement of the basis set and then examination of the convergence, which are straight-forward in the case of the plane-wave basis set, are not feasible. Thus, in order to assure the reliability of the Gaussian-orbital basis set, it is imperative to compare the results from the Gaussian-orbital basis set with those from a well-converged plane-wave basis set, at least for key materials. This is what we have actually done, as will be described below.

We have determined Gaussian exponents in the basis set by fitting numerically obtained pseudo-orbitals. It is found that the following exponents provide satisfactory fitting with a mean deviation from the numerical orbitals less than 10^{-3} in atomic units: 0.299 and 2.908 for carbon s, 0.362 and 2.372 for carbon p, 0.06, 0.101 and 0.202 for s of alkali atoms, 0.06 and 0.101 for p of alkali atoms, 0.06, 0.61 and 0.74 for bromine s, and 0.33 and 1.96 for bromine p orbitals in atomic unit, i.e. a_0^{-2}. Brillouin zone (BZ) integration is performed using $2k$ points in an irre-ducible wedge (32 k points in the whole zone) in a self-consistent calculation for doped fullerides. Γ point sampling is found to be valid for pristine C_{60}. The interval of the real-space mesh in the fast-Fourier-transformation is set to be 0.15 Å to express compact C valence orbitals.

We have examined the validity of the present scheme of the calculation by comparing results for diamond and graphite with those from a well-converged plane-wave (PW) basis set and/or with the experimental results. The calculated bond length in diamond is 1.548 Å (experimental value: 1.545 Å), and the cohesive energy is 8.4 eV (PW basis-set value: 8.63 eV, experimental value: 7.37 eV). For graphite, the calculated bond length is 1.442 Å (experimental value: 1.418 Å), and the cohesive energy is 8.1 eV. This agreement for the structural constants in both sp^3-bonded and sp^2-bonded materials assures the validity of the present scheme of the calculation for fullerenes in which sp^3 and sp^2 bonds coexist. How-ever, the calculated interlayer distance of graphite is 6% larger than the experimental value. To resolve the reason for this deviation, we have performed a completely-converged PW basis-set calculation for graphite using the conjugate-gradient minimization technique [18]. Using the PW basis set with a cut-off kinetic energy of 80 Ry, the interlayer distance of graphite is calculated to be more than 10% larger than the experimental value. Graphite layers are regarded as being bound via an induced dipole–dipole interaction (van der Waals force) which is unlikely to be properly described by LDA theory. We thus ascribe the deviation in the interlayer distance of graphite to the limitation of LDA.

3. COHESION AND ENERGY BANDS OF C_{60}

In face-centered cubic (fcc) C_{60}, each C_{60} cluster is centered at the fcc lattice points [8] and rotates freely [19] at relatively low temperature. X-ray diffraction [20] and the NMR [21] measurements show that there is a transition at 250–260 K from a phase in which the clusters rotate to a phase in which they jump between symmetry-equivalent orientations and the lattice has

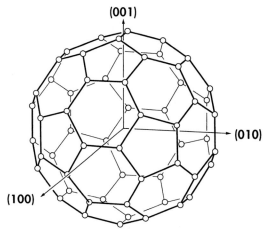

Fig. 1. C$_{60}$ cluster placed in a fcc lattice. Each crystal axis crosses a double bond shared by two hexagons.

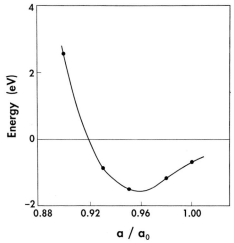

Fig. 2. Total energy per C$_{60}$ cluster in fcc C$_{60}$ as a function of the lattice constant a. The energy is measured from the total energy of an isolated C$_{60}$ cluster. a_0 is the experimental lattice constant. The solid line is a guide for the eye.

simple cubic symmetry. The orientational disorder at higher temperature affects the electronic structure to some extent, as will be seen below, but the characteristics of solid C$_{60}$ are nevertheless well-represented by a fcc structure in which each cluster has a fixed and equivalent orientation [22].

On each fcc lattice point, we place the C$_{60}$ cluster so that the point group attains the highest possible symmetry, T_h (Fig. 1). It is known that in C$_{60}$ there are two different bonds, one which constitutes regular pentagons (with length r_1) and the other which is shared by two hexagons (with length r_2). In fact, the LDA calculation [19] using the plane-wave basis-set with a 26 Ry cutoff gives values of $r_1 = 1.45$ Å and $r_2 = 1.40$ Å. In the present calculation, we

use $r_1 = 1.46$ Å and $r_2 = 1.40$ Å, which have been determined by NMR measurements [23].

Figure 2 is the calculated total energy of fcc C$_{60}$, measured from the total energy of an isolated C$_{60}$ cluster, as a function of the ratio of the fcc lattice constant a with the experimental value ($a_0 = 14.20$ Å [8]). The isolated cluster has been simulated by a large-enough lattice-constant calculation ($a = 1.2a_0$), and is found to have a binding energy of 7.40 eV per atom. Of note is the fact that the calculated binding energy [24] for isolated C$_{70}$, which is less abundant than C$_{60}$ in the mass spectroscopy measurements, is 7.42 eV, suggesting the importance of dynamic and

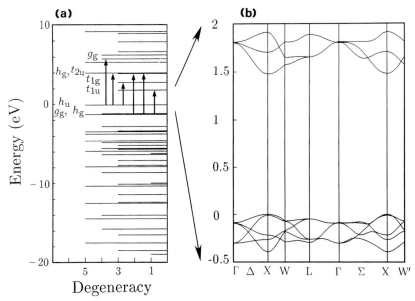

Fig. 3. Electronic energy levels of an isolated C$_{60}$ (a) and the energy band structure of fcc C$_{60}$ (b). The valence-band top at the X point is defined as the zero energy. The optically allowed transitions with excitation energies less than 6 eV are indicated by arrows.

kinetic aspects in the formation of cage-structure clusters. As seen from Fig. 2, C_{60} clusters are condensed via rather weak van der Waals-like forces. The calculated cohesive energy of 1.6 eV per C_{60}, which has been corroborated by recent experiments [25], is much smaller than the C–C bond energy but somewhat larger compared to the interlayer binding energy of graphite. This suggests that the chemical hybridization of π-like orbitals belonging to the different fullerenes plays a role to some extent in the condensation of C_{60} clusters. The theoretical lattice constant of fcc C_{60} is smaller than the experimental value by about 5% (Fig. 2). We provisionally regard this deviation as a limitation of the LDA.

In Fig. 3, the calculated energy levels of C_{60} and the band structure of fcc C_{60} having the experimental lattice constant are shown. The isolated C_{60} has icosahedral symmetry and its energy levels have up to fivefold degeneracies. The levels below -6 eV and above 7 eV in Fig. 3a are the bonding and antibonding σ-like states, whereas the rest are mostly the π-like states. The highest-occupied state is the h_u state, which is completely occupied by 10 electrons. The energy gap between the h_u and the lowest unoccupied state t_{1u} is about 1.9 eV. If the system has perfect spherical symmetry, the energy levels are characterized by an angular momentum number l, and the optically-allowed transitions are from l to $l \pm 1$ states ($\Delta l = 1$). The energy levels of the C_{60} cluster indeed have a certain correspondence to the l state: e.g. the second and the third highest occupied states, g_g and h_g, correspond to the $l = 4$ states for spherical symmetry, and the highest occupied state h_u is one of the $l = 5$ states. The optical transitions in C_{60} corresponding to $\Delta l = 1$ transitions are thus expected to have strong oscillator strengths. Such transitions are shown by arrows in Fig. 3a: $h_u \rightarrow t_{1g}$, $h_g \rightarrow t_{1u}$, $h_u \rightarrow h_g$, $g_g \rightarrow t_{2u}$, $h_g \rightarrow t_{2u}$ and $h_u \rightarrow g_g$. The calculated excitation energies for these optically-allowed transitions are 2.9 eV, 3.1 eV, 4.1 eV, 5.1 eV, 5.2 eV and 5.9 eV, respectively. Since there is a clear correspondence between the energy levels of the C_{60} cluster and the energy bands of fcc C_{60}, we can expect that the optical spectra of fcc C_{60} will be similar to those of the C_{60} cluster. The calculated excitation energies above are in qualitative agreement with the observed peaks in photoabsorption measurements [6], although more effort is necessary for quantitative comparison.

The energy gap of the C_{60} cluster remains finite even in solid C_{60}, with a value for the energy gap of 1.5 eV in the present calculation. Gaussian-orbital basis sets overestimate the energy gap compared with the well-converged plane-wave basis set [26], whereas the LDA underestimates it compared with experiment [27]. Thus the value of 1.5 eV may be a good

estimate for the energy gap of fcc C_{60} [28]. Both the valence-band top and the conduction-band bottom are located at the X point (Fig. 3b). The transition is, however, optically forbidden since both the conduction and the valence bands have *ungerade* symmetry under inversion operation. Another unique feature of the energy bands lies in the inequivalent Z lines (i.e. the XW and XW′ lines in Fig. 3b). This inequivalence is related to the internal degrees of freedom of fullerenes. As shown in Fig. 1, the two XW lines, e.g. from X ($[200]\pi/a$) to W ($[201]\pi/a$) and from X ($[200]\pi/a$) to W′ ($[210]\pi/a$), respectively, are inequivalent, due to the presence of an internal bond network in the fullerene. The fullerene molecule lacks perfect spherical symmetry, and consequently the energy bands exhibit different dispersion along the

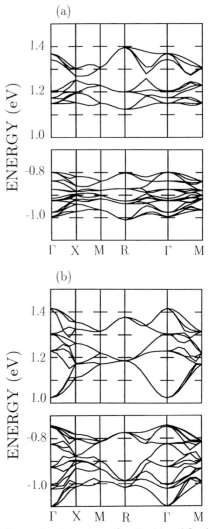

Fig. 4. Energy band structure of sc C_{60} (a) and fcc C_{60} (b). The upper panels are the conduction bands, and the lower panels the valence bands. The calculations have been performed with a tight-binding model [30]. In (b) the energy bands are folded in the small BZ of the sc lattice. The X, M and R points correspond to the X/2, 2K/3 and L points of the BZ of fcc C_{60}.

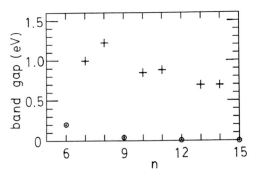

Fig. 17. Calculated energy band gap as a function of the number of hexagons on the circumference for the $B(1,0)n$ tubule. Symbols, $\odot$ and $+$, represent the classes of narrow-gap and moderate-gap semiconductors, respectively.

for the band gap of graphitic micro-tubules. A tubule $A(n_1,n_2)$ is (1) a metal for $n_1 - 2n_2 = 0$, (2) a narrow-gap semiconductor for $n_1 - 2n_2 = 3m (m = 1,2,\ldots)$, or (3) a moderate-gap semiconductor otherwise [77]. We thus find that the band-gap is tunable by choosing the tubule structure. According to the HRTEM observation [13], the micro-tubules are assembled co-axially, and it might therefore be possible to design multi-tubule semiconductor–metal atom-scale interfaces.

8. SUMMARY

In this paper we have discussed the cohesion mechanism, electronic structures and the mechanism of the superconductivity of solid C_{60} and related materials, on the basis of microscopic total-energy band-structure calculations within the local density approximation (LDA). We have found that the C_{60}

clusters are condensed via van der Waals-like forces and the resulting fcc C_{60} is a semiconductor with an energy gap of ~ 1.5 eV. Both the valence-band top and the conduction-band bottom are at the zone boundary (X point), but rotation of the C_{60} clusters modifies the energy-band dispersion. For the alkali-atom doped C_{60} fullerides, we have found that electrons are transferred from alkali s orbitals to the t_{1u} conduction bands of the pristine C_{60}. Alkali-atom doping transforms the C_{60} solid into a strongly-bonded ionic metal in which both Madelung and band energies contribute to its cohesion. The overall structures of the calculated density of states are in excellent agreement with the photoemission and inverse photoemission data for both pristine and alkali-doped fullerides. A unique linear relation between the superconducting transition temperature and the Fermi-level density-of-states over a wide range of temperature has been found for potassium-doped and rubidium-doped C_{60}. It is predicted that the high electronegativity of C_{60} hinders hole injection by halogen-atom doping. Systematic calculations for carbon micro-tubules have been performed using a realistic tight-binding model. We have found that the micro-tubule exhibits a striking variation in electronic conduction, from a metal to narrow-gap and wide-gap semiconductors, depending on the diameter and the degree of helical arrangement of the tubule. This drastic variation has been explained in terms of the band structure of graphite and the curvature of the micro-tubules. The present calculation demonstrates that C_{60}-related materials provide exciting topics in condensed matter physics, and have great potential as a novel unit for the design of new materials.

Acknowledgements—We wish to thank S. Sawada for collaborating in part of the present work and for valuable discussions. We are also grateful to O. Sugino, S. Iijima, J. S. Tsai, T. W. Ebbesen and K. Tanigaki for informative conversations, and to J. H. Weaver, R. L. Whetten and J. E. Fisher for sending their experimental data prior to publication.

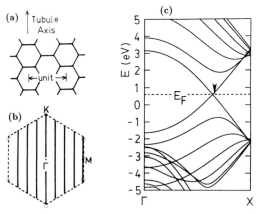

Fig. 18. (a) Axis and the construction unit on the circumference of the tubule $B(2,1)n$. (b) First BZ of a graphite sheet (dashed line) and the wavevectors allowed by the periodic boundary condition along the circumference of the $B(2,1)4$ tubule (solid line). (c) Calculated band structures of $B(2,1)6$ tubule. Here the X point has a wavenumber near $\sqrt{3}\pi/a$. The points at which the two π-like bands are degenerate are shown by open circles in (b). The arrow in (c) denotes the K point.

REFERENCES

1. Osawa E., *Kagaku (Kyoto)* **25**, 854 (1970).
2. Iijima S., *J. Cryst. Growth* **50**, 675 (1980).
3. Kroto H. W., Heath J. R., O'Brien S. C., Curl R. F. and Smalley R. E., *Nature* **318**, 162 (1985).
4. For a review of earlier work, see Kroto H. W., *Science* **242**, 1139 (1988).
5. Krätschmer W., Lamb L. D., Fostiropoulos K. and Huffman D. R., *Nature (London)* **347**, 354 (1990).
6. Ajie H. *et al.*, *J. Phys. Chem.* **94**, 8630 (1990).
7. Taylor R., Hare J. P., Abdul-Sada A. K. and Kroto H. W., *J. Chem. Soc., Chem. Commun.* **20**, 1423 (1990); Yannoni C. S., Johnson R. D., Meijer G., Bethune D. S. and Salem J. R., *J. Phys. Chem.* **95**, 9 (1991).

8. Fleming R. *et al.*, *MRS Late News Session on Buckyballs: New Materials Made from Carbon Soot*, videotape. Materials Research Society, Pittsburgh (1990).
9. Hebard A. *et al.*, *Nature* **350**, 600 (1991).
10. Rosseinsky M. J. *et al.*, *Phys. Rev. Lett.* **66**, 2830 (1991); Holczer K. *et al.*, *Science* **252**, 1154 (1991).
11. Tanigaki K. *et al.*, *Nature* **352**, 222 (1991).
12. For a recent review, Huffman D., *Physics Today*, November, p. 22 (1991).
13. Iijima S., *Nature* **354**, 56 (1991).
14. Oshiyama A. and Saito M., *J. Phys. Soc. Jpn.* **56**, 2104 (1987); Miyamoto Y. and Oshiyama A., *Phys. Rev.* **B41**, 12,680 (1990).
15. Hamann D. R., Schlüter M. and Chiang C., *Phys. Rev. Lett.* **43**, 1494 (1979); Bachelet G. B., Hamann D. R. and Schlüter M., *Phys. Rev. B* **26**, 4199 (1982).
16. Ceperley D. M. and Alder B. J., *Phys. Rev. Lett.* **45**, 566 (1980).
17. For example, see *Theory of the Inhomogeneous Electron Gas* (Edited by S. Lundqvist and N. H. March). Plenum, New York (1983).
18. Sugino O. and Oshiyama A., unpublished. The conjugate gradient minimization technique combined with LDA is described by Sugino O. and Oshiyama A., *Proc. Int. Conf. Defects in Semiconductors* (Lehigh Univ. 1991, Trans Tech), p. 469; *Phys. Rev. Lett.* **68**, 1858 (1992).
19. Zhang Q.-M., Yi J.-Y. and Bernholc J., *Phys. Rev. Lett.* **66**, 2633 (1991).
20. Heiney P. A. *et al.*, *Phys. Rev. Lett.* **66**, 2911 (1991).
21. Tycko R. *et al.*, *Phys. Rev. Lett.* **67**, 1886 (1991).
22. Saito S. and Oshiyama A., *Phys. Rev. Lett.* **66**, 2637 (1991).
23. Johnson R. D., Yannoni N., Meijer G. and Bethune D. S., in [8].
24. Saito S. and Oshiyama A., *Phys. Rev.* **B44**, 11532 (1991).
25. Pan C., Sampson M. P., Chai Y., Hauge R. H. and Margrave J. L., *J. Phys. Chem.* **95**, 2944 (1991).
26. An LDA calculation using soft pseudopotentials (core radii: 0.79 Å) and a plane-wave basis set with a cut-off of 49 Ry gives an energy gap of 1.18 eV: Troullier N. and Martins J. L., *Phys. Rev.*, submitted.
27. For example, Perdew J. P. and Levy M., *Phys. Rev. Lett.* **51**, 1884 (1983); Sham L. J. and Schlüter M., *ibid.* **51**, 1888 (1983).
28. Measured values of the energy gap of fcc C_{60} are reported by: Skumanich A., *Chem. Phys. Lett.* **182**, 486 (1991); Mort J. *et al.*, *Chem. Phys. Lett.* **186**, 281 (1991); Cohen H., Kolodney E., Maniv T. and Folman M., *Solid State Commun.* **81**, 183 (1992); Saito Y. *et al.*, *Chem. Phys. Lett.* **189**, 236 (1992); Gensterblum G. *et al.*, *Phys. Rev. Lett.* **67**, 2171 (1991).
29. Ching W. Y. *et al.*, *Phys. Rev. Lett.* **67**, 2045 (1991).
30. The calculation has been performed with a newly-developed realistic tight-binding model which satisfactorily reproduces the LDA results for graphite and fcc C_{60} (Hamada N. and Sawada S., to be published in *Proc. Int. Workshop on Electronic Properties and Mechanisms in High T_c Superconductors*, Tsukuba (1991) Elsevier, New York). In this model, the carbon $2s$ and $2p$ orbitals are used to express the tight-binding hamiltonian, and the non-orthogonality between atomic orbitals of neighboring sites is taken into account. The transfer and the overlap integrals have a suitable distance dependence in order to describe the band structures in various atomic configurations.
31. Saito S., Hamada N. and Sawada S., preprint.
32. David W. I. F. *et al.*, *Nature* **353**, 147 (1991).
33. Penn D. R., *Phys. Rev.* **128**, 2093 (1962).
34. Weaver J. H. *et al.*, *Phys. Rev. Lett.* **66**, 1741 (1991).
35. Jost M. B. *et al.*, *Phys. Rev. B* **44**, 1966 (1991).
36. Saito Y. *et al.*, *J. Phys. Soc. Jpn* **60**, 2518 (1991).
37. Saito S. and Oshiyama A., *Phys. Rev.* **B44**, 11536 (1991).
38. Stephens P. W. *et al.*, *Nature* **351**, 632 (1991).
39. Fleming R. M. *et al.*, *Nature* **352**, 701 (1991).
40. Zhou O. *et al.*, *Nature* **351**, 462 (1991).
41. The calculated values for the cohesive energy contain some ambiguities, since the calculated lattice constants for K_xC_{60} are smaller than the experimental values by about 5%.
42. In general, ionic materials could have cesium-chloride, sodium-chloride or zinc-blende structures. According to the rigid-sphere model, when the ionic radius of an anion is larger to some extent than that of a cation, the material is expected to have the zincblende structure. See e.g., Cotton F. A. and Wilkinson G., *Advanced Inorganic Chemistry*. John Wiley, New York (1972).
43. Saito S. and Oshiyama A., *Physica C* **185–189**, 421 (1991).
44. Oshiyama A. and Saito S., *Solid State Commun.* **82**, 41 (1992). In the calculation of DOS, $12\,k$ points per irreducible wedge ($256\,k$ points in the whole Brillouin zone) are adopted as sampling points, and the eigenvalues are interpolated using the calculated values at those sampling points. The calculated DOS at each energy value is broadened with a Gaussian width of 0.1 eV.
45. Hamada N., Saito S., Miyamoto Y. and Oshiyama A., *Jpn J. appl. Phys.* **30**, L2036 (1991).
46. Benning P. J., Martins J. L., Weaver J. H., Chbante L. P. F. and Smalley R. E., *Science* **252**, 1417 (1991).
47. Wertheim G. K. *et al.*, *Science* **252**, 1419 (1991).
48. Tycko R. *et al.*, *Science* **253**, 884 (1991).
49. Erwin S. C. and Pickett W. E., *Science* **254**, 842 (1991).
50. Fleming R. M. *et al.*, *Nature* **352**, 787 (1991).
51. Sparn G. *et al.*, *Science* **252**, 1829 (1991).
52. Schirber J. E. *et al.* *Physica C* **178**, 137 (1991).
53. Sparn G. *et al.*, *Phys. Rev. Lett.* **68**, 1221 (1992).
54. Zhou O. *et al.*, *Science* **255**, 833 (1992).
55. The present calculation within LDA underestimates the lattice constants of fcc C_{60} and K_3C_{60} by about 5% so that the calculated bulk moduli are considerably larger than the experimental values. We thus multiply the calculated bulk moduli by this factor, which is determined by scaling theory and experiment for pristine C_{60}.
56. Fischer J. E. *et al.*, *Science* **252**, 136 (1991).
57. Duclos S. J. *et al.*, *Nature* **351**, 380 (1991).
58. Takahashi T. *et al.*, *Phys. Rev. Lett.* **68**, 1232 (1992); Fink J. *et al.*, *Proc. First Italian Workshop on Fullerenes*, Bologna, February (1992).
59. Lof R. W., Veenendaal M. A., Koopmans B., Jonkman H. T. and Sawatzky G. A., preprint.
60. de Coulon V., Martins J. L. and Reuse F., preprint.
61. The value for α, where $T_c \propto M^{-\alpha}$, M being the ionic mass, varies from 0.37 (Ramirez A. P. *et al.*, *Phys. Rev. Lett.* **68**, 1058 (1992)) to 1.4 (Ebbesen T. W. *et al.*, *Nature* **355**, 620 (1992)).
62. Possible mechanisms in which vibrational or phonon modes are crucial for superconductivity have been proposed on the basis of McMillan's equation: Martins J. L., Troullier N. and Schabel M., preprint; Varma C. M., Zaanen J. and Raghavachari K., *Science* **254**, 989 (1991); Schlüter M., Lannoo M., Needels N., Baraff G. A. and Tomanek D., *Phys. Rev. Lett.* **68**, 526 (1992).
63. Bethune D. S. *et al.*, *Chem. Phys. Lett.* **174**, 219 (1990).
64. For a review, see Fisher J. E. and Thompson T. E., *Physics Today* July, p. 36 (1978); Kamimura H., *Physics Today* December, p. 64 (1987).
65. Saito S., *Clusters and Cluster Assembled Materials* (Edited by Averback R. S., Nelson D. L. and Bernholc J., MRS Proceedings, Materials Research Society, Pittsburgh. **206**, 115 (1990).

66. Lichtenberger D. L. *et al.*, p. 673 in [65].
67. Miyamoto Y., Oshiyama A. and Saito S., *Solid State Commun.* **82,** 437 (1992).
68. Kobayashi M. *et al.*, *Solid State Commun.* **81,** 93 (1992); *ibid.*, to be published.
69. Zhu Q. *et al.*, *Nature* **355,** 712 (1992).
70. The cage-structure BC$_{59}$ has been observed in a cluster beam; see Guo T., Jin C. and Smalley R. E., *J. Phys. Chem.* **95,** 4948 (1991).
71. Miyamoto Y., Hamada N., Oshiyama A. and Saito S., *Phys. Rev. B.*, submitted.
72. Hamada N., Sawada S. and Oshiyama A., *Phys. Rev. Lett.* **68,** 1579 (1992).
73. Tersoff J., *Phys. Rev. Lett.* **61,** 2879 (1988).
74. A detailed description of the total energy minimization within the Tersoff model will be published elsewhere; Sawada S. and Hamada N., preprint.
75. For example, Painter G. S. and Ellis D. E., *Phys. Rev.* **B1,** 4747 (1970).
76. The lengths of the bonds parallel and normal to the tubule axis are different. However, the calculated difference in the bond length is less than 0.1%, so that the variation in the overlap and the transfer integrals due to the bond length difference is insignificant.
77. Recently a LDA calculation has been performed for the tubule $B(2,1)5$, showing metallic energy bands; Mintmire J. W., Dunlap B. I. and White C. T., *Phys. Rev. Lett.* **68,** 631 (1992). Also, after completion of the present work, we have become aware of tight-binding calculations which contain results similar to those in the present calculations; Saito R., Fujita M., Dresselhaus G. and Dresselhaus M. S., preprint.

SUPERCONDUCTIVITY IN ALKALI INTERCALATED C_{60}

M. SCHLÜTER,† M. LANNOO,†‡ M. NEEDELS,† G. A. BARAFF† and D. TOMÁNEK§

†AT&T Bell Laboratories, Murray Hill, NJ 07974, U.S.A.

§Department of Physics and Astronomy, Michigan State University, East Lansing, MI 48824-1116, U.S.A.

(*Received* 4 *March* 1992)

Abstract—Superconductivity observed in alkali intercalated C_{60} solid can be explained on the basis of conventional BCS theory. Intra-molecular Jahn–Teller type vibrations with high frequencies couple to conduction electrons in C_{60} π-orbitals with strength V. The density of these states (N) is determined by the relatively weak intermolecular coupling. This results in a real space factorization of the coupling parameter $\lambda = NV$ which has several experimental consequences. We present detailed calculations that lead to this picture and compare with existing experiments.

Keywords: BCS, superconductivity, Jahn–Teller distortions.

1. INTRODUCTION

The discovery of superconductivity [1] in fcc alkali intercalated A_3C_{60} compounds (A = K, Rb, Cs) with T_c-values exceeding 30 K has created considerable excitement. The question arises whether superconductivity in these compounds can be explained in standard BCS terms using electron–phonon coupling or whether electron–correlation based couplings are operative. Also at issue is the appropriateness of BCS type coupling vs the condensation of strongly coupled, preformed bosons. Numerous models falling into these various categories have been proposed [2–11]. We here present the results of detailed studies of the electronic and vibronic properties of A_3C_{60} and their coupling to each other [3]. We find that standard BCS-type coupling to molecular vibrations can well account for all known observations. The uniqueness of the present situation arises from the particular molecular, chemical nature of the compounds rather than from unusual superconductivity mechanisms.

In section 2 we will discuss the basic electronic structure of A_3C_{60}, details relevant to superconductivity and a variety of approximations used to describe the electron states near the Fermi-level. In section 3 we will focus on the vibrational properties of A_3C_{60}, subdivide those into several groups of vibrations according to the nature of the molecule, and present numerical results for several models. In section 4 we study the electron-phonon coupling in detail and relate it to the Jahn–Teller coupling problem of an isolated C_{60} molecule. Finally, in section 5 we describe

our results for superconductivity, including a discussion of Coulomb repulsion effects, the isotope effect, changes upon alkali substitution, effects of pressure, etc. We will compare with experiments where available and conclude in section 6 with a discussion of limiting circumstances for further enhancement of superconductivity in A_3C_{60}.

2. ELECTRONIC STATES IN A_3C_{60}

It is convenient to view the electronic structure of A_3C_{60} as a result of a stepwise refinement of energy scales. In this context we begin with the largest scale, that of an isolated C_{60} molecule. Each of the 60 atoms is three-fold coordinated at the vertex of two hexagons and one pentagon. There are a total of 12 pentagons, 20 hexagons and 90 bonds. While each atom is equivalent, the bonds are not: 60 (long) bonds exist separating pentagons and hexagons, while 30 (short) bonds separate hexagons. Experimentally [12, 13], the bond lengths are determined to be about 1.4 Å and 1.45 Å. The electronic structure of the isolated C_{60} molecule has been studied early by empirical methods [14]. The states can be classified into σ- and π-like revealing an insulating gap (1–2 eV) between the highest occupied, mostly π-like level (HOMO) and the lowest unoccupied mostly π^*-like level (LUMO). In the icosahedral group I_h, the HOMO is classified as a five-fold degenerate h_u level, while the LUMO is represented by a three-fold degenerate t_{1u} level. Both are part of a 11-fold degenerate $\ell = 5$ manifold of π-states (on a sphere) that is split by the iscosahedral symmetry of C_{60}. Since intercalation with alkali atoms will result in electron donation to C_{60}, the LUMO is of particular interest. Careful

---

‡ Permanent address: ISEN, Lille, France.

analysis shows that the phases of the three LUMO t_{1u}-states are bonding along the short bonds and antibonding along the long (pentagon) bonds. When placed into an fcc lattice with the measured lattice constant of $a_0 = a\sqrt{2} = 14.2\,\text{Å}$, the closest carbon atoms of neighboring molecules are separated by $d \simeq 3.1\,\text{Å}$, which is smaller than the interlayer separation in graphite (3.45 Å). While the electrons remain largely localized to the individual C_{60} molecules, small inter-molecular overlap exists.

The distribution of electronic charges is illustrated in Fig. 1 where the total valence electron density (top) is compared to the hypothetical charge of an electron added to the LUMO (bottom). The overlap between these LUMO states of two molecules is emphasized in Fig. 2. The calculations were done in the Density Functional (LDA) approach using a plane wave expansion and pseudopotentials yielding results in close agreement with other studies [15–19]. While the electronic structure of an isolated C_{60} is well understood, and near the gap given by the large intramolecular π–π hopping energy scale, details of the weak inter-molecular hopping and band formation are less well established. In particular, these details depend on the relative orientation of C_{60} molecules [20, 21], NMR [22, 23] and X-ray [24] experiments show that in pure, undoped C_{60} the molecules rotate freely at room temperature undergoing an ordering transition only below 260 K. At these temperatures the molecules still show correlated rotations which are only frozen out below $\sim 140\,\text{K}$, as indicated by NMR studies [25]. Pulsed neutron studies show considerable orientational disorder to remain [26]. For intercalated material the structure is already frozen at room temperature [27]. The degree of order is not well established. The influence of molecular orientation on the electronic structure of the LUMO complex has recently been studied [20]. Ignoring details in the electronic structure (which would be washed out because of residual disorder) the main results of these studies are the formation of a $\sim 0.5\,\text{eV}$ wide LUMO band, separated from the next higher band by $\sim 0.5\,\text{eV}$. The two energy scales of the $\sim 10\,\text{eV}$ wide π–π^* complex and the $\sim 0.5\,\text{eV}$ wide inter-molecular band dispersion are clearly separated, emphasizing the strong molecular nature of fcc C_{60}.

Solid C_{60}, intercalated with alkali atoms exhibits a rich phase diagram with both semiconducting and metallic phases [28]. We have concentrated on the (only) metallic phases of fcc A_3C_{60} or A_2BC_{60}. In

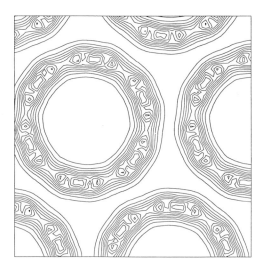

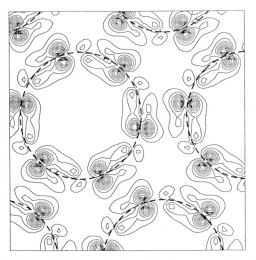

Fig. 1. Calculated electronic charge density contour plots of fcc C_{60}. The total valence charge density (top) is compared to the hypothetical charge density of an electron added to the LUMO (bottom). The densities are shown in a (111) plane intersecting the C_{60} molecules in an equatorial plane through bond, but not individual atoms. The calculations were done using pseudopotentials and plane waves.

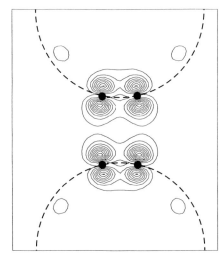

Fig. 2. Calculated charge density distribution of t_{1u} LUMO electrons, shown in a plane (close to (110)) containing two pairs of atoms on adjacent molecules. Overlap of this kind is responsible for the formation of a narrow conduction band in A_3C_{60}.

these phases alkali atoms occupy the two tetrahedral (A) and the one octahedral (B) interstitials. The C$_{60}$ molecules are thought to be in their 'standard' T_h^3 configurations, with hexagons pointing in the (111) directions. There are two such orientations which presumably remain disordered. The key question regarding the electronic structure of A$_3$C$_{60}$ is whether the alkali atoms merely act as donors of electrons into the mostly rigid C$_{60}$ LUMO complex, or whether significant hybridization with these states takes place. Several independent calculations [3, 20, 30] show that the rigid band donor picture is essentially correct. The first alkali derived states are found $\sim 2\,\text{eV}$ above the t_{1u} LUMO band. The states near E_F in A$_3$C$_{60}$ have then the approximate spatial distribution shown in Fig. 1 (bottom). The bandwidth W and the density of states $N(E_F)$ to first order depend only on the inter-molecular π-electron overlap, illustrated in Fig. 2. With this picture in mind, and with the purpose to conduct a study of electron–phonon inter-actions, we consider a simple empirical tight-binding Hamiltonian [3, 31] of the form:

$$H_{\text{TB}} = \sum_{i\alpha} \epsilon_{i\alpha} C_{i\alpha}^+ C_{i\alpha} + \sum_{i<j} \sum_{\alpha\alpha'}^{nn} t_{i\alpha j\alpha'} C_{i\alpha}^+ C_{i\alpha'},$$

where the on-site energies $\epsilon_{i\alpha}$ and the nearest neighbor hopping integrals $t_{i\alpha,j\alpha'}$ are empirically determined parameters. We use a four state $(\alpha = s, p_x p_y p_x)$ basis set. The Slater–Koster matrix elements (ϵ, t) are obtained from weighted fits to first-principles, LDA-type band structures of C$_2$, graphite and diamond at different interatomic separations. Two representative sets of parameters, as obtained by us [31] (TB1) and by Goodwin [32] (TB2) are listed in Table 1. The parameters are normalized to the intra-atomic distance d_0 in diamond. The scaling of these parameters with distance is an important and more subtle question. One can use the traditional d^{-2} dependence proposed for p-electron overlap [33]. Our fits to LDA results indicate a d^{-3} dependence of $t_{\text{pp}\pi}$ and a d^{-2} dependence of all other hopping integrals. Goodwin [32] proposes $d^{-2.8}$ for all hoppings. When calculating the electron–phonon coupling strength for intra-molecular vibrations we shall compare all different

Table 1. Tight binding parameters for carbon (in eV) for an inter-atomic separation of $d_0 = 1.55\,\text{Å}$ (diamond). TB1, TB2 refer to different models given in Refs 31 and 32, respectively

	TB1	TB2
ϵ_s	-7.3	-7.45
ϵ_p	0	0
$t_{ss\sigma}$	-3.63	-4.43
$t_{sp\sigma}$	$+4.20$	$+3.79$
$t_{pp\sigma}$	$+5.38$	$+5.66$
$t_{pp\pi}$	-2.24	-1.83

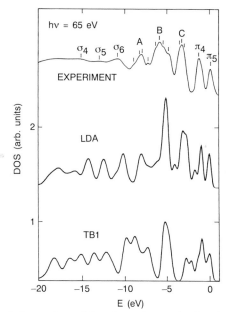

Fig. 3. Comparison of the valence electron density of states of C$_{60}$ as measured by photoemission (top) (Ref. 15) and as calculated by the LDA pseudopotential scheme (middle) and the TB1 empirical tight binding scheme (bottom). The theoretical spectra are convoluted with an energy dependent Gaussian 'self-energy' function with width $0.23 + 0.02|\epsilon|$ (Ref. 19).

models. Inter-molecular distance dependence of t is empirically determined from the superconductivity results to be about $d^{-2.7}$ (see section 6). A comparison of the overall valence electron spectrum of C$_{60}$ between LDA, TB1 and an experimental photoemission spectrum is shown in Fig. 3. The close similarity of the molecular features confirms the overall correctness of the theoretical models.

3. VIBRATIONAL STATES IN A$_3$C$_{60}$

As for the electronic states it is also instructive to subdivide the vibrations of A$_3$C$_{60}$ into individual groups reflecting the molecular nature of the compound. In Fig. 4 a sketch of the full vibrational spectrum is shown. The highest frequency band (A) is due to intra-molecular vibrations of C$_{60}$. The highest modes are mostly tangential in character, while the modes at the lower end of the spectrum have mostly radial character. There are 174 modes grouped into one, three, four and five-fold degenerate representations. Experimentally, the modes have been studied by neutron scattering [34, 35], Raman scattering [36–39] and infrared absorption [37]. The Raman studies are of particular interest to us, since the symmetry selection rules of this process select the same identical modes that couple to the conduction electrons in the t_{1u} LUMO. These are two symmetric, one-fold degenerate A_{1g} modes and eight five-fold

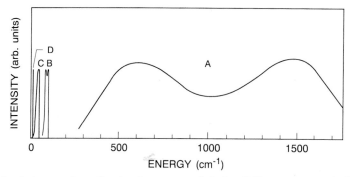

Fig. 4. Sketch of the complete vibrational spectrum of A_3C_{60}. Different groups of vibrations are emphasized. Intra-molecular vibrations are highest in energy (A), optic alkali vibrations are lower (B), followed by acoustic inter-molecular C_{60} vibrations (C) and C_{60} librations at very low energy (D).

degenerate H_g modes. The two A_g modes correspond to the (relatively low energy, $\sim 500\ cm^{-1}$) overall radial breathing mode and the (relatively high energy, $\sim 1500\ cm^{-1}$) tangential double-bond stretching mode. This latter mode can also be viewed as a tangential breathing of the 12 pentagons. The displacement patterns of the five-fold degenerate (d-like quadru-polar) H_g modes are more complex. They are depicted in Fig. 5, as calculated from a bond charge model. Detailed discussion of these modes can be found in several earlier studies [40–42]. In our work we have employed four different, approximate descriptions of these vibrations. These are a Keating-type model [43] with two parameters describing nearest neighbor bond–stretch (α) and bond–bend (β) forces. We used two different β/α ratios (0.1 and 0.3) with the overall scale adjusted to experiment. The value of 0.1 is close to what has been used previously to describe the vibrations of benzene [42] and the in-plane modes of graphite. However, Keating-type models, well suited for diamond-type sp^3 networks, work less well for planar sp^2 graphite. In fact, with only nearest neighbor interactions they do not stabilize any out-of-plane motion in planar graphite. In C_{60}, because of finite curvature, they are somewhat better suited. A further empirical model we use is the so-called bond–charge model, developed by Weber [44] for diamond and extended to graphite and C_{60} by Onida and Benedek [45]. In this model, Keating-like bond–stretch and bond–bend potentials are augmented by a (long-range) screened Coulomb interaction between adiabatic charges localized at the atoms and about midway between the atoms in the bonds. The strength of this interaction is one further empirical parameter. The fourth vibrational model we used is based on the MNDO empirical electronic structure method, with the results described in Ref. 40. In this well estab-lished method, vibrational eigenmodes are found from the second derivative of a parameterized electronic Hamiltonian. The parameters are fit to a large set of

molecular energies and not specifically adjusted to C_{60}. In Table 2 we compare [46] the 10 Raman active mode frequencies ($2A_g + 8H_g$) of the four employed models with experimental results. The average devi-ations from experiment range from $\sim 3\%$ to $\sim 10\%$ with the empirical bond charge model giving the best overall agreement.

When C_{60} is condensed into a A_3C_{60} solid additional vibrational degrees of freedom appear of finite frequencies. Since the mass of alkali atoms is significantly smaller than that of C_{60} the vibrations of the ionic A_3C_{60} solid are well separated into three acoustic branches (C) of mostly C_{60} character at low frequencies ($\approx 50\ cm^{-1}$) and nine optic branches (B) of mostly alkali character at higher frequencies ($\approx 100\ cm^{-1}$). All frequencies are generally low because of the weak inter-molecular interactions, but they are increased in the intercalated material because of ionic Coulomb contributions. Experimentally, alkali induced, strongly dipole active modes are found near $80\ cm^{-1}$ using electron energy loss spectroscopy [47]. Finally, there is another class of very low frequency modes (D), corresponding to the librational motion of C_{60} molecules in their angular potential wells.

Table 2. Comparison of the measured eight H_g and two A_g Raman active modes of C_{60} to various theoretical models described in the text. The average deviation from experiment is given in the bottom row. All energies are in cm^{-1}

Mode	ω_p^{exp}	Keating $\beta/\alpha = 0.1$	Keating $\beta/\alpha = 0.3$	Bond charge	MNDO
$H_g(1)$	273	250	298	271	263
(2)	437	347	411	410	447
(3)	710	444	621	718	771
(4)	744	774	766	793	924
(5)	1099	1145	1162	1157	1261
(6)	1250	1299	1226	1218	1407
(7)	1428	1662	1500	1452	1596
(8)	1575	1718	1718	1691	1722
$A_g(1)$	497	492	476	499	610
(2)	1469	1678	1452	1455	1667
$\Delta\%$		11.9	5.8	3.2	10.7

Hg modes: Raman Active

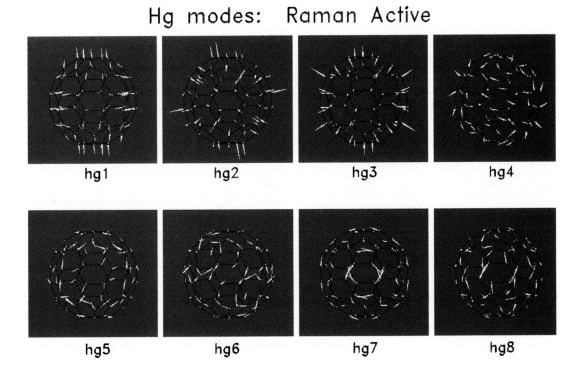

Fig. 5. Eigenvector representation of representative members of each of the eight (five-fold degenerate) H_g vibrational modes. These modes are both Raman active and couple to electrons in the t_{1u} LUMO (plot courtesy M. Grabow).

These modes have recently been identified in pure C_{60} itself. This observation is extremely useful in a by quasi-elastic neutron scattering [48] and by specific heat measurements [49]. They are in the 10–20 cm^{-1} regime and presumably at somewhat higher frequencies for the intercalated materials. In the context of our present investigations we make no particular effort to quantitatively describe vibrations other than the intra-molecular modes. As will become evident in the next section, for superconductivity coupling to electrons in the t_{1u} LUMO derived bands is dominated by these intra-molecular modes. Coupling to the low frequency modes may affect the normal state resistivity.

molecular crystal with strongly different energy scales. It permits to neglect in (4) all contributions except those that modulate the strong *intra*-molecular π–π overlap of the t_{1u} states. In this limit of $t_{inter}/t_{intra} \rightarrow 0$, the sum in (4) can be further simplified [51] and one obtains:

$$\lambda = N(0)V = N(0) \sum_p V_p = N(0) \sum_{p,\mu} \frac{\text{Trace}(I^2)_{p,\mu}}{9M\omega_p^2}, \quad (5)$$

where the trace corresponds to the (3×3) matrix in the t_{1u} subspace and summations of p and μ are over the normal modes p of an isolated C_{60} with their degeneracy index μ. These modes are normalized per C_{60} molecule. The quantity V_p in eqn (5) is just the average coupling energy in the (3×3) t_{1u} subspace.

4. ELECTRON–PHONON COUPLING IN A_3C_{60}

In the BCS theory of superconductivity the dimensionless electron–phonon coupling constant is given by:

$$\lambda = \frac{2}{N(0)} \sum_{p,q} \frac{1}{2K_p(q)} \sum_{\substack{nk \\ n'k'}} |I_{nk,n'k'}(p,q)|^2 \delta(\epsilon_{nk}) \delta(\epsilon_{n'k'}), \quad (2)$$

where ϵ_{nk} is the energy of the electronic Bloch state of band n with wavevector $\mathbf{k}$. The delta functions ensure the sum to be restricted to the Fermi surface. $K_p(q) = M\omega_p^2(q)$ is the force constant of the pth photon with wavevector $\mathbf{q}$ and $I_{nk,n'k'}(p,q)$ is the electron–phonon matrix element, linear in the phonon normal mode amplitudes. $N(0)$ is the density of states at the Fermi level per spin orientation. We now use a tight binding representation for the Bloch states [50]

$$\psi_{nk}(\mathbf{r}) = \frac{1}{\sqrt{N}} \sum_{\substack{\tau\ell \\ \mathbf{R}}} c_{\tau\ell}(n\mathbf{k}) e^{i\mathbf{k}\mathbf{R}} \phi_\ell(\mathbf{r} - \mathbf{R} - \tau) \quad (3)$$

with the sum running over all sites τ in the unit cell, all cells $\mathbf{R}$ and all orbitals ℓ, contributing to the t_{1u} manifold near E_F. Inserting (3) into (2), together with the phonon eigenmode representation in terms of individual atomic displacements $\mathbf{u}_\tau(p,q)$ we obtain:

$$
\begin{aligned}
I_{nk,n'k'} = & \sum_{\substack{\tau\ell \\ \tau'\ell'}} C^*_{\tau\ell}(n\mathbf{k}) C_{\tau'\ell'}(n'\mathbf{k}') \\
& \times \sum_{\mathbf{R}} \{ \nabla_\tau \langle \phi_\ell(\mathbf{r} - \tau) | H | \phi_{\ell'}(\mathbf{r} - \mathbf{R} - \tau') \rangle \} \\
& \times \{ \mathbf{u}_\tau(p,q) e^{i\mathbf{k}'\mathbf{R}} \\
& - \mathbf{u}_{\tau'}(p,q) e^{i\mathbf{k}\mathbf{R}} \} \delta(\mathbf{k} - \mathbf{k}' - \mathbf{q}).
\end{aligned} \quad (4)
$$

The crucial term in (4) is the first parenthesis, where the derivative of an atomic hopping matrix-element is calculated with respect to the motion of an atom at site τ. As pointed out early on, this gradient matrix element is proportional to the original matrix element

The beauty of this result is that it directly relates to the adiabatic Jahn–Teller problem of an electron in the three-fold degenerate t_{1u} manifold coupled to distortions of the isolated C_{60} molecule. This relationship has recently been shown by Lannoo et al. [51] by revisiting the classic Jahn–Teller problem [52] of a three-fold degenerate electronic state coupled to five-fold degenerate distortions ($t_{1u} \times H_g$). Interestingly, the problem can be expressed as a special case of a tetrahedral defect of T_d symmetry coupled to modes of T and E symmetry (like the vacancy in silicon). The resulting static Jahn–Teller coupling energy is given in terms of the coupling matrix elements I as:

$$E_p(H_g) = \sum_\mu \frac{\text{Trace}(I^2)_{p,\mu}}{15M\omega_p^2}. \quad (6)$$

Therefore the contribution V_p to λ in eqn (5) is

$$V_p = \tfrac{5}{3} E_p(H_g). \quad (7)$$

for H_g modes. Analogously one finds for the totally symmetric A_g modes

$$V_p = \tfrac{2}{3} E_p(A_g). \quad (8)$$

Using the different electronic structure models and the different phonon models we can now numerically evaluate $V = \Sigma_p V_p$. The results are compiled in Tables 3–5. Looking first at Table 5 we see that, for a given electronic structure model (e.g. column 3, TB1 with $n_{pp\pi} = 3$), the values for V are reasonably constant, except for the MNDO model. For a given phonon model (e.g. row 3, the bond charge model) the strength of the coupling depends of course strongly ($\sim n^2$) on the distance exponent n. Comparing with the LDA results an exponent near $n \approx 2.5$ seems to be appropriate for the present situation. Inspecting the results in more detail (Tables 3, 4) one

Table 3. Electron–phonon coupling constants $V = \Sigma_p V_p$ (in meV) for individual modes p and the logarithmically averaged frequency $\omega_{\log}$ (in cm^{-1}) used for estimates of T_c (see eqn (10)). Here the results are for different phonon models and the same TB1 electronic tight-binding model with a $n_{pp\pi} = 3$ distance dependence of the matrix elements

Mode	Keating $\beta/\alpha = 0.1$	Keating $\beta/\alpha = 0.3$	Bond charge	MNDO
$H_g(1)$	1.1	11.0	9.4	1.4
(2)	3.2	5.1	3.2	1.8
(3)	26.6	16.0	7.6	6.3
(4)	3.7	11.6	10.8	0.7
(5)	0.1	0.2	0.2	0.8
(6)	0.0	0.0	1.1	1.1
(7)	2.2	11.2	18.5	18.3
(8)	35.9	27.1	22.2	16.3
V	72.8	82.2	73.0	46.7
$\omega_{\log}$	786	887	982	1320

Table 5. Summary of calculated electron–phonon matrix elements V (in meV). The columns correspond to electronic structure models, the rows to phonon models. Details are in the text. The result labelled (*) is extracted from Ref. 2

	LDA	TB1 $n = 2$	TB1 $n_{pp\pi} = 3$	TB2 $n = 2.8$	MNDO
Keating $\beta/\alpha = 0.1$		38.1	72.8	57.4	
Keating $\beta/\alpha = 0.3$		37.6	82.2	58.0	
Bond charge	52.0	32.2	73.0	49.1	
MNDO		21.6	46.7	32.8	56.0*

can see that the spectral distribution of V_p is rather similar in all cases, except for the phonon calculations with the Keating ($\beta/\alpha = 0.1$) model and the MNDO model. In all other models roughly one-half of the oscillator strength comes from the lowest four (radial) modes and one half from the highest four (tangential) modes. These general results are in contrast with the MNDO results of Ref. 2 which report about 80% of the contributions to V to originate from the highest two modes ($H_g(7)$, $H_g(8)$). We shall, in the next section discuss experiments that do measure quantities related to these coupling strengths.

In the limit $t_{inter}/t_{intra} \to 0$, which we discussed so far, no $q = k - k'$ dependence of the scattering exists. For the Jahn–Teller types symmetry lowering H_g modes, the scattering is thus inter-band (off-diagonal in the t_{1u} LUMO manifold) on an individual C_{60} molecule. The coupling to the A_g symmetric modes merits further discussion. The mode coupling is diagonal in the t_{1u} space and therefore does not scatter on an individual molecule. The $q = 0$ limit

Table 4. Electron–phonon coupling constants $V = \Sigma_p V_p$ (in meV) for individual modes p and the logarithmically averaged frequency $\omega_{\log}$ (in cm^{-1} used for estimates of T_c (see eqn (10)). Here the results are for different electronic models and the same bond-charge phonon model. The LDA results are obtained by the 'frozen phonon' method (Ref. 3)

Mode	TB1 $n = 2$	TB1 $n_{pp\pi} = 3$	TB2 $n = 2.8$	LDA
$H_g(1)$	3.0	9.4	4.7	8.0
(2)	2.4	3.2	2.9	7.0
(3)	6.0	7.6	8.5	4.0
(4)	4.8	10.8	7.7	7.0
(5)	0.0	0.2	0.1	1.0
(6)	0.6	1.1	0.6	3.0
(7)	7.0	18.5	11.1	13.0
(8)	8.4	22.2	13.4	9.0
V	32.2	73.0	49.1	52.0
$\omega_{\log}$	939	982	946	950

corresponds simply to an overall shift, as also pointed out in Ref. 2. The modes can, however, scatter between molecules for a finite t_{inter} and a finite q. In this case the scattering strength would again be given by the strong t_{intra}. For finite doping, such as in A_3C_{60}, however, the inter-molecular potential produced by a $q \neq 0$ A_g mode is likely to be screened out, effectively eliminating the contributions of A_g modes to the electron–phonon coupling parameter in the metallic compound. Screening also eliminates the coupling to the alkali modes. Since the LUMO wavefunctions have vanishingly small amplitudes at the alkali sites, only long-range potentials could scatter [4]. These are, however, efficiently screened in the metallic compounds. This is in contrast to on-ball Jahn–Teller modes which directly modulate nearest neighbor wavefunction overlap. Here screening has only a quantitative effect.

Equation (5) shows that the dimensionless coupling constant is proportional to the density of states $N(0)$ of electrons at E_F. Although this quantity should in principle be available from calculations and/or experiments, in practice the situation is still unclear. On the theoretical side the problems are mainly due to uncertainties and disorder in the orientational arrangement of C_{60} molecules. Intra-molecular bonding merely determines that $N(0)$ derives from a well-isolated three-fold degenerate t_{1u} LUMO. Although this is rather important, the weak inter-molecular interactions determine the value of $N(0)$. On an approximate scale, $N(0) = 3/W$ per spin, where W is the t_{1u} conduction band width. Most calculations [3, 19, 20], in reasonable agreement with each other show $W \approx 0.5$ eV resulting in an average $N(0) \approx 6$ states/eV–spin–C_{60}. However, the inter-molecular density of states has considerable structure [5], some of which may be washed out by disorder which can drastically change this value. A recent model calculation [53] supports this view. On the experimental side, large variations exist in reported estimates for $N(0)$. Photoemission data [54] are interpreted to give small values, $N(0) \approx 1$–2. Difficulties here are associated with hole lifetime effects and/or surface sensitivity. Susceptibility measurements [55] suggest values in the range of $N(0) \approx 10$–15, while NMR data [56] suggest even larger values $N(0) \gtrsim 20$. The difficulties here are associated with extracting 'bare' density of states values in the presence of interactions and disorder. The questions are largely unsettled at this time and we have to consider reasonable ranges of $N(0)$.

5. SUPERCONDUCTIVITY

The complete theory of superconductivity, given by Eliashberg allows one to calculate T_c, given the interaction strengths and their respective frequency distributions. For the attractive electron–phonon interaction, these are the coupling parameters $\lambda_p = N(0) \cdot V_p$ and the vibrational frequencies ω_p. Less detail is known for the repulsive Coulomb interaction μ, which we shall discuss below. An approximate, explicit formula for T_c has been given by McMillan [57], which works remarkably well for not too strong coupling. For the present case, we have tested the validity of McMillan's formula by solving Eliashberg's equation numerically [58] for a variety of scenarios. We have found that for the vibrational intra-molecular coupling of C$_{60}$, McMillan's formula gives excellent results. Then T_c is given by

$$T_c = \frac{\hbar \omega_{log}}{1.2 K_B} \exp\left[\frac{-1.04(1+\lambda)}{\lambda - \mu^* - 0.62\lambda\mu^*} \right], \quad (9)$$

where the logarithmically averaged phonon frequency ω_{log} is given by

$$\ln \omega_{log} = \frac{1}{\lambda} \sum_p \lambda_p \ln \omega_p, \quad (10)$$

with $\lambda = \Sigma_p \lambda_p$. Values for ω_{log} have been calculated for the various models and some are indicated in Tables 3 and 4. They are remarkably constant and typically of order 800–1000 cm^{-1} or 1150–1450 K (with the exception of the MNDO results of Ref. 2).

The effective Coulomb interaction μ^* is reduced from the full Coulomb repulsion μ by retardation via the approximate relationship [59]

$$\left(\frac{1}{\mu^*}\right) = \left(\frac{1}{\mu}\right) + \ln\left(\frac{\omega_{el}}{\omega_{ph}}\right). \quad (11)$$

Here ω_{el} is the characteristic cut-off frequency for Coulomb interactions while $\omega_{ph} \approx \omega_{log}$. The effectiveness of retardation has been questioned [60] for C$_{60}$ molecular solids on grounds that the bandwidth W for inter-molecular hopping is comparable to $\hbar\omega_{log}$. It is, however, important to realize that the electronic structure of A$_3$C$_{60}$ is strongly molecular only for the

valence states and the first few conduction bands. Higher lying states intermix with the alkali states and are truly extended throughout the solid. Coulomb scattering into these higher lying states allows electrons therefore to hop off the C$_{60}$ molecules at a much faster rate than that given by W. The correct frequency to use in eqn (11) is therefore a characteristic A$_3$C$_{60}$ plasma frequency of order 10 eV or higher. Recent calculations [61] and electron energy loss experiments [62] show strong structures in the scattering form factor $S(q, \omega)$ near 6 eV and 25 eV, due to C$_{60}$, π and σ plasmons, respectively. We therefore believe that the Coulomb repulsion in A$_3$C$_{60}$ is rather standard, i.e. a value of $U \approx$ few eV for carbon orbitals yields a $\mu = N(0) \cdot U$ of order 0.5–1.0 for the molecule which is then renormalized via retardation to values of $\mu^* \approx 0.1$–0.2. Isotope measurements, discussed below are consistent with these values.

It is clear that with all the uncertainties in $N(0)$, V and μ^*, T_c cannot be reliably calculated. Inversely, however, the observed T_c values can well be explained with parameters within the discussed range. For instance, for a $\hbar\omega_{log} \approx 1400$ K as obtained from the bond charge model, a $V \approx 50$ meV which is about the calculated LDA value (see Table 5), an average $N(0) \approx 14$ and a $\mu^* \approx 0.2$ one obtains $T_c \approx 20$ K which is the observed T_c for K$_3$C$_{60}$. The important question is whether these estimates sensibly explain observed trends and whether the overall picture is consistent with all experiments.

It was noted earlier [63] that T_c scales monotonically with the A$_3$C$_{60}$ lattice constants upon chemical alkali substitution. This observation is beautifully confirmed by the present scenario. The molecular nature of A$_3$C$_{60}$ 'factors' all quantities in real space. The electron–phonon coupling matrix element V, and the prefactor $\hbar\omega_{log}$ are intra-molecular quantities of C$_{60}$ and should be invariant. If we also assume μ^* to be constant to first order, only the density of states $N(0)$ varies from compound to compound. To first-order, it scales with inter-molecular distance d as

$$N(0) \sim \frac{1}{W} \sim \frac{1}{t_{inter}^0}\left(\frac{d}{d_0}\right)^n. \quad (12)$$

In Fig. 6 and Table 6 we show how this simple argument used with the selected values given above and a distance scaling of $n = 2.7$ beautifully explains all observed trends. This value of n empirically adjusted is within the range expected for inter-molecular interactions [20].

The effect of chemical pressure is fully equivalent to mechanical pressure. Also shown in Fig. 6 and Table 6 are experimental results [64–66] obtained by applying hydrostatic pressure to K$_3$C$_{60}$ ($\bigcirc$) and to contradicts the model for superconductivity proposed in Ref. 4.

The situation is drastically different for carbon isotope substitution. According to our model for superconductivity we expect $\hbar\omega_{\log}$ to change $\sim 1/\sqrt{M}$. Since $\hbar\omega_{\log}$ enters T_c as prefactor and in the reduction of μ^*, we expect from McMillan's formula

$$T_c \sim M^{-\alpha} \qquad (13)$$

with

$$\alpha = \frac{1}{2}\left\{1 - \left(\mu^* \ln \frac{\hbar\omega_{\log}}{1.2kT_c}\right)^2 \frac{1 + 0.62\lambda}{1 + \lambda}\right\}.$$

In Table 7 we illustrate the variation of α for two scenarios. One corresponding to our model with $\hbar\omega_{\log} \approx 1400$ K and one for a hypothetical low energy phonon model with $\hbar\omega_{\log} \approx 200$ K. In both cases λ and μ^* have been chosen to reproduce the $T_c \approx 29$ K for Rb_3C_{60}. For the model developed here and with $\mu^* = 0.2$ we expect a reduction of α from 0.5 to 0.29. This is in excellent agreement with recent measurements by Ramirez et al. [68] which indicate $\alpha = 0.37 \pm 0.05$ and by Chen and Lieber [69] who find $\alpha = 0.30 \pm 0.06$ for K_3C_{60}. Very different results ($\alpha > 1$) have been reported by Ebbesen et al. [70], but for rather incomplete isotope substitution.

The strong coupling scenario does not yield any appreciable reduction of α, unless one goes beyond the range of validity of McMillan's theory [67]. While this agreement seems to confirm our model we have to caution that sizeable isotope effects can in principle also occur for different reasons. Inspection of Fig. 6 shows that the observed ≈ 0.5 K change of T_c upon C^{13}/C^{12} substitution could also be accounted for if the lattice parameter was reduced by ≈ 0.01 Å.

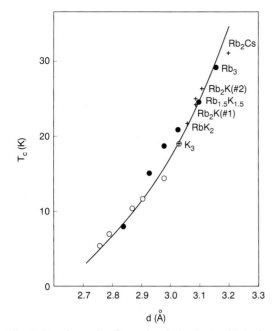

Fig. 6. Experimentally observed variation in T_c with lattice constant variations, converted here into an approximate surface distance between C_{60} molecules. Shown are the data for chemically substituted compounds (+), Ref. 63, and those obtained from K_3C_{60} (○) and Rb_3C_{60} (●) via mechanical pressure (Ref. 66). The solid line is the calculated variation of T_c using McMillan's formula and the parameters discussed in the text.

Rb_3C_{60} (●). The results follow the same curve. This shows unambiguously that there is no alkali isotope effect. Rb_3C_{60} reduced to the volume of K_3C_{60} has virtually the same T_c as K_3C_{60}. This

Table 6. Compilations of experimental T_c values vs lattice constants a. The chemical substitution data are from Ref. 63, the Rb_3 and K_3 pressure data are from Ref. 66. The inter-molecular distance d is approximated using a C_{60} radius of 3.52 Å. The calculated values are obtained from scaling the K_3C_{60} λ-value using a $n = 2.7$ distance variation of $N(0)$ (see eqn (12) in text). T_c is calculated with $\mu^* = 0.2$. The table is illustrated in Fig. 6

	a	d	$T_c^{\exp}$	T_c^{calc}	λ^{calc}
Rb_2Cs	14.49	3.20	31.3	34.9	0.85
Rb_3	14.44	3.16	29.4	31.0	0.82
Rb_2K	14.36	3.11	26.4	25.9	0.78
$Rb_{1.5}K_{1.5}$	14.34	3.09	25.2	24.7	0.77
RbK_2	14.30	3.06	21.8	22.3	0.75
K_3	14.25	3.03	19.3	19.9	0.73
Rb_3	14.44	3.16	29.4	34.9	0.85
	14.35	3.10	24.8		
	14.27	3.03	21.0		
	14.18	2.98	18.9		
	14.11	2.93	15.2		
	13.98	2.84	8.1	8.0	0.61
K_3	14.25	3.03	19.3	19.9	0.73
	14.18	2.98	14.5		
	14.09	2.91	11.9		
	14.03	2.87	10.6		
	13.92	2.79	7.1		
	13.87	2.76	5.5	5.1	0.57

Table 7. Calculated isotope shift exponents for a weak coupling model (top) and a strong coupling model (bottom). The values of λ, μ^* are chosen to reproduce the T value of 29 K (Rb_3C_{60}) using McMillan's formula [57]. The model developed in this paper corresponds to $\lambda = 0.81$ which together with $\mu^* = 0.2$ predicts $\alpha = 0.29$. The measured value is 0.32 ± 0.05 (Ref. 68)

$\hbar\omega_{\log} = 1400$ K		
λ	μ^*	α
0.39	0.0	0.50
0.58	0.1	0.45
0.81	0.2	0.29
1.09	0.3	0.04

$\hbar\omega_{\log} = 200$ K		
λ	μ^*	α
1.5	0.0	0.5
2.1	0.1	0.49
2.9	0.2	0.46
4.2	0.3	0.41

Anharmonic zero-point motion effects [71] could in principle account for this. Precision lattice parameter measurements are needed to clear up this ambiguity. For models where superconductivity derives from intra-molecular correlations [5, 6] isotope effects can also be obtained by similar arguments from on-ball zero point fluctuations.

The coupling of the Jahn–Teller type modes to conduction electrons modifies these modes themselves. As proposed by Allen [72], these mode self-energy effects should be observable. The self-energy can be expressed in a phonon linewidth (Lorentzian full width)

$$\gamma_p = \pi \omega_p^2 N(0)^2 \frac{V_p}{d_p} \tag{14}$$

and a frequency shift

$$|\Delta\omega_p| = \omega_p N(0) \frac{V_p}{d_p},$$

where the coupling strength V_p is given in eqn (5) and where d_p is the degeneracy of mode p. In Table 8 we list these phonon self-energy parameters for the case of the bond charge phonon model and the LDA frozen phonon couplings (see Table 4). A value of $N = 14$ states/eV–spin–C_{60} was used, as appropriate for K_3C_{60}. We see significant phonon linewidth broadening. Particularly affected are the higher frequency modes due to the ω_p^2 factor in eqn (14). These results can be compared to Raman [36, 38, 39] and neutron [35] scattering data, contrasting the insulating C_{60} and A_6C_{60} phases with the metallic A_3C_{60} phase.

In the case of Raman scattering the (dispersionless) intra-molecular modes are probed near $q \approx 0$. In a perfect crystal the continuum of t_{1u}-derived electronic states near E_F couples to these modes for $E(k) = \omega_{ph}$ near k_F, which is much larger. There is, however, as stated earlier, considerable disorder (in particular orientational) in these materials which does not only give a short mean free path but which also should

Table 8. Phonon self-energies calculated (eqn (14)) from the parameters of the LDA/bondcharge model. A value of $N = 14$ states/eV–spin–C_{60} and the values of V_p in Table 4 were used. All energies are given in cm^{-1}. The line width γ is a Lorentzian full width at half maximum

	ω_p^{calc}	ω_p^{exp}	γ_p	$|\Delta\omega_p|$
$H_g(1)$	271	273	7	6
(2)	410	437	15	8
(3)	718	710	23	7
(4)	793	744	56	15
(5)	1157	1099	15	3
(6)	1218	1250	50	9
(7)	1452	1428	307	45
(8)	1691	1575	192	24

destroy q-conservation and allow for Raman scattering to probe the electron–phonon coupling. This was studied experimentally by several groups [36, 38, 39] who have identified substantial coupling of the H_g modes in metallic A_3C_{60} to a continuum which is absent in the insulating phases (C_{60}, A_6C_{60}). Coupling is observed in all cases to both low energy (radial) and high energy (tangential) modes, in general agreement with the present calculations. The coupling to the low energy H_g modes, in particular $H_g(2)$ is also clearly observed by neutron scattering [35]. Line frequency shifts are a more subtle problem, since other effects occur in going from C_{60} to A_3C_{60}, e.g. crystal field shifts due to the A^+ ions, weakening of the C_{60} bonds due to occupying the antibonding t_{1u} level, etc. Detailed comparisons between theory and experiment should, however, become possible.

We summarize the scenario developed here for superconductivity in A_3C_{60}. The parameters are as follows: relatively high energy phonons with $\hbar\omega_{log} \approx 1400$ K couple with weak to intermediate strength of $\lambda \approx 0.73$, subject to a reasonably large Coulomb repulsion of $\mu^* \approx 0.2$ to yield T_c near 20 K for K_3C_{60}. We are thus not in the strong coupling limit and expect the BCS value of $2\Delta \approx 3.5kT_c$ for the superconducting gap and a coherence length of $\xi_0 \approx 130$ Å. NMR [73] and optical [74] data seem to indicate a BCS gap value, while point contact tunneling [75] yields larger values. Coherence lengths of ≈ 150 Å have been inferred from H_{c2} measurements [76] on granular films; much smaller values are typically quoted [75] using a clean limit interpretation.

Can we use what we learned here and extrapolate to new, hypothetical materials? The obvious first choice is C_{60} itself with different doping levels. If one could hole-dope C_{60} the conduction electrons would be situated in a band derived from a five-fold degenerate h_u level. Calculations [20] show the width of this band also to be about 0.5 eV, so $N(0)$ could be larger by a factor ~ 1.7 over the electron doped case. The coupling V can be calculated analogously to the electron case and is found to be smaller by a factor ~ 1.4, such that the overall coupling strength λ becomes only somewhat larger. Although different modes couple, $\hbar\omega_{log}$ is about the same, so we expect a T_c in the same range as for the electron doped case. Similar studies for the next higher conduction band complexes (three-fold t_{1g} and three-fold t_{2u}) yield coupling strengths V, increased by $\sim 10\%$ and $\sim 40\%$ over the t_{1u} LUMO value, respectively. These findings may be of relevance for the recently discovered Ca intercalated materials [78].

Coupling V derives some of its contributions from coupling to lower energy radial or transverse modes. These modes couple to π-like electrons only for curved geometries. In fact, in graphite there is no first order coupling to transverse modes for symmetry reasons, which we suggested [3] to be the reason why T_c is much lower in intercalculated graphite. (The density of states there is comparable [79] to A_3C_{60}.) Reversing the argument we may increase T_c if we find a highly symmetric molecule (with high electronic degeneracies) exhibiting a larger curvature than C_{60}. We have studied the hypothetical C_{20} molecule (only pentagons) which is insulating in its $2+$ charge state. The symmetry of the LUMO is four-fold degenerate G_u which couples to A_g, G_g and H_g modes. The coupling V is found to be indeed about 1.5 times stronger than in C_{60}. Chemically, however, the atoms have near perfect sp^3 bond angles and the dangling bonds make C_{20} probably highly reactive.

6. CONCLUSIONS

We have examined the electronic and vibrational structure of A_3C_{60} compounds in detail and found by direct calculations that the Jahn–Teller type intramolecular modes do efficiently couple to the conduction electrons induced by alkali intercalation. Since the frequencies of these modes are high, they should be efficient for superconductivity. The coupling (V) is largely an isolated molecule property and can be derived from Jahn–Teller type studies. The hopping between the molecules is the second important ingredient in that it determines the kinetic energy or the density of states (N) of conduction electrons. Details are probably washed out by orientational disorder. The dimensionless BCS coupling parameter $\lambda = NV$ is then factorized in real space, a picture which is beautifully confirmed by several experiments, such as studies of chemical or mechanical pressure affecting T_c and vibrational linewidth studies in Raman or neutron experiments. Moreover, the absence of any alkali isotope effect and the observation of a strong effect upon carbon isotope can be quantitatively explained.

In spite of all these obvious confirmations the scenario developed here puts A_3C_{60} close to edge of validity of the underlying BCS model. The electron kinetic energy is only a few times larger than the average phonon energy $\hbar\omega_{log}$ and Migdal's approximation used in BCS theory becomes less appropriate [80]. However, the factorization in real space may be helpful here too. Furthermore, estimates show that the kinetic energy is probably not much larger than the intra-molecular Coulomb repulsion [21, 81]. Further reduction of the kinetic energy, by e.g. increasing the lattice constant should ultimately lead to magnetic instabilities [82] limiting a further increase in T_c.

Acknowledgements—We thank W. Zhong and Y. Wang for assistance with numerical calculations and many of our colleagues for discussions of their results.

REFERENCES

1. Hebard A. F., Rosseinsky M. J., Haddon R. C., Murphy D. W., Glarum S. H., Palstra T. T. M., Ramirez A. P. and Kortan A. R., *Nature* **350**, 600 (1991).
2. Varma C. M., Zaanen J. and Raghavachari K., *Science* **254**, 989 (1991).
3. Schlüter M. A., Lannoo M., Needels M., Baraff G. A. and Tomanek D., *Phys. Rev. Lett.* **68**, 526 (1992).
4. Zhang F. C., Ogato M. and Rice T. M., *Phys. Rev. Lett.* **67**, 3452 (1991).
5. Charkavarty S., Gelfand M. P. and Kivelson S., *Science* **254**, 970 (1991).
6. Baskaran G. and Tossati E., *Current Science* **61**, 33 (1991).
7. Johnson K. H., McHenry M. E. and Cloughtery D. P., *Physica C* (1992) submitted.
8. Mazin I. I., Rashkeev S. N., Antropov V. P., Jepsen O., Liechtenstein A. I. and Andersen O. K., *Phys. Rev.* (1992) submitted.
9. Stollhoff G., *Phys. Rev.* **B44**, 10,998–11,000 (1991).
10. Pietronero L., preprint.
11. Martins J. L. and Troullier N., preprint.
12. Yannoni C. S., Bernier R. P., Bethune D. S., Meijer G. and Salem J. R., *J. Am. Chem. Soc.* **113**, 3190 (1991).
13. Fleming R. M., Siegrist T., Marsh P. M., Hessen B., Kortan A. R., Murphy D. W., Haddon R. C., Tycko R., Dabbagh G., Mujsce A. M., Kaplan M. L. and Zahurak S. M., *Mater. Res. Soc. Symp. Proc.* **206**, 691 (1991).
14. Hadon R. C., Brus L. E. and Raghavachari K., *Chem. Phys. Lett.* **125**, 459 (1986).
15. Weaver J. H., Martins J. L., Komeda T., Chen Y., Ohno T. R., Kroll G. H., Troullier T., Hauffer R. E. and Smalley R. E., *Phys. Rev. Lett.* **66**, 1741 (1991).
16. Zhang Q., Yi J. Y. and Berhnolc J., *Phys. Rev. Lett.* **66**, 2633 (1991).
17. Saito S. and Oshiyama A., *Phys. Rev. Lett.* **66**, 2637 (1991).
18. Feuston B. P., Andreoni W., Parrinello M. and Clementi E., *Phys. Rev.* **B44**, 4056 (1991).
19. Troullier N. and Martins J. L., to be published.
20. Satpathy S., Antropov V. P., Andersen O. K., Jepsen O., Gunnarsson O. and Liechtenstein A. I., *Phys. Rev.* submitted (1992).
21. Gunnarsson O., Satpathy S., Jepsen O. and Andersen O. K., *Phys. Rev. Lett.* **67**, 3002 (1991).
22. Yannoni C. S., Johnson R. D., Meijer G., Bethune D. S. and Salem J. R., *J. Phys. Chem.* **95**, 9 (1991).
23. Tycko R., Haddon R. C., Dabbagh G., Glarum S. H., Douglass D. C. and Mujsce A. M., *J. Phys. Chem.* **95**, 518 (1991).
24. Heiney P. A., Fischer J. E., McGhie A. R., Romanow W. J., Denenstein A. M., McCauley J. P., Smith A. B. and Cox D. E., *Phys. Rev. Lett.* **66**, 2911 (1991).
25. Tycko R., Dabbagh G., Fleming R. M., Haddon R. C., Makhija A. V. and Zahurak S. M., *Phys. Rev. Lett.* **67**, 1886 (1991).
26. Hu R., Egami T., Li F. and Lannin J. S., preprint.
27. Stephens P. W., Mihaly L., Lee P. L., Whetten R. L., Huang S.-M., Kaner R., Diedrich F. and Holezer K., *Nature* **351**, 632 (1991).
28. Murphy D. W., Rosseinsky M. J., Fleming R. M., Tycko R., Ramirez A. P., Haddon R. C., Siegrist T., Dabbagh G., Tully J. C. and Walstedt R. E., *J. Phys. Chem. Solids* **53**, 1321 (1992).

29. Erwin S. C. and Pickett W. E., *Science* **254**, 842 (1991).
30. Martins J. L. and Troullier N., *Phys. Rev.* (1992) submitted.
31. Tomanek D. and Schlüter M., *Phys. Rev. Lett.* **67**, 2331 (1991).
32. Goodwin L., *J. Phys. C* **3**, 3869 (1991).
33. Harrison W. A., *Phys. Rev. B* **27**, 3592 (1991).
34. Cappelletti R. L., Copley J. R. D., Kamitakahara W. A., Li F., Lannin J. S. and Ramage D., *Phys. Rev. Lett.* **66**, 3261 (1991).
35. Prassides K., Tomkinson J., Christides C., Rosseinsky M. J., Murphy D. W. and Haddon R. C., *Nature* **354**, 462 (1991).
36. Duclos S. J., Haddon R. C., Glarum S. H., Hebard A. F. and Lyons K. B., *Science* **254**, 1625 (1991).
37. Bethune D. S., Meijer G., Tang W. C., Rosen H. J., Golden W. G., Seki H., Brown C. A. and de Vries M. S., *Chem. Phys. Lett.* **179**, 181 (1991).
38. Mitch M. G., Chase S. J. and Lannin J. S., *Phys. Rev. Lett.* **68**, 883 (1992).
39. Zhou P., Wang K., Rao A. M., Eklund P. C., Dresselhaus G. and Dresselhaus M. S., *Phys. Rev.* submitted (1992).
40. Stanton R. E. and Newton M. P., *J. Phys. Chem.* **92**, 2141 (1988).
41. Brendsdal E., *Spec. Lett.* **24**, 319 (1988).
42. Weeks D. E. and Harter W. G., *J. Chem. Phys.* **90**, 4744 (1989).
43. Martin R., *Phys. Rev. B* **4**, 4005 (1970).
44. Weber W., *Phys. Rev. B.* **15**, 4789 (1977).
45. Onida G. and Benedek G., *Europhys. Lett.* submitted (1992).
46. We thank K. Raghavachati for providing us with the MNDO C_{60} vibrational eigenvectors and G. Onida for his bond charge model results.
47. Rowe J., Malic R. A., Chaban E. E. and Haddon R. C., to be published.
48. Neuman D. A., Copley J. R. D., Kamitakahara W. A., Rush J. J., Cappelletti R. L., Coustel N., McCauley J. P., Fisher J. E., Smith III A. B., Creegan K. M. and Cox D. M., preprint.
49. Beyermann W. P., Hundley M. F., Thompson J. D., Diederich F. N. and Grüner G., preprint.
50. Varma C. M., Blount E. I., Vashista P. and Weber W., *Phys. Rev. B* **19**, 6130 (1979).
51. Lannoo M., Baraff G. A., Schlüter M. and Tomanek D., *Phys. Rev. B* **44**, 1210 (1991).
52. O'Brien M. C., *J. Phys. C* **4**, 2524 (1971).
53. Gelfand M. P. and Lu J. P., *Phys. Rev. Lett.* **68**, 1050 (1992).
54. Chen C. T., Tjeng L. H., Rudolf P., Meigs G., Rowe J. E., Chen J., McCauley J. P., Smith A. B., McGhie A. R., Romanow W. J. and Plummer E., *Nature* **352**, 603 (1991).
55. Ramirez A. P., Rosseinsky M. J., Murphy D. W. and Haddon R. C., to be published.
56. Tycko R., Dabbagh G., Rosseinsky M. J., Murphy D. W., Fleming R. M., Ramirez A. P. and Tully J. C., *Science* **253**, 884 (1991).
57. McMillan W. L., *Phys. Rev.* **167**, 331 (1968).
58. Eliashberg G. M., *JETP* **11**, 696 (1960). We thank P. B. Allen for providing us with his Eliashberg code.
59. Morel P. and Anderson P. W., *Phys. Rev.* **125**, 1263 (1962).
60. Anderson P. W., preprint (1991).
61. Bertsch G. F., Bulgac A., Tomanek D. and Wang Y., preprint (1991).
62. Gensterblum G., Pineaux J. J., Thiry P., Candano R., Vigneron J. P., Lambin P. and Lucas A. A., *Phys. Rev. Lett.* **67**, 2171 (1991).
63. Fleming R. M., Ramirez A. P., Rosseinsky M. J., Murphy D. W., Haddon R. C., Zahurak S. M. and Makhija A. V., *Nature* **352**, 787 (1991).
64. Schirber J. E., Overmyer D. L., Wang H. H., Williams J. M., Carlson D. D., Kini A. M., Pellin M. J., Welp U. and Kwok W.-K., *Physica C* **178**, 137 (1991).
65. Sparn G., Thompson J. D., Huang S.-M., Kaner R. B., Diederich F., Whetten R. L., Gruner G. and Holczer K., *Science* **252**, 1829 (1991).
66. Zhou O., Vaughn G. B. M., Zhu Q., Fischer J. E., Heiney P. A., Coustel N., McCauley J. P. and Smith A. B., preprint.
67. Kresin V. Z., *Phys. Lett.* **A122**, 434 (1987).
68. Ramirez A. P., Kortan A. R., Rosseinsky M. J., Duclos S. J., Mujsce A. M., Haddon R. C., Murphy D. W., Makhija A. V., Zahurak S. M. and Lyons K. B., *Phys. Rev. Lett.* **68**, 1058 (1992).
69. Chen C. C. and Lieber C. M., preprint.
70. Ebbesen T. W., Tsai J. S., Tanigaki K., Tabuchi J., Shimakawa Y., Kubo Y., Hirosawa I. and Mizuki J., *Nature* **355**, 620 (1992).
71. Fisher D. S., Millis A. J., Shraiman B. and Bhatt R. N., *Phys. Rev. Lett.* **61**, 482 (1988).
72. Allen P. B., *Phys. Rev. B* **6**, 2577 (1972).
73. Tycko R., Dabbagh G., Rosseinsky M. J., Murphy D. W., Ramirez A. P. and Fleming R. M., preprint.
74. Rotter L. D., Schlesinger Z., McCauley J. P., Jr., Coustel N., Fischer J. E. and Smith A. B., III, preprint.
75. Zhang Z., Chen C.-C., Kelty S. P., Dai H. and Lieber C. M., *Nature* **253**, 333 (1991).
76. Palstra T. T. M., Haddon R. C., Hebard A. F. and Zaanen, *J. Phys. Rev. Lett.* **68**, 1054 (1992).
77. Sparn G., Thompson J. D., Whetten R. L., Huang S. M., Kaner R. B., Diederich F., Grüner G. and Holczer K., *Phys. Rev. Lett.* **68**, 1228 (1992).
78. Kortan A. R., Kopylov N., Glarum S. H., Gyorgy E. M., Ramirez A. P., Fleming R. M., Thiel F. A. and Haddon R. C., preprint.
79. Fisher J. E., in *Intercalculated Layer Materials*, pp. 481–532. D. Reidel, Dordrecht (1979).
80. Grabowski M. and Sham L. J., *Phys. Rev. B* **29**, 6132 (1984).
81. Miller B., Rosamilia J. M., Dabbagh G., Muller A. and Haddon R. C., preprint.
82. Allemand P. M., Khemani K. C., Koch A., Wudl F., Holczer K., Donovan S., Gruner G. and Thompson J. D., *Science* **253**, 301 (1991).

SUBJECT INDEX

AUTHOR INDEX